The Production and Distribution of Roman Military Equipment

Proceedings of the Second Roman Military Equipment Research Seminar

edited by

M. C. Bishop

BAR International Series 275
1985

B.A.R.

5, Centremead, Osney Mead, Oxford OX2 0ES, England.

B.A.R.-S275, 1985: 'The Production and Distribution of Roman Military Equipment'

ISBN 9780860543473 paperback
ISBN 9781407343792 e-book
DOI https://doi.org/10.30861/9780860543473
A catalogue record for this book is available from the British Library
This book is available at www.barpublishing.com

CONTENTS

III: A Case Study

LIST OF ILLUSTRATIONS

		Page
WEBSTER		
Fig.1	Sketches of the British dagger scabbards	215
COULSTON		
Fig.1	Bow terminology	329
Fig.2	Yrzi bow	330
Fig.3	Belmesa bow ear	331
Fig.4	Qum Darya bow fragments	331
Fig.5	Stages of stave construction	332
Fig.6	Ch'engtu workshop plan	332
Fig.7	Stave positions. I Yrzi bow; II 'Sassanid' bow	333
Fig.8	Arrow terminology	333
Fig.9	Bar Hill ear laths. Convex faces	334
Fig.10	Bar Hill ear laths. Flat faces	334
Fig.11	Caerleon ear laths. Flat faces	335
Fig.12	Caerleon ear laths. Concave faces	335
Fig.13	Caerleon grip laths. Convex faces	336
Fig.14	Caerleon grip laths. Flat faces	336
Fig.15	Belmesa bow ear. Sinister face	337
Fig.16	Belmesa bow ear. Back face	337
Fig.17	Belmesa bow ear. Dexter face	338
Fig.18	Belmesa bow ear. Belly face	338
Fig.19	Trajan's Column. Archer in Scene XXIV	339
Fig.20	Trajan's Column. Archers in Scene CXV	339
Fig.21	Trajan's Column. Barbarian bow on the north-east face of the pedestal	339
Fig.22	Trajan's Column. Archer in Scene CVIII	340
Fig.23	Trajan's Column. Detail of bow in Scene CVIII	340
Fig.24	Marcus Column. Archer in Scene LXXVIII	340
Fig.25	Marcus Column. Horse-archer in Scene LVII	340
Fig.26	Archer's tombstone from Housesteads	341
Fig.27	Detail of Housesteads archer's bow	341
Fig.28	Tombstone of Monimus, *cohors I Ituraeorum*, Mainz, West Germany	342
Fig.29	Tombstone of an *eques singularis Augusti*, Mainz, West Germany	342
Fig.30	Tombstone from Gÿor, Hungary, *ala I Augusta Ituraeorum*	342
Fig.31	Tombstone from Mainz, West Germany, *ala Parthorum et Araborum*	343
Fig.32	Tombstone from Tipasa, Algeria, *ala I Augusta Ituraeorum*	343
Fig.33	Relief of Palmyrene rider-gods, Abgal and Ašar, from Palmyra. National Museum in Damascus, dated to A.D.154	343
Fig.34	Detail of bow on a Palmyrene page relief. Located south-east of the 'Funerary Temple', Palmyra.	344
Fig.35	Detail of statue of Atys. London, British Museum.	344

LIST OF CONTRIBUTORS

*Lindsay Allason-Jones	Museum of Antiquities, The University, Newcastle upon Tyne, NE1 7RU, England
Julian Bennett	Central Excavation Unit, H.B.M.C., Fort Cumberland, Eastney, Portsmouth, England
*M.C. Bishop	Department of Archaeology, The University, Newcastle upon Tyne, NE1 7RU, England
*J.C. Coulston	Department of Archaeology, The University, Newcastle upon Tyne, NE1 7RU, England
*Mrs C. van Driel-Murray	I.P.P., Singel 453, 1012 WP Amsterdam, The Netherlands
Michel Feugère	C.N.R.S., Centre Camille Jullian, Université de Provence, 29 Av. Robert Schuman, 13621 Aix-en-Provence, France
*Dr J. Oldenstein	Institut für Vor- u. Frühgeschichte, Saarstraße 21, 6500 Mainz, West Germany
*John Paddock	Institute of Archaeology, 31-34 Gordon Square, London, WC1 0PY, England
Ian R. Scott	Department of Archaeology, University College, P.O. Box 78, Cardiff, CF1 1XL, Wales
Dr Graham Webster	The Old School House, Chesterton, Harbury, Warks., CV33 9LF, England

N.B. Contributors marked (*) read their papers at the original seminar in March 1984

PREFACE

The first Roman military equipment research seminar, held in the spring of 1983, sought to define the interests of workers in this field.[1] Such was the enthusiasm of all the participants, that it was decided to meet again in 1984, but this time a specific theme was to be discussed: the production of military equipment. This volume contains many of the papers given at that second meeting, along with some additional relevant contributions.

Works of synthesis on military equipment have been comparatively few and far between, although those of MacMullen,[2] Robinson,[3] and Oldenstein[4] are exceptions (and the value of these is attested by the number of references to them in this volume). It is a curious fact that arms, the raison d'être of the Roman army, have largely been neglected in discussions of that institution, beyond their purely antiquarian interest. The fact is that the army needed weapons and had to procure them from somewhere. This meant that an industry (of whatever form) had to exist to supply the estimated 300,000 troops of the frontier forces. All this has, however, received scant interest from historians and archaeologists[5].

Clearly, no one book can hope to provide the definitive answer to the question "how did the Roman army equip itself?", and it was never intended that the seminar would produce a single comprehensive answer, but a number of plausible solutions are proposed here. The structure of this book comprises three elements: first, a largely theoretical section considers production and distribution in three general papers which seek to define models for the operation of the system; second, specific types of object are discussed in relation to their manufacture; third, Jon Coulston's important examination of archery equipment demonstrates the need to relate wider questions of organisation and matériel to the fundamental matter of production and distribution.

I should like to take this opportunity to thank all of the contributors, including those who spoke at the seminar (namely Peter Connolly, Nicholas Fuentes, and Peter Price) but who have not included papers within this volume for various reasons. Foremost amongst these exclusions must be Simon James, who gave a paper based upon his (forthcoming) article on the fabricae of the later Roman army[6] - essential reading for anyone interested in that period or in seeing the whole question of production in context. The Ermine Street Guard's participation and informed discussion was especially welcome. Prof. D.J. Mosely and Dr D.L. Kennedy of the Department of Ancient History and Classical

Archaeology at the University of Sheffield provided help and encouragement during the seminar itself and Dr Kennedy has been a useful source of advice during the preparation of this volume. Finally, I must thank my wife, Martha Andrews, who helped type the text.

The Roman military equipment seminars are an annual event that attract both scholars and amateurs. Further details of future meetings may be obtained from the editor.

NOTES

1. BISHOP, 1983.

2. MACMULLEN, 1960.

3. ROBINSON, 1975.

4. OLDENSTEIN, 1977.

5. But see now BREEZE, 1984, esp. 269-77.

6. JAMES, forthcoming.

BIBLIOGRAPHY

BISHOP 1983: M.C. Bishop (ed.), *Roman Military Equipment. Proceedings of a Seminar held in the Department of Ancient History and Classical Archaeology at the University of Sheffield, 21st March 1983*, (Sheffield 1983)

BREEZE 1984: D.J. Breeze, 'Demand and supply on the Northern Frontier' in R. Miket & C. Burgess (eds.), *Between and Beyond the Walls: Essays on the Prehistory and History of North Britain in Honour of George Jobey*, (Edinburgh 1984), 264-86

JAMES forthcoming: S.T. James, 'The *fabricae*, state arms factories of the later Roman Empire' in R. Jackson & M.J. Stone (eds.), *Roman Military Studies I*, (London forthcoming)

MACMULLEN 1960: R. MacMullen, 'Inscriptions on armor and the supply of arms in the Roman Empire', *American Journal of Archaeology*, 64, 1960, 23-40

OLDENSTEIN 1977: J. Oldenstein, 'Zur Ausrüstung römischer Auxiliareinheiten. Studien zu Beschlagen und Zierat an der Ausrüstung der römischen Auxiliareinheiten der obergermanisch-raetischen Limesgebietes aus dem zwieten und dritten Jahrhundert n. Chr.', *Bericht der römisch -germanischen Kommission*, 57, 1976 (1977), 49-366

ROBINSON 1975: H.R. Robinson, *The Armour of Imperial Rome*, (London 1975)

THE MILITARY FABRICA AND THE PRODUCTION OF ARMS IN THE EARLY PRINCIPATE

M.C. Bishop

> "The main source of supply of arms in the earlier Empire was small shops and dealers. Fine armor beyond the call of duty could be ordered by the military swell from local artists, or was hawked about in the camp."
>
> MACMULLEN, 1960, 25

> "It is not known for certain whether the fabri, or specialist craftsmen, of a legion actually made armour or only repaired it. No doubt many of them were capable of making it in times of emergency, or when a legion was stationed in newly conquered territory far away from production centres, but had manufacture been their full-time occupation there surely could not have been any need for private firms to manufacture arms of any kind.
>
> "I would suggest that maintenance of equipment and the production of javelin heads, arrows and the like were the main tasks assigned to the average military workshop."
>
> ROBINSON, 1975, 8

These two opinions might be held to represent a school of thought that has gained considerable, but by no means universal, credence amongst students of the Roman army. The nature of production of military equipment in the early imperial period is an extremely complex problem and it must be stressed that there is no one easy answer. Nevertheless, there is sufficient evidence to allow us not only to study certain areas in a reasonable amount of detail (and it is the purpose of this paper to examine the role of the fabricae in this process with particular reference to metalworking, although other areas of manufacture will necessarily be brought into the discussion), but also to consider some of the wider implications of this matter.

Fabricae themselves have received detailed treatment on a number of occasions[1] and there have been attempts to study the personnel who worked within them,[2] but in order to understand their role in the production of military equipment, it is necessary to examine critically the archaeological, literary, and sub-literary source material. Moreover, it will prove necessary to be particular about the contemporaneity of this material, so the evidence used, wherever possible, will be drawn from the 1st century A.D.

THE SOURCES

1) Literary

Our most important literary sources are the De Re Militari of Vegetius and a fragment of Taruttienus Paternus[3] preserved in Justinian's Digest. Obviously, neither of these are, on the face of it, first century in date, but they will still prove to be admissable once we study their background in some detail.

The main passage of relevance in Vegetius is II,11 where he is describing the duties of the praefectus fabrum (Appendix No.1). He relates how each legion had a number of workshops producing a variety of weapons (*'omniaque armorum genera formabantur'*) and, most importantly from our point of view, stresses the self-sufficiency of the legion (*'haec enim erat cura praecipua, ut quicquid exercitui necessarium uidebatur numquam deesset in castris'*) and that they met their needs by manufacture.

It is generally accepted that Vegetius compiled his work during the 4th century A.D.,[4] using authors from a variety of periods, ranging from the Republic up to his own day.[5] There has been considerable debate as to which source Vegetius used for this particular section, the chief candidates being Taruttienus Paternus, Iulius Frontinus, and Cornelius Celsus (who was, in turn, using Cato, it is presumed).[6] Now, whilst both Frontinus and Celsus were writing in the 1st century A.D.,[7] Cato was quite clearly a Republican source and so of doubtful relevance to the changed circumstances of the imperial army.[8] The office of praefectus fabrum certainly existed under the Republic, but it does not appear to have had anything to do with military workshops and was, in fact, largely an honorary title.[9]

What appears to have happened is that Vegetius has become confused over the posts of praefectus fabrum and praefectus castrorum, the latter first being attested at the end of the 1st century B.C.[10] Many of the duties which he says are the responsibility of the former (II,11) are also to be found associated with the latter (II,10).[11] Therefore, if these duties of the 'praefectus fabrorum' belong in the imperial army, the description is unlikely to come from Cato. The most likely origin would seem to be either Celsus himself, writing in the Julio-Claudian period, or Paternus.[12]

P. Taruttienus Paternus, writing towards the end of the 2nd

century A.D.,[13] included a list of _immunes_ who, by the nature of their work, were exempted from ordinary fatigues (Appendix No.2).[14] Whilst the post of _immunis_ is not known before the 2nd century, it is strongly suspected that a similar system was in operation well before that.[15] Paternus' list is more interesting than it may at first seem, for all of the posts included would have fallen within the area of responsibility of the _praefectus castrorum_; not surprisingly, men who could be considered to be _fabrica_ staff are included.[16]

2) Sub-Literary

Our two fundamental literary sources for the activities of the military _fabricae_ would both seem to belong within the period of the principate, but we can be more certain of the dating of our sub-literary evidence. Here we are dealing with three main texts, one an undated papyrus from Egypt, the other two being closely-dated writing tablets from one of the pre-Hadrianic forts at Vindolanda.[17]

The Berlin papyrus _inv._ 6765, which probably dates to the 2nd or 3rd centuries A.D., preserves a record of two days' activity within a legionary _fabrica_, probably that of _legio II Traiana Fortis_ (Appendix No.3).[18] The salient points include the facts that 100 men were employed in the workshop on one day, including _immunes_, _cohortales_, civilians, and a group - _galliari_ - who may be slaves.[19] Lists of items being produced are given and these include _spathae_, two sorts of shield, iron plates, bows, and catapult fittings.[20] An interesting distinction is also made between items made (_fabricatus_) and those completed (_peractus_).[21] The value of this text in demonstrating the volume of production possible and the numbers of personnel involved is immediately obvious.[22]

Two of the wooden writing tablets found at Vindolanda - _Tab. Vind._ 1 and 3 - provide us with evidence which demonstrates that a similar situation was pertaining in the early empire, the period that concerns us here. _Tab. Vind._ 1 (Appendix No.4) lists men being sent to work in the workshops on April 25th. The figure given is 343 men and this includes cobblers, and various assignments of a constructional nature, the precise details of which are lost to us.[23] _Tab. Vind._ 3 (Appendix No.5), on the other hand, is just a fragmentary list of names of men, identified by century, but it almost certainly includes _gladiarii_, _fabri_, and _scutarii_. The first tablet supports the idea that large contingents of troops could be assigned to the _fabricae_, whilst the second shows that weapons manufacture may have been one of the tasks in which they would be employed.

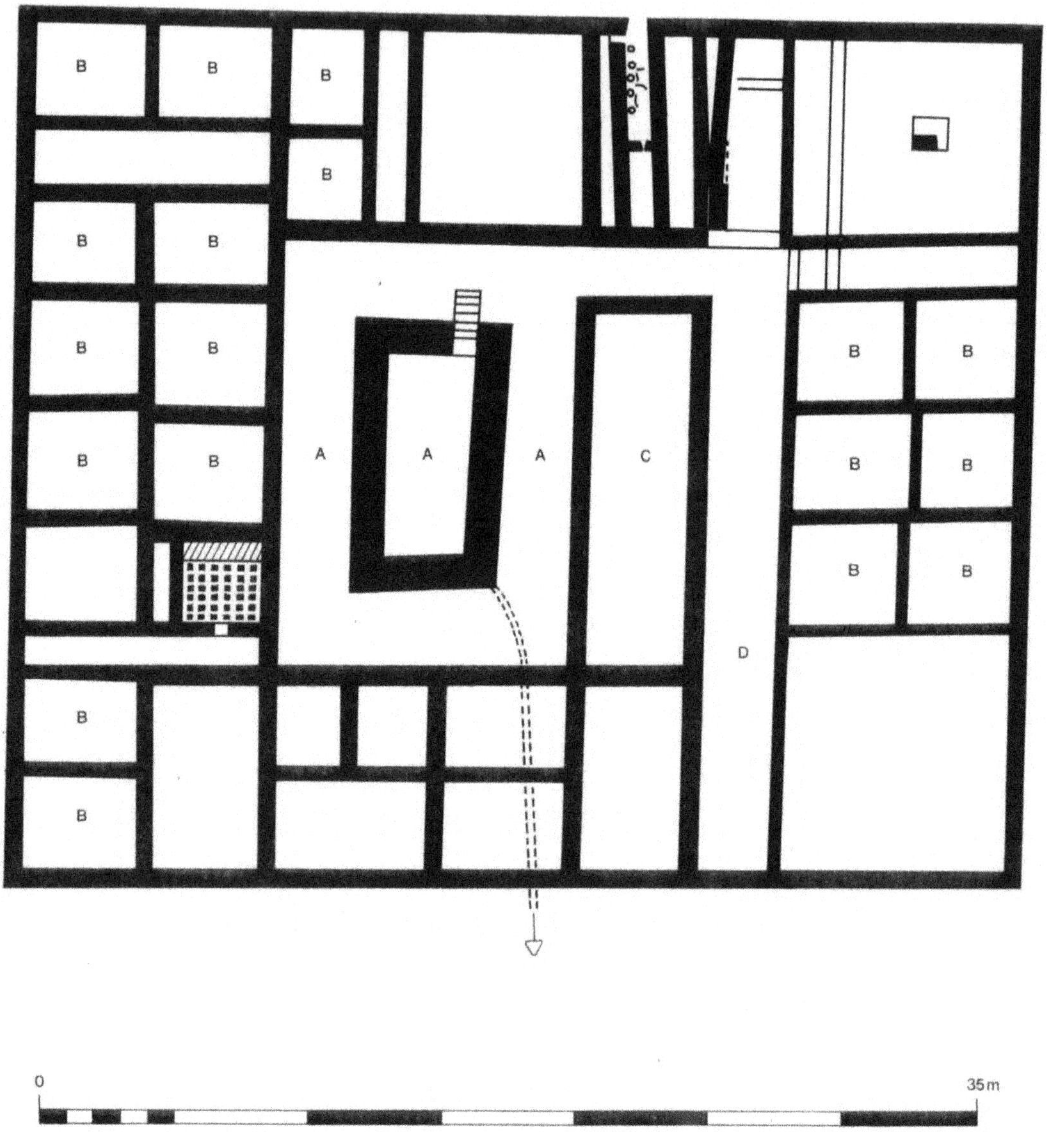

Fig.1: The stone _fabrica_ building at Wiesbaden (after ORL Nr.31, Taf.V,1). There is a courtyard with a large watertank (A) at the centre of the complex. Around the periphery are pairs of rooms (B) that may be workshops with their associated storage space. The large, centrally located room (C) may be an assembly hall. Direct access to the courtyard of a _fabrica_ is usually afforded by a corridor (D).

To these important documents we can add an address on a writing tablet found in the Schutthügel at Vindonissa and probably dating to the mid-first century (Appendix No.6), which mentions a scutarius called Valerius, of the eighth cohort.[24] Vindonissa was also the findplace of a bronze votive inscription (Appendix No.7) which, appropriately enough as it turns out, was set up to Mars by Tiberius Iulius Aquila, describing himself as a gladiarius.[25]

3) Archaeological

The literary and sub-literary sources would therefore appear to provide sufficient evidence for us to postulate direct participation by the army in the production of military equipment during the early principate, but we must now turn to the archaeological evidence to test this hypothesis.

Fabricae have been identified - with more or less certainty - at a number of sites of the 1st and early 2nd centuries A.D. Sites such as Haltern, Hofheim, Valkenburg, Oberstimm, Inchtuthil, Wiesbaden, and Vindolanda may be included amongst these, if only because they are the most familiar.[26] At these sites, the typical workshop building took the form of a courtyard structure, usually with a large central water tank fed by the establishment's water supply (Fig.1). It is interesting to compare the layout of these fabrica buildings with more recent parallels, such as the Abbeydale industrial hamlet in Sheffield (Fig.2).[27] Characteristically, the Roman buildings feature a comparatively large number of rooms, the purpose of which is not always clear.[28] By comparison with such sites, a number of other buildings have been suggested as fabricae, but the most interesting of these is building G at Vetera I. If this was indeed a workshop, then it is the largest building of this type known to us (it would have been designed to serve a double-legionary garrison).[29]

It is worth noting in passing that no building has yet been conclusively identified as a fabrica - industrial evidence, although present, is seldom overwhelming - and that many structures which clearly had another primary function, could produce the sort of evidence associated with industrial processes.[30] That being said, a number of buildings which we are, for argument's sake, accepting as fabricae have produced evidence suggestive of the production of military equipment.

Here we may discern a number of classes of evidence that may be of some use. First there is direct evidence of metalworking (but equally wood-, bone-, and leatherworking) in the form of waste products, raw materials, or material intimately bound-up

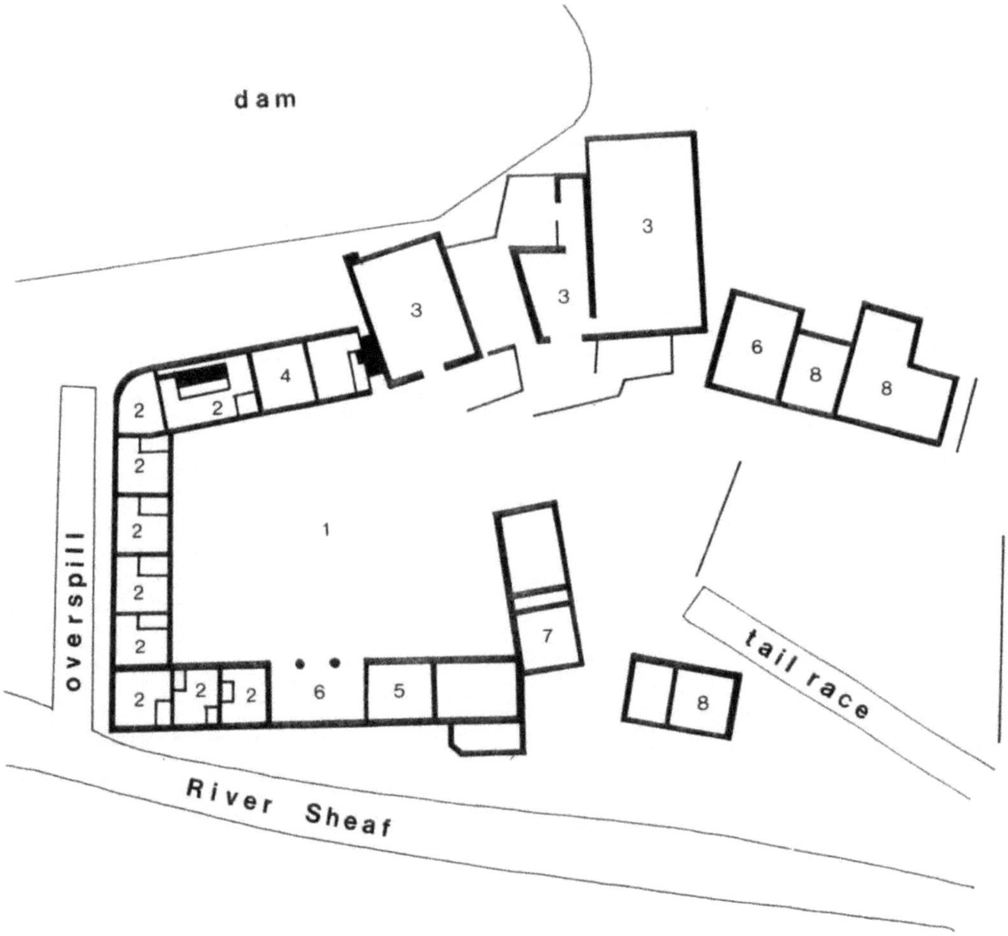

(not to scale)

Fig.2: The Abbeydale Industrial Hamlet (after PEATMAN, 1981, 2 Fig.). The plan shows Abbeydale as it is today. Its main components are the courtyard (1), the rooms (2) associated with primary production processes (crucible furnace, forges), those (3) devoted to secondary processes (finishing), final assembly (4), where the products were painted to prevent corrosion (5). In addition there was storage space (6), an adminstrative area (7), and accommodation for the workers and the manager (8). The main products were two types of scythe and all processes necessary for their manufacture (with the exception of rolling the steel, done at a nearby mill) were performed on-site.

with the production process (such as crucibles). The waste products of iron and copper alloy working have been recorded from Hofheim,[31] Oberstimm,[32] Wiesbaden,[33] Rheingönheim,[34] and Exeter[35] to name but a few. Second, there is the indirect evidence of tools, particularly those of the smith or armourer in the case of metalworking. It is said that a number of tools were found on the site of the fabrica at Haltern when it was excavated.[36] Finally, there is the equipment itself; but this class, for reasons which will soon become apparent, is not as straightforward as it might at first seem.

Dr Jürgen Oldenstein has demonstrated that military equipment was produced on the 2nd and 3rd century limes in Germany and has pointed to the important role of scrap metal as a resource in this scheme.[37] Scrap is, by definition, a waste product that is recycled and the process of recycling has long been an important part of the production of metal goods.[38] There is every reason to suspect that it was equally, if not more, important to the Roman army.

Sir Ian Richmond suggested that the hoard of iron nails and wheel tyres found in a pit in the fabrica at Inchtuthil was deliberately placed there by the Romans when they abandoned the site, in order that this precious resource would not fall into the hands of their enemies.[39] Twelve tons of nails were, for whatever reason, deposited in a pit, but they were not all used nails from the demolition of the site.[40] Excavation of other military sites has not yet revealed a comparable hoard of nails, but other artefacts have been recovered in considerable quantities.

James Curle's excavations at Newstead produced a rich haul of military equipment of all kinds, but chiefly metalwork. Deposited in pits and wells, this was long thought to be the result of some sort of disaster;[41] at the same time, a ritual motive was sometimes suggested as at least a partial cause.[42] The 'disaster' theory has now been discounted on good archaeological grounds,[43] and, seen in the context of the Inchtuthil find, an alternative explanation has been offered for the Newstead material: namely that it was not the debris from a battle or the spoils of a triumphant foe, but rather (part, at least, of) the stock of scrap metal belonging to the fort's workshop.[44]

In this instance, military equipment found in the archaeological record was scrap. It is understandable, if we imagine the considerable amount of organisation involved in the abandonment and systematic demolition of a Roman military site, that not all waste material could be moved with the army.[45] Thus the logical move for a praefectus castrorum would be to remove all trace of its existence by burying it.[46] But it is at this point that we must ask how military equipment came to be in the archaeological record under 'normal' circumstances: the reason that has come to be tacitly accepted is what might be termed 'accidental loss',[47] but whilst this is acceptable for very small items, it becomes increasingly unlikely the larger the item involved.[48] Indeed examining military equipment more closely, a pattern is clearly discernible: even single finds are usually damaged in some way.[49] Examination of the finds from any military site will inevitably reveal a long list of defects. Confirmatory evidence comes in the form of the bent scraps of metal, particularly copper alloy, that are so common on these sites.[50]

The evidence would seem to suggest, then, that most of the military equipment that is excavated on Roman military sites was discarded as scrap by the time it entered the archaeological record. If this supposition is linked together with the idea that a fabrica would keep a hoard of stock, taking with it as much as was practicable when it moved on, then we are beginning to uncover the basic framework of the system of military equipment production. Hoards of military metalwork, similar to those from Inchtuthil and Newstead, are known from other sites. A large quantity of material, both iron and copper alloy, was found at Augsburg-Oberhausen.[51] Although mooted as the site of a major military base, this hoard, or something similar, might be seen as the sort of 'core hoard' a military unit would take with it when it moved.[52]

Yet more information about military equipment can be derived from its location within a Roman fort or fortress.[53] The archaeological context of military equipment is usually pits and ditches, such finds far outweighing the 'casual' find of an item in a floor surface or such like.[54] The deposition of military equipment in the archaeological record was inextricably linked with the abandonment and demolition of a site.[55] The process of demolition can be reconstructed with some accuracy:[56] there would be an intensive search for anything re-usable that was transportable, but once the fabrica carts (or even boats) were full, everything else had to be discarded where it would no longer be accessible - throwing odd pieces of military equipment into pits and ditches along with the other rubbish would ensure their complete burial once the demolition process was completed.[57] Some pieces, such as small numbers of 'lorica segmentata' fittings, may simply not have seemed worth saving.[58]

Virtually every military site of the 1st century A.D. has produced military equipment and this might be taken to imply widespread manufacture.[59] It also allows the thought that whenever a military site was abandoned, military equipment would be deposited, an idea that is exemplified in the clearest possible fashion at Vindonissa.[60]

The analysis of copper alloy artefacts has shown that several different forms of brass (or gilding metal) were in use in the army,[61] but that these classes were extremely homogeneous within their groupings, something that, given the level of Roman military technology, was only possible with the recycling or large amounts of scrap metal and its segregation into the particular categories of metal, where appropriate.[62]

THE FABRICA SYSTEM

Theory

Once it has been established that non-archaeological and archaeological sources both imply that production of equipment was undertaken by the army on a fairly large scale (indeed, to the point of self-sufficiency, if we are to believe Vegetius), then it is possible to begin to attempt a reconstruction of the system, based upon the fundamentals discussed above, and some models are presented here to demonstrate how the system could have operated.

It is well-known that military equipment could pass into the hands of a number of owners, a fact most eloquently attested by items with several owner-inscriptions.[63] Moreover, the means by which this took place have already been studied and it is now generally accepted that, with a few exceptions,[64] equipment passed back to the army upon the retirement or death of the former owner in exchange for a (nominal?) sum of money.[65] This process of recyling items, along with the inevitable repair of damaged equipment, would have meant that the army's actual production needs at any given time would have been comparatively low and the life of an object could reasonably be expected to be above the notional 20 man-years period (but only, note, in times of peace; it is impossible to estimate loss-rates under campaign conditions).[66] Keeping a legion (and, presumably, its associated auxiliaries[67]) equipped was thus not as difficult as might first appear. The use of scrap metal would further ensure that the demand for raw materials was very low, so a campaigning army did not need rapidly to locate sources or constantly replenish its stocks from the rear.[68] The picture presented by P. Berlin inv.6765 and the levels of production that were clearly attainable (although it must be remembered that we do not know the circumstances in which the document was written) are therefore not implausible.

That document, along with the two writing tablets from

Vindolanda, shows us how the <u>fabrica</u> system must have worked. Accurate modern reconstructions of Roman military equipment have usually been undertaken by an individual craftsman, responsible for all phases of the work, from the simplest to the most skilled.[69] This is not, of course, the most efficient way of doing things: it is far better to apportion menial tasks to unskilled labourers, who are in turn supervised by skilled craftsmen who can finish a job where necessary.[70] If this sort of system operated in the Roman army, then the legionaries could clearly provide the labour-force, and the <u>immunes</u> (as they were later to be called) the skill and supervision.[71] In its way, this goes some way to explaining the status and the function of the <u>immunis</u> in the army.[72] The principle of close supervision of technically difficult tasks was well-established, particularly in the construction of fortifications.[73] The use of ordinary soldiers in the workshops might also help to explain the widely varying quality of military equipment - the '<u>lorica segmentata</u>' fitting produced by the hasty or inexperienced man cannot help but be inferior to one produced by a diligent or experienced worker.[74]

We can, perhaps, see this system in operation to its best advantage if we submit an item of military equipment to an analysis of the personnel needed to produce its component parts. First, a simple object, like a spear:

1) A woodman, to supervise gathering of wood of the required shape and quality; unskilled porters
2) A smith (semi-skilled?[75]) to forge the blade, rivets, and butt; unskilled men as porters (raw material, fuel, water, finished object), bellows-operator.
3) A carpenter to shape and finish the wood; unskilled porters
4) An expert to perform or supervise assembly and to check standard of work; unskilled porters to store finished items

All of these four principal jobs <u>could</u> be performed by one experienced spear-maker, but it would be more efficient to divide the work in this way.

In many ways, a simple object is not a true test of the system, so an examination of a complex object, such as a '<u>lorica segmentata</u>' is called for:

1) One or more smiths to prepare the iron plates from ingots; unskilled men as porters (raw material, fuel, water, finished object), bellows-operator.
2) One or more copper smiths to prepare copper-alloy sheet and rivets from ingots; unskilled men as porters (raw material, fuel, water, finished object), bellows-operator.
3) Leatherworker(s) to produce (tan, cut, stitch) strapping for the armour; unskilled (or semi-skilled) workers for tanning, cutting, stitching, and as porters
4) Unskilled men to cut out sheet copper-alloy components,

assemble and rivet them together where necessary; porters to transfer components for final assembly

5) An expert to perform or supervise assembly and to check standard of work; unskilled porters to store finished items

There is no reason why the same porters could not be used at every stage of production, providing there were enough of them.

Hierarchy

Vegetius tells us, in a garbled fashion, that the officer with ultimate responsibility for the *fabrica* was the *praefectus castrorum*.[76] Each such *praefectus* would appear to have had his own *officium*,[77] who no doubt dealt with the administrative side of the workshops, doubtless producing documents like *P. Berlin* inv.6765 or *Tab. Vind.* 1 and 3. The *praefectus castrorum* seems to have taken on these duties from the *quaestor* in a Republican army.[78]

Beneath the *praefectus* were the staff of the workshops, the most senior of whom may well have been the *optio fabricae*, a post which is only definitely recorded in the Digest passage.[79] His precise status is ambiguous,[80] but it seems likely that he was the *de facto* head of the *fabrica*, whether an administrator or a professional.[81]

The main body of the staff was provided by the men we have already seen referred to as *immunes*. There has been a tendency to view these specialists as a separate entity within their unit,[82] perhaps even living near the workshops[83] or in the accommodation of the first cohort of a legion,[84] but the available evidence seems to suggest that they were held on century strength[85] - a fact apparently confirmed by *Tab. Vind.* 3. The range of skills embodied amongst these men is hinted at, but surely not exhaustively listed, in the surviving fragment of Taruttienus Paternus.[86] Taking the singular and plural nouns in the list seriously, we might expect at least 56 staff attached to the *praefectus castrorum* (and, possibly, to the *fabrica*) and more likely over 60.

The bulk of the workforce was provided from the ranks, with men being assigned, probably on a daily basis,[87] according to the needs of the workshop at any given time. Over 340 men were working at Vindolanda on one particular day,[88] whilst the *fabrica* described in *P. Berlin* inv. 6765 could easily have had over 200 men working in it, which explains the productivity attested by that document.

Versatility was clearly the key to the successful operation of the *fabrica* and it would have been a system that was well adapted to the unpredictable demands that alternating periods of campaigning and rest would have placed upon it.

Modus operandi

Many military sites of the early principate were, on the basis of the archaeological evidence, engaged in the production of military equipment.[89] Scrap material was collected, stored, and recycled; an important aspect of this process is obviously the collection - how was it organised? If we discard the accidental loss hypothesis as largely irrelevant,[90] then we may suppose that soldiers were fully aware of damage to their equipment and would seek to remedy it. A broken '*lorica segmentata*' fitting would not be casually left where it fell,[91] but would be retrieved in case it could be repaired - we must bear in mind that the cost of replacement would have to be born by the soldier,[92] so there was a strong incentive to retain even the most insignificant scrap of material. Given a certain length of time spent around barracks, such odds and ends were bound to accumulate to the point where it would be worth the soldier's while to trade them in to help defray the cost of his deductions for new kit.[93]

This hypothesis may well be behind the pattern of distribution of finds within military sites,[94] since when it was time for a fort to be abandoned, the *fabrica* would surely take its central stock of scrap with it, rather than gather odd pieces remaining in the barracks and awaiting delivery to it. These would then be tidied away by the demolition parties as close to the point where they were found as was possible - hence the distribution pattern.[95] However, the organization of the scrap cycle may well have been more organised than we have so far suggested: building VI at Carnuntum was found to have various items of equipment within its rooms and these were separated according to their type.[96] Segregation of scrap is, of course, of fundamental importance to its re-use for the production of military equipment.[97] We may well have, in such a building, an example of the enigmatic *armamentarium*;[98] if this were indeed so, we might go even further and suggest a link with the *custos armorum*, an *immunis* found on the strength of every *centuria* or *turma*.[99] The *custos armorum* would then become a convenient agent for the collection, storage, and general administration of the flow of scrap metal within his company,[100] but this must, for the time being, remain pure speculation.[101]

The main workshops were most likely to be found in those sites termed *hibernae* or winter quarters.[102] The ideal conditions would then prevail - a semi-permanent site with the pick of local resources, available manpower, the ability to work on long-term projects[103] - whereas on campaign, when the army was presumably normally working from what we call 'temporary' or 'marching' camps,[104] a permanent workshop would be far too cumbersome, and the troops would be needed for combat, rather than making equipment.[105] Here we must envisage the use of a field-forge, largely concerned with the repair of damaged equipment and manufacture of simple weapons.[106] The archaeological remains of such field-forges would almost certainly be negligible.[107] Such a system is briefly mentioned by Josephus[108] and is supported, albeit in a very negative sense, by the rarity of military equipment from temporary military sites.

Analysis of the fabrica-system of the 1st and early 2nd centuries A.D. would therefore suggest that it operated much the same as in later periods.[109] The main production was by the army, although there may have been room for specialist craftsmen working outside the army (possibly ex-fabrica workers), but they would necessarily be geared to a much lower level of production, since such a person had to perform all stages of the job himself,[110] so that private production must have been responsible for a very small proportion of the equipment made at any given time.[111] Moreover, that sort of production was better geared to catering for the luxury end of the market and there is a little evidence that suggests this is indeed what it did - producing items which only the senatorial or equestrian officers (or perhaps some of the top centurions) of a legion could have afforded.[112] In short, there was no large-scale private industry in the west.

The model suggested by the various strands of evidence (Fig.3) shows just how little raw material the army needed at any given time. When it was necessary to issue new equipment, then it could rely partly on recycled intact items, and partly on new artefacts manufactured from scrap. Since manufacture is kept on the legionary level, supply and demand could be monitored closely by the fabrica and the production programme regulated accordingly. The two great virtues of the system are, therefore, efficiency and flexibility.

EFFECTS OF FABRICA PRODUCTION

The craftsmen working in the fabricae of the frontier legions in the west were almost certainly free of any central control.[113] This meant they could pursue their own particular tastes, and those of the market they served, and this in turn meant that there was great scope for variety and individuality in Roman military equipment of the early imperial period.

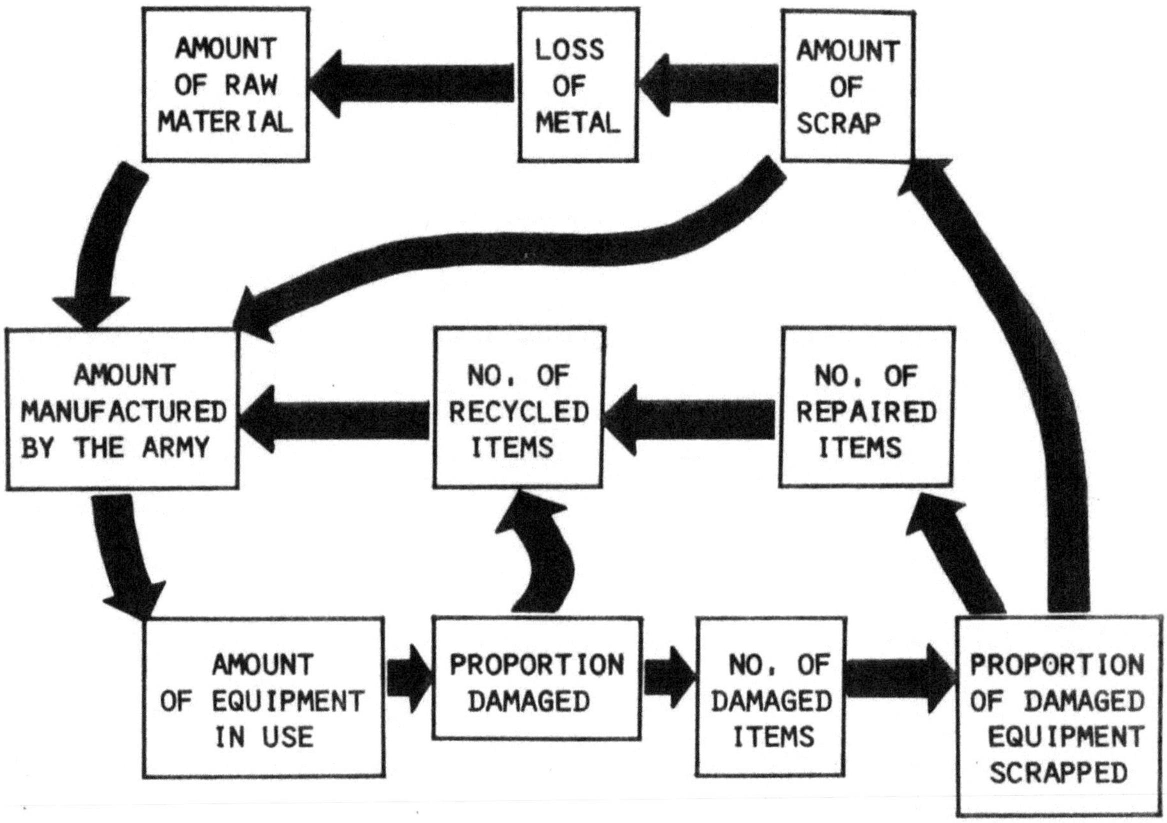

Fig.3: Model for the production of military equipment in military _fabricae_.

Craftsmen working within a legionary <u>fabrica</u> would have a set repertoire of designs for items like belt-plates or pendants. They would have learnt elements of this when they were first introduced to the job, but would have added to it from their own experience, both by what they saw during their career, and what the soldiers wanted: this made for a very eclectic system.[114] Thus, in conditions of isolation, a legion's equipment would become very noticeably distinct from that of other legions - however, the fact that army groups campaigned together most summers[115] meant that craftsmen would have the opportunity to meet and exchange ideas, so the common pool of ideas would belong to the army-group, rather than the legion (although peculiarities may still have been apparent).[116]

The movement of units between army-groups meant that the influences on any given legion's craftsmen were rendered even more diverse and could in turn be passed on further through the Roman army as a whole, as and when the opportunity presented itself.[117] The craftsman thus had a rich profusion of sources of inspiration upon which he could draw.

Examples of these craft traditions are obviously going to be very difficult to isolate in the archaeological record, given this diversity of influences, but it may nevertheless be possible to indicate some likely candidates. Prime amongst these is what might be termed the Upper German tradition, a style of equipment decoration and design manifested in the region of that province in the Tiberio-Claudian period and then dispersed in the decades that followed.[118]

Some examples that may be cited include a certain type of belt-plate, made of beaten copper alloy and depicting a range of mythological scenes and devices.[119] The most common motifs include the wolf and twins, lotus blossoms, and a human bust with <u>cornucopiae</u>.[120] The distribution of these objects, many of them reasonably securely dated to the Tiberio-Claudian period, show a bias towards Upper Germany, with a few from areas where legions from that army group were subsequently transferred (such as Britain and Pannonia).[121] Likewise, belt-plates inlaid with niello show a similar, but more widespread distribution - the finds of these items in Britain are particularly interesting and again suggest an Upper German connection.[122]

One particular style of scabbard fitting, associated with the <u>legio</u> XIII <u>Gemina</u> by its presence in the Schutthügel at Vindonissa,[123] is also found at Carnuntum, Magdalensberg, and Mainz.[124] These cases are exceptional in the comparative ease with which army-groups, or even legions, can be traced, but it becomes much more difficult once units are shifted round to accommodate the invasion of Britain.[125] Usually, only broad trends can be suggested, such as the use of lobate decoration on Danubian '_lorica segmentata_' fittings,[126] the prevalence of lunate pendants in Lower Germany,[127] or the concentration of Imperial-Gallic helmets in the Rhineland.[128]

THE CITY-STATE

The _fabrica_-system is here proposed as the means of producing military equipment in certain parts of the Roman world, but it was by no means the only way. The archaeological evidence, which would seem to be a reliable indicator of military production in the west, is largely absent in the eastern half of the empire, a fact usually ascribed to the state of archaeological knowledge in the respective areas.[129] A few items are known, but the large amounts of damaged pieces are lacking, despite the fact that the Middle East, at least, was the scene of some of the most intense campaigning by the Roman army in the early principate.[130]

The solution to this state of affairs lies, it might be suggested, in two well-known and often-cited references to the production of equipment in the East: both Tacitus and Cassius Dio mention equipment being produced for the army in eastern cities, both implying an element of quality control on the part of the Romans.[131] The tradition of cities (or, more properly, city-states) producing equipment to demand had a respectable pedigree: in 399 B.C., Dionysius I converted Syracuse into one large workshop in order to equip his army.[132] Scipio turned New Carthage into a vast munitions factory in 210 B.C.,[133] and during the third Punic war, the city of Carthage itself went into mass-production of equipment after having surrendered all its arms to the Romans and then changing its mind.[134] Many more examples can be found from late Hellenistic and Republican times to confirm this picture.[135]

The characteristics of such enterprises were their makeshift nature, the speed with which they were set up, and the volume of production of which they were capable.[136] This system was obviously ideally suited to equipping large armies quickly.[137] The most important common factor was, however, the classical city - wherever such settlements were found, it seems to have been possible to use this peculiar facet of their nature.[138] This gives us our first hint that the difference lay not between east and west, but between city-state and military _fabrica_. So long as the Roman army was operating within the sphere of influence of the classical city, then it was possible to organise its supplies as and when they were needed. Expansion outside of this essentially Mediterranean zone (most notably into 'Celtic' areas), however, meant the army was forced to choose between establishing long lines of communication with its centres of production, or meeting its own needs: it chose the latter course.[139]

This use of the _polis_ to provide large amounts of equipment quickly may be one reason why most new legions were recruited within the Mediterranean zone, rather than in 'frontier' areas.[140] We may even see a manifestation of this system in the Imperial-Italic class of helmets,[141] generally thought to be of a lower standard of manufacture than those of Imperial-Gallic manufacture.

CONCLUSIONS

Vegetius certainly seems to have been right when he described the legion as self-sufficient in its needs. There is, however, a more important side to this than purely the logistical aspect: army-groups in the north and west of the empire, restricted by the weather outside the campaigning season, needed to maintain their morale.[142] The everyday routine of sentry-go or patrols could not occupy any more than a small proportion of a force,[143] so employing troops in the workshops or on production-related tasks was a valuable source of variety in what must have been an extremely monotonous existence at times. Our evidence seems to suggest that it was the legionaries who undertook various manual tasks, in the early empire at least, and that the bulk of patrolling and police-work would have been left to the auxiliary forces.[144] Manufacture could therefore be left to the legions and their troops (but we cannot totally exclude auxiliary involvement).[145] It is clear, however, that the production of equipment was not so much a necessary chore as an important factor in the continued efficiency of the army.

The strength of the mechanism of arms-supply in the early empire was undoubtedly its flexibility. Adopting the Greek _polis-system_ of production and then supplementing it with the _fabrica-system_ meant that the Romans were capable of adapting to the prevailing local conditions. Changes in the nature and structure of the empire under Diocletian meant that it was necessary to evolve the chain of arms factories attested in the _Notitia Dignitatum_ and elsewhere, but there can be no doubting that they looked back, at least in part, to military _fabricae_ of the earlier period, rather than any wholly contractual civilian scheme.[146]

Finally, this study would also seem to suggest that we can apply a number of rules relating to the presence of Roman military equipment in the archaeological record.

1. Military equipment discovered on land is most likely to have

been deliberately discarded by the Roman army and - by implication - that a military presence of more than temporary duration can be deduced.[147]

2. Such discarded material is inherently characteristic of a particular unit or army group.
3. This material is normally only discarded when a site is abandoned and is thus linked to the demolition process.
4. Areas where the military situation was complex and sites were frequently abandoned will tend to produce more finds of military material than those with a stable situation and long periods of occupation.
5. The relative distributions of military equipment in the west and east of the empire is not simply a result of differing levels of archaeological investigation.

ACKNOWLEDGEMENTS

I should like to thank my wife Martha Andrews, Mr P.W.M. Freeman, and Dr D.L. Kennedy for reading this paper in draft form and discussing it with me. Whilst I have benefitted greatly from their advice, they are in no way responsible for any mistakes contained within it.

NOTES

1. VON PETRIKOVITS, 1970; 1974a; 1974b; 1975, 88-98; cf. JOHNSON, 1983, 183-8.

2. SANDER, 1962. Cf. VON PETRIKOVITS, 1970, 246-7.

3. Taruttienus is to be preferred over Tarruntenus - cf. AE 1971, 534; I am grateful to Dr D.L. Kennedy for this reference.

4. For a discussion of Vegetius' work in its contemporary setting, see GOFFART, 1977 and BARNES, 1979.

5. On the sources for Vegetius work, see SCHENK, 1930; PARKER, 1932; SANDER, 1929; 1932; 1939.

6. These sources are named by Vegetius I,8. Paternus: SCHENK, 1930, 8-26; Frontinus: ibid, 39-83; Celsus: ibid., 26-39; Cato: SANDER, 1932, 374.

7. Celsus: SCHENK, 1930, 28 with further refs.; Frontinus: ibid., 39-41.

8. A fragment of Cato's military writings is preserved in Fronto, Ad Verum Imp., II,1,20.

9. DOBSON, 1965, 62-3.

10. KEPPIE, 1984, 176 with n.5.

11. II,9-11 seems to be the result of confusion over the posts of praefectus castrorum, praefectus fabrum, and praefectus legionis. All of the duties listed in this section fell within the sphere of the praefectus castrorum.

12. SCHENK, 1930, 29 discusses originality in Celsus' works.

13. Paternus' contribution to military writing is considered in SCHENK, 1930, 23-4; cf. RE 'Tarruntenus Paternus'.

14. Digest L,6,7.

15. WATSON, 1969, 75-7.

16. These are: artifices qui fossam faciunt, architectus, naupegi, specularii, fabri, sagittarii, aerarii, bucularum structores, carpentarii, scandularii, gladiatores, aquilices, tubarii, cornuarii, arcuarii, plumbarii,

ferrarii, lapidarii, 'qui calcem cocunt', 'qui siluam infindunt', 'qui carbonem caedunt', optio fabricae. This assumes that tubarii and cornuarii are horn-makers, rather than musicians.
Vegetius summarises the responsibilities of the praefectus castrorum in II,10.

17. Naturally, these documents also benefit from being original and not transmitted.

18. See the brief discussion in BRUCKNER & MARICHAL, 1979, 6-7. The two days are the 18th and 19th of April.

19. But note that these were almost certainly personal slaves, calones, and not any sort of organised slave labour (loc. cit.). Cf. MACMULLEN, 1984, 44 & nn.27-8.

20. Spathae: Col.I,11; scuta planata: II,13; scuta talaria: II,9; lamnae levisatae: I,10; II,10; arcus: II,14; capitula ballistaria: II,15.

21. Fabricatus: Col.I,11; I,12; peractus: I,14; II,14.

22. It is not clear whether the figure of 100 refers to all workers, or just to legionaries - i.e. milites legion]ari n(umero) C (Col.I,4-5) - although we may suspect the latter to be the case.

23. It is noteworthy that the list is not dealing exclusively with staff carrying out building work, but includes sedentary workers (i.e. the cobblers), so is likely to be reasonably representative of the sort of list normally drawn up.

24. The form IIX, instead of the more usual VIII, may be early, although SANDYS, 1927, 55-6 disagrees.

25. Almost certainly not a civilian as MACMULLEN, 1960, 25-6 implies; his name suggests (but does not demand) a Julio-Claudian date.

26. Haltern: VON SCHNURBEIN, 1974, 65 - cf. id., 1982, 59; Hofheim: RITTERLING, 1904, 8-14 and 1913, 59-65; Valkenburg: SCHÖNBERGER, 1979; Oberstimm: SCHÖNBERGER, 1978, 30ff; Inchtuthil: TAYLOR & WILSON, 1961, 160; Wiesbaden: ORL Nr.31, 32-6; Vindolanda: BIRLEY, 1977, 111-5.

27. The groundplan of the fabrica was certainly ergonomically efficient, since it provided each part with easy access to any other and to the main water tank. At the same time, the

open yard could be used for temporary storage or open-air processes.
For a convenient discussion of the Abbeydale industrial hamlet, see PEATMAN, 1981.

28. Many of these smaller rooms have been identified as accommodation for men associated with the respective buildings (VON PETRIKOVITS, 1975, 49; cf. below, n.29).

29. LEHNER, 1928, 20-1 with Abb.2.

30. Such as the tribunes' houses at Colchester - FRERE, 1983, 309 with Fig.15.

31. RITTERLING, 1904, 9.

32. SCHÖNBERGER, 1978, 304.

33. ORL Nr.31, 34.

34. ULBERT, 1969, 113.

35. BIDWELL, 1980, 31-5.

36. VON SCHNURBEIN, 1974, 65.

37. OLDENSTEIN, 1977a; 1977b.

38. Hoards of scrap from the prehistoric period are well-known - cf. MEGAW & SIMPSON, 1979, 297-8.

39. TAYLOR & WILSON, 1961, 160.

40. ANGUS _et al_., 1962.

41. CURLE, 1911, 113-5; FRERE, 1978, 143-4.

42. ROSS & FEACHEM, 1976.

43. MANNING, 1972, 243-6.

44. _Ibid_., 246.

45. The maximum capacity of a wagon is open to debate; Diocletian's edict on prices includes a 1200lb wagon load (_De pretiis_. XVII), whilst the Theodosian Code sets a limit of 1075lb (WHITE, 1984, 129). Even in the case of the former, at least 20 wagons would have been needed to haul the nails deposited at Inchtuthil.

46. Since he was in charge of logistical matters, such a decision must have been his responsibility.

47. Cf. The report by Dr G. Webster on the Chichester gladius in DOWN, 1981, 173; LEAHY, 1980, 84 draws a similar conclusion about a pendant from Normanby.

48. The accidental loss of very small fittings may seem acceptable, but larger objects like swords are not just 'lost' in this way and certainly not in the quantities that the archaeological record would seem to suggest.

49. The military equipment from any early imperial site will show convincing evidence of this phenomenon, but a few examples will suffice: FRERE & ST.JOSEPH, 1974, Fig.30,58 & 59 (damaged harness junction-loops); 60 (fragment of a box latch); CRUMMY, 1983, Fig.151,4211 (bent belt-plate); NIBLETT, 1985, Fig.64,37 (broken helmet browguard - I follow FITZPATRICK, forthcoming in identifying Sheepen with a large pre-Flavian military base).

50. As at Rheingönheim (ULBERT, 1969, Taf.59), but most sites where military equipment is found produce such scrap material - cf. Oberstimm (SCHÖNBERGER, 1978, Tafn.38-41).

51. HÜBENER, 1973.

52. Similar such core hoards may include the Künzing (HERRMANN, 1969), Straubing (KEIM & KLUMBACH, 1951), and Corbridge (DANIELS, 1968) examples.

53. On this subject, see BISHOP, forthcoming.

54. Longthorpe: FRERE & ST.JOSEPH, 1974, 46-78 & Fig.22 (pits & ditches); Newstead: CURLE, 1911, 116-39 & 419 (for full references to ditch-finds; pits & ditches); Sheepen: NIBLETT, 1985, 112 (pits); Nijmegen: BOGAERS & YPEY, 1963 (pit)

55. Hence its frequent association with destruction debris - one of the original reasons for the 'disaster hypothesis' at Newstead: CURLE, 1911, 113-4 reasoned that intact quernstones were unlikely to be discarded normally, but their weight may have precluded their being taken with the departing army. In connection with the Antonine abandonment, he notes the association of military equipment and building debris in Pit I (ibid., 115).

56. The order in which things were done may have varied to some extent, but the basic principle of the recovery of re-usable

items, discarding of unwanted material, and demolition of structures is common sense.

57. It was, of course, standard practice for the abandoning force to cast the ramparts into the ditches - cf. Longthorpe (FRERE & ST.JOSEPH, 1974, 11), Strageath (WILSON, 1974, Fig.3), or Colchester (CRUMMY, 1984, 93 & Sheet 6b, Sx65).

58. Such items were probably kept in ones or twos in barracks by the troops against the time when they would be collected.

59. There are notable exceptions to this: temporary camps seldom produce anything of this nature, whilst the fort of Pen Llystyn was comparatively poor in finds of this nature - HOGG, 1969, 181-5.

60. Where the 'Schutthügel' was added to each time a legion left (cf. RE 'Vindonissa' 103-4). It is unlikely that it was used for dumping military equipment when the site was in occupation.

61. This is demonstrated by the work of a number of researchers over the past few years - Gowland in FOX & ST.JOHN HOPE, 1901, 245-6; CRADDOCK *et al.*, 1973, 15; reports by Justine Bayley in HURST, 1985, 30 and NIBLETT, 1985, 115; BISHOP, unpublished.

62. On scrap segregation: BAYLEY & BUTCHER, 1981. For a practical example of this, see RLÖ II, 41-2.

63. MACMULLEN, 1960, 33-6.

64. See BREEZE *et al.*, 1976, 77-81.

65. *Ibid.*, 93-5; GILLIAM, 1967, 237-8.

66. There are too many unknown factors, such as the complete loss of equipment (implied in Tacitus, *Ann.*, II,5) or the capture of enemy material (thus supplementing the scrap stock).

67. Assuming that legionary workshops supplied a legion's associated auxiliaries; on the attachment of auxiliary units to a legion, see SADDINGTON, 1982, 183-4.

68. In his calculations of the Roman demand for iron, WHITE, 1984, App.4 completely disregards the role played by the recycling of scrap, as well as the value of coppiced wood as a fuel source (RACKHAM, 1976, 23 & 51).

69. See, for example, GARLICK, 1980.

70. As suggested by ANSTEE, 1953, 202.

71. Immunes are mentioned in P. Berlin inv.6765 (Appendix No.3) and men who fall within this class are mentioned in Tab. Vind. 3 (Appendix No.5).

72. Paternus says these men achieved their status 'by the conditions of their service' (Appendix No.2) - in other words responsibility by virtue of their expertise. Cf. WATSON, 1969, 76.

73. The construction of camps was supervised and checked by the centurions (Vegetius, III,8; Polybius, VI,34).

74. ROBINSON, 1975, 181-2 with Fig.182 thought that the variety in lobate lorica hinges was due to evolution, but it is easily explained by varying degrees of competence in execution.

75. Whilst skill at blacksmithing takes a long time to achieve, functional objects could be produced with only a minimum amount of practice (particularly under supervision); the basic beating out of ingots could have been undertaken by ordinary legionaries (who might, if they showed a talent, have been upgraded to immunes).

76. On the praefectus castrorum: DOBSON, 1978, 68-74; WEBSTER, 1979, 117; KEPPIE, 1984, 176-7; MAXFIELD, 1981, 204-5.

77. WATSON, 1969, 85; cf. VON DOMASZEWSKI, 1908, 48.

78. For the duties of the quaestor in the Republican army see Polybius VI,39; cf. HARMAND, 1967, 172-9 and 366-8.

79. There are a few suggested references - MACMULLEN, 1960, 28 n.49; cf. BREEZE, 1976, 128 who only cites the Digest.

80. Loc. cit., using the optio valetudinarium to say that "the duties of the post were not medical but administrative".

81. Ibid., 132.

82. VON PETRIKOVITS, 1975, 122-3; SANDER, 1962, 147.

83. VON PETRIKOVITS, 1975, 49.

84. BREEZE, 1969, discounts this idea, pointing out that the laterculi from Lambaesis illustrate the fact that immunes

could come from any of the cohorts of a legion.

85. We find an armorum custos, carrarius, librarius, and a cerarius in P. Gen. Lat. I iv b verso being recorded as exempt from duty in the century ('opera vacantes'); cf. WATSON, 1969, 73-4. P. Gen. Lat. I verso v and verso iv I indicate that only 31 (with an additional 5) remain available for fatigues from a theoretical century strength of 80 - FINK, 1971, 210.

86. The passage makes no claim to be an exhaustive list, only listing a sample of such posts (see Appendix No.2).

87. This is clearly indicated by Tab. Vind. 3 and P. Berlin inv. 6765, which record specific dates - two consecutive days in the last case.

88. The fact that this, may have been an unusually large work detail is irrelevant, but it clearly demonstrates the principle of using the troop body as labour for the workshop.

89. As we have seen, finds of military equipment from nearly all first century A.D. sites would seem to support this assertion.

90. That is to say that items were accidentally lost, but such material forms an insignificant proportion of military equipment recovered from the archaeological record.

91. 'Lorica segmentata' fittings that failed would in fact be more likely to break and remain in situ, rather than detach themselves completely from the cuirass.

92. Since soldiers were expected to purchase their equipment, they must also have been responsible for lost or damaged material. It is possible, but not likely, that a regular deduction, such as 'caligas fascias' in P. Gen. Lat. I recto i, could cover such an eventuality.

93. This is suggested by the concentration of equipment in the barrack area at Oberstimm (BISHOP, forthcoming); cf. KLUMBACH & BAATZ, 1970. For equipment in pits in the barracks at Longthorpe, see FRERE & ST.JOSEPH, 1974, 32.

94. It is a pattern also observed by the excavators of Dangstetten - FINGERLIN, 1981, 422.

95. Ditch terminals and pits being the favoured repositories for such material - BISHOP, forthcoming.

96. RLÖ II, 41-2.

97. On segregation of scrap, see BAYLEY & BUTCHER, 1981.

98. The assumption that because weapons were found in the principia of a fort then this was the location of the armamentarium does not necessarily follow (cf. JOHNSON, 1983, 108-9).

99. On the custos armorum, see VON DOMASZWESKI, 1908, 44 (legionary) and 55 (ala).

100. The post seems to have been on a par with the tesserarius (BREEZE, 1974, 267), in being the next stage up from the librarius (ibid., 268; the situation was different in the auxilia - ibid., 281), so we should perhaps envisage responsibility of the sort enjoyed by the tesserarius (Vegetius II,7).

101. It has been shown that the custos armorum could not have been the guardian of a company's weapons whilst they were locked away (ROBINSON, 1975, 9). It would seem logical to have one man in each company whose specific task was to monitor the state of his men's weapons.

102. Vegetius tells us (II,11 - Appendix No.1) that the fabrica staff included men who constructed buildings for hibernae, permanent structures obviously being one of the main differences between the hiberna and aestiva. A workshop of the kind envisaged here would be a difficult thing to move about under normal circumstances, quite apart from the fact that the labour force would be on active duty in the summer months.

103. The fabrica presumably strove to build up a surplus of equipment during the winter months and may have embarked upon new ventures (as with the introduction of the 'lorica segmentata'?).

104. The relationship between modern and ancient terminology is difficult here; whether the castra aestiva (e.g. Tacitus Ann.I,37) was capable of supporting a full fabrica is unknown, although Hyginus 35 gives us an idea of the size and location of the fabrica within it.

105. This would imply that the field workshop was staffed largely by the immunes when on campaign.

106. Such forges have been used by many armies through the ages;

I am grateful to Mr J. Turner for identifying a number of similarities between the proposed operation of the fabrica outlined in this paper and REME workshops during the second world war.

107. Traces of any sort of occupation within temporary sites are ephemeral - RICHMOND, 1933, 58-69 could identify ovens/hearths, 'dugouts', and low turf mounds in camp B at Cawthorn. Even in permanent sites, the evidence for their original purpose from fabricae is surprisingly minimal once they have been cleared by the army, but then it is extremely difficult to prove the original purpose of any Roman military building purely by using the archaeological record.

108. Josephus, Bell. Iud., III,83; cf. Hyginus, 4 & 35.

109. The finds from the first, second, and early third centuries consistently seem to indicate that this was so: the range of finds from Antonine Newstead matches that from the Trajanic abandonment (CURLE, 1911, esp. 113 with 116-39). The Künzing (HERRMANN, 1969) and Straubing (KEIM & KLUMBACH, 1951) hoards are both apparently third-century, as is the cavalry sports helmet from one of the pits in the barracks at Echzell (KLUMBACH & BAATZ, 1970).

110. Even with a few slaves to help him, he could not achieve volume production.

111. Evidence for private production is slight, but is summarised in MACMULLEN, 1960, 25-6, which is largely followed by OLDENSTEIN, 1977a, 79-83.

112. See, for example, the description of luxury weapons contained in P. Giss. 47 where the Greek terms used are for officers arms, rather than the more usual graecisized Latin terms normally found (e.g. balteum in P. Mich. VIII 474,8-9).

113. Apart from the fact that such control was impractical in the Roman world (MILLAR, 1982, 7-11 for difficulties of communication), the variety of equipment design confirms this.

114. Vegetius 1,8 stresses the importance of tradition in the early imperial legion.

115. Tacitus Ann. I,16 for 3 legions of the Pannonian army in an aestiva together; I,31 for 4 legions of the Lower German army in a similar camp.

116. Such peculiarities would probably depend upon the whim of individual craftsmen.

117. Empire-wide army movements, such as major campaigns, were a good opportunity (e.g. RE 'legio', 1250-1), whilst the civil wars of A.D.68-70 brought many legions into contact for the first time in years (ibid., 1265-6).

118. By 'Upper Germany', the Upper German army-group is meant: the province, as such, did not exist before A.D.82 (RAEPSAET-CHARLIER, 1973, 161.

119. VON GONZENBACH, 1966, 184-9.

120. Cf. SCHÖNBERGER, 1978, Abb.76.

121. Britain: Chichester, Hod Hill; Pannonia: Magdalensburg (see n.120).

122. Although not many sites of the Pannonian legion (VIIII Hispana) have been investigated as thoroughly as those of legio II Augusta.

123. VON GONZENBACH, 1965, 6.

124. Carnuntum: RLÖ XXXII, Taf.66,13; Magdalensburg: VON GONZENBACH, 1965, 20; Mainz: loc. cit.

125. See above n.117.

126. For lobate hinged-strap and -buckle fittings from the 'lorica segmentata', see RLÖ II, Taf.XVIII & XIX.

127. ZADOKS-JOSEPHUS JITTA & WITTEVEEN, 1977 includes many examples from the Netherlands.

128. ROBINSON, 1977, 557-9 saw Imperial-Gallic helmets as being the work of Gallic smiths supplying the German armies, but the evidence for them is weak.

129. This imbalance is shown up clearly in the proportions of helmets surviving from the west and east respectively - ROBINSON, 1975, 13-139.

130. Such as the campaigns of Corbulo in Armenia and Vespasian in Judaea.

131. Tacitus Hist. II,82; Dio LXIX,12,2.

132. Diodorus Siculus, XIV,41-3.

133. Livy XXVI,47.

134. Appian VIII,93.

135. In 68 B.C., Mithridates manufacturing weapons in his towns in preparation for war against Lucullus (Appian XII,87); inhabitants of Massillia set up _officina_ for production of arms after they have shut their gates against Caesar (Caesar _Bell. Civ._ I,34).

136. Characterstically utilising all available space within a city (Diodorus Siculus XIV,41,6; Appian VIII,93), all available personnel (specialised and non-specialised - Diodorus Siculus XIV,43,1; Appian VIII,93), and working very fast (_loc. cit._).

137. Dionysius was evidently thinking in terms of an army of 140,000 men (Diodorus Siculus XIV,43,2).

138. Presumably because they possessed a) expert craftsmen who could turn their hand to weapons production; b) a large, able, (and usually) willing workforce; and c) the organizational ability to harness all of this potential. Note that Dionysius (Diodorus Siculus XIV,41,3) and Sertorius (SCHLESINGER & GEER, 1959, 191) had to import specialist craftsmen to their production centres.

139. Long-distance transport (both land and sea) was difficult and not something to which sensitive bulk shipments of arms would be entrusted.

140. On new legions being raised in Italy, see MANN, 1963. Caesar raised _legio V Alaudae_ in Transalpine Gaul, where there were classical cities (Suetonius _Div. Iul._ 24).

141. Cf. ROBINSON, 1975, 62.

142. It may be no coincidence that eastern armies, largely freed of the need to manufacture their equipment, had a reputation for slovenliness (Tacitus, _Ann._ XIII,35; Fronto, _Ad Verum Imp._, II,1,9).

143. _P. Gen. Lat._ I _verso_ v is the best known record of such duties and relates to only 36 men from a century (WATSON, 1969, 73-4) available for fatigues. Of these, never more than half-a-dozen are occupied in guard duties on any one day.

144. Rightly or wrongly, this is certainly the impression given

by Trajan's Column.

145. On the face of it, the troops in _Tab. Vind._ 1 and 3 (Appendix Nos.4 & 5) are auxiliary, but it must be remembered that these documents do not specifically state this fact. Otherwise, references to _cohortales_ in _P. Berlin_ inv.6765 (Appendix No.3 Col.ii,5) may be taken to indicate the use of auxiliaries (BRUCKNER & MARICHAL, 1979, 7).

146. Late _fabricae_ are discussed in detail in JAMES, forthcoming.

147. Finds of equipment from water are a different matter. I hope to discuss this subject in greater detail elsewhere.

APPENDIX: SUBLITERARY AND LITERARY SOURCE MATERIAL

No.1: **Vegetius II,11**

habet praeterea legio fabros tignarios structores carpentarios ferrarios, pictores reliquosque artifices ad hibernorum aedificia fabricanda, ad machinas turres ligneas ceteraque, quibus uel expugnantur aduersariorum ciuitates uel defenduntur propriae, praeparatos, qui arma vehicula ceteraque genera tormentorum uel noua facerunt uel quassata repararent. habebant etiam fabricas scutarias loricarias arcuarias, in quibus saggitae missibilia cassides omniaque armorum genera formabantur. haec enim erat cura praecipua, ut quicquid exercitui necessarium uidebantur numquam deesset in castris.

"A legion had workmen, joiners, masons, wagon-makers, smiths, painters, and other craftsmen for the construction of buildings in the winter-quarters and for the preparation of machines, wooden towers (which are either for assaulting enemy cities or defending their own) and so forth, who will either make from scratch or repair wagons, vehicles, and other sorts of engines. They even had workshops for shields, cuirasses, and bows, in which they fashioned arrows, missiles, helmets, and all sorts of weapons. For this was a principal concern: that whatever they saw was necessary for the army was never missing from the camp."

No.2: **Digest 50,6,7**

quibusdam aliquam uacationem munerum grauiorum condicia tribuit, ut sunt mensores, optio ualetudinarii, medici, capsarii, et artificies qui fossam faciunt, ueterinarii, architectus, gubernatores, naupegi, ballistarii, specularii, fabri, sagittarii, aerarii, bucularum structores, carpentarii, scandularii, gladiatores, aquilices, tubarii, cornuarii, arcuarii, plumbarii, ferrarii, lapidarii, et qui calcem cocunt, et qui siluam infindunt, qui carbonem caedunt ac torrent. in eodem numero haberi solent lani, uenatores, uictimarii, et optio fabricae, et qui aegris praesto sunt, librarii quoque qui docere possint, et horreorum librarii, et librarii depositorum, et librarii caducorum et adiutores cornicularioum, et stratores, et polliones, et custodes armorum, et praeco, et bucinator. hi igitur omnes inter immunes habentur.

"Certain soldiers are granted by their conditions of service some exemption from the heavier fatigues. These are men such as surveyors, the medical sergeant, medical orderlies and dressers, ditchers, farriers, the architect, pilots, shipwrights, artillerymen, glaziers, workmen, arrowsmiths, coppersmiths, helmet-makers, waqon-makers, roof-tile-cutters, swordcutlers, water engineers, trumpet-makers, horn-makers, bow-makers, lead-workers, blacksmiths, stonecutters, lime-burners, woodmen, and charcoal burners. In the same category there are usually

included butchers, huntsmen, keepers of sacrificial animals, the workshop sergeant... armoury sergeants, the herald, and the trumpeter. These are all included amongst the <u>immunes</u>."
Based on WATSON, 1969, 76.

No.3: P. Berlin inv. 6765 (ChLA 409)

XIIII] Kal(endas) Ma[ias]
[op]erati]sunt in f.
milit]es legion abricam legionis
immun]es
[ ]ari n(umero) C
cohorta]les
galliar]ii f[ic
]...es
] VIII
] ' IIII
]lamnae leuisatares X
]spathar[u]m fabricatae X
]. fabricatae VI
]elaria V
]peractae CXXV
]....... .
c]lauis carrum ..

XIII Ka[l(endas) Maias
[op]erati sun[t in fabricam legionis
milites leg[ionari
immunes [
cohortales [
galliarii fic[
pagani [
custodiae [
scuta talari[a
leuesatae la[
item breues[
..........[
[s]cuta planat[a
arcus peracti[
capitula ball[istaria
seruo in p[
.[

Published in BRUCKNER & MARICHAL, 1979, 6-7

No.4: **Tab. Vind. 1**

VII K(alendas) Maias fabricis h(omines) CCCXXXXIII
ex eis sutores ***vacat*** *XII*
s[tr]uctores ad balneum *XVIII*
a]d plumbum [
a]d ar [
[ ] a[ ]ualetudinar[
ad furnaces [
ad lutum [
tectores [
m rpil [
ad cae[
] b[
ad p [
cum[

Published in BOWMAN & THOMAS, 1983, 77-9

No.5: **Tab. Vind. 3**

Side A

[[ ] em[ ]]
[[ ] riu [ ]]
[[ ]utarius[ ]]
]arium [
]us adiu [
[[ ]arium [ ]]
]s faber[
] s 7 u[
] um gladi [
]usas.7 fru[
[[ ] ralem[ ]]
[[ ]is faber [ ]]
[[ ]rrum[ ]]
]t qp e [

Side B

[[ ] eni a[ ]]
[[ ]ius 7 [ ]]
[[ ] us ma ri [ ]]
[[ ] lius faber [ ]]
] am [
]s . rufini [
] a ean [

traces
] . . *faber* [
[[ ]*atorium* [ ]]
[[ ] . *frumen*[ ]]
]*a* . . . *am*[
]? *frum*[
]*m* [
trace

Published in BOWMAN & THOMAS, 1983, 81-3

No.6: **AE 1926, No.3**

ualerio scu[ta]rio
IIX

No.7: **CIL XIII 11504**

marti uotum
tib(erius) iul(ius) aquil(a)
gladia[ri]us
s(oluit) l(aetus) l(ibens) m(erito)

ABBREVIATIONS

AE L'Année *Epigraphique*

ChLA *Chartae Latinae Antiquiores*

CIL *Corpus Inscriptionum Latinarum*

ORL *Der obergermanisch-raetische Limes*

RE *Paulys Realencyclopädie der classischen Altertumswissenschaft*

RLO *Der römische Limes in Österreich*

BIBLIOGRAPHY

ANGUS *et al.* 1962: N.S. Angus, G.T. Brown, H.F. Cleere, 'The iron nails from the Roman legionary fortress at Inchtuthil, Perthshire', *Journal of the Iron and Steel Institute*, 200, 1962, 956-68

ANSTEE 1953: J.W. Anstee, 'Fragments of Roman "bronze" scale armour from Corbridge', *Museums Journal*, 53, 1953, 200-2

BARNES 1979: T.D. Barnes, 'The date of Vegetius', *Phoenix*, 33, 1979, 254-7

BAYLEY & BUTCHER 1981: J. Bayley & S. Butcher, 'Variations in alloy composition of Roman brooches', *Revue d'Archéométrie*, suppl.29, 1981, 29-36

BIDWELL 1980: P.T. Bidwell, *Roman Exeter: Fortress and Town*, (Exeter 1980)

BIRLEY 1977: R. Birley, *Vindolanda: a Roman Frontier Post on Hadrian's Wall*, (London 1977)

BISHOP forthcoming: M.C. Bishop, 'The distribution of military equipment within Roman forts of the first century A.D.' in PLANCK & UNZ, forthcoming

BISHOP unpublished: M.C. Bishop, *Report on the Analysis of Some Military Equipment from the Roman Fortress at Longthorpe Using Atomic Absorption Spectroscopy*, unpublished report prepared for Peterborough City Museum and Art Gallery, (Sheffield 1984)

BOGAERS & YPEY 1963: J.E. Bogaers & J. Ypey, 'Ein neuer römischer

Dolch mit silbertauschierte und emailverzierter Scheide aus dem Legionslager Nijmegen', Berichten van de Rijksdienst voor het oudheidkundig bodemondezoek in Nederland, 12-13, 1962-3, 87-98

BOWMAN & THOMAS 1983: A.K. Bowman & J.D. Thomas, Vindolanda: The Latin Writing Tablets, Britannia Monograph Series No.4, (London 1983)

BREEZE 1969: D.J. Breeze, 'The organization of the legion: the first cohort and the equites legionis', Journal of Roman Studies, 59, 1969, 50-55

BREEZE 1974: D.J. Breeze, 'The organisation of the career structure of the immunes and principales of the Roman army', Bonner Jahrbücher, 174, 1974, 245-92

BREEZE 1976: D.J. Breeze, 'A note on the use of the titles optio and magister below the centurionate during the principate', Britannia, 7, 1976, 127-33

BREEZE et al. 1976: D.J. Breeze, J. Close-Brooks, J.N.G. Ritchie, 'Soldiers' burials at Camelon, Stirlingshire, 1922 and 1975', Britannia, 7, 1976, 73-95

BRUCKNER & MARICHAL 1979: A. Bruckner & R. Marichal (eds.), Chartae Latinae Antiquiores, Part X, Germany I, (Zürich 1979)

CRADDOCK et al. 1973: P.T. Craddock, J. Lang, K.S. Painter, 'Roman horse-trappings from Fremington Hagg, Reeth, Yorkshire, N.R.', British Museum Quarterly, 37, 1973, 9-17

CRUMMY 1983: N. Crummy, Colchester Archaeological Report 2: The Roman Small Finds from Excavations in Colchester 1971-9, (Colchester 1983)

CRUMMY 1984: P. Crummy, Colchester Archaeological Report 3: Excavations at Lion Walk, Balkerne Lane, and Middleborough, Colchester, Essex, (Colchester 1984)

CURLE 1911: J. Curle, A Roman Frontier Post and its People. the Fort of Newstead in the Parish of Melrose, (Glasgow 1911)

DANIELS 1968: C.M. Daniels, 'A hoard of iron and other materials from Corbridge', Archaeologia Aeliana, ser.4, XLVI, 1968, 115-26

DOBSON 1965: B. Dobson, 'The praefectus fabrum in the early principate' in JARRETT & DOBSON, 1965, 61-84

DOBSON 1978: B. Dobson, Die Primipilares, (Köln-Bonn 1978)

VON DOMASZEWSKI 1908: A. von Domaszewski, 'Die Rangordnung des römischen Heeres', Bonner Jahrbücher, 97, 1908, 1-278

DOWN 1981: A. Down, Chichester Excavations Vol.5, (Chichester 1981)

FINGERLIN 1981: G. Fingerlin, 'Eberzahnanhänger aus Dangstetten', Fundberichte aus Baden-Württemburg, 6, 1981, 417-32

FINK 1971: R.O. Fink, Roman Military Records on Papyrus, American Philological Association Monograph 26, (Cleveland 1971)

FITZPATRICK forthcoming: A. Fitzpatrick, 'Camulodunum: a reassessment' in PLANCK & UNZ, forthcoming

FOX & ST.JOHN HOPE 1901: G.E. Fox & W.H. St.John Hope, 'Excavations on the site of the Roman city at Silchester, Hants., in 1900', Archaeologia, 57, 1901, 229-56

FRERE 1978: S.S. Frere, Britannia. A History of Roman Britain, ed.2, (London 1978)

FRERE 1983: S.S. Frere, 'Roman Britain in 1982. I Sites explored', Britannia, 14, 1983, 280-335

FRERE & ST.JOSEPH 1974: S.S. Frere & J.K. St.Joseph, 'The Roman fortress at Longthorpe', Britannia, 5, 1974, 1-129

GARLICK 1980: M. Garlick, 'The reconstruction and use of a lorica hamata', Exercitus, 1:1, Summer 1980, 6-8

GILLIAM 1967: J.F. Gilliam, 'The deposita of an auxiliary soldier (P. Columbia inv.325)', Bonner Jahrbücher, 167, 1967, 233-43

GOFFART 1977: W. Goffart, 'The date and purpose of Vegetius' "De Re Militari"', Traditio, 33, 1977, 65-100

VON GONZENBACH 1965: V. von Gonzenbach, 'Schwertscheidenbleche von Vindonissa aus der Zeit der 13. Legion', Jahresbericht der Gesellschaft Pro Vindonissa, 1965, 5-35

VON GONZENBACH 1966: V. von Gonzenbach, 'Tiberische Gürtel- und Schwertscheidenbeschläge mit figürlichen Reliefs', Helvetia Antiqua. Festschrift Emil Vogt, Beitr. z. Prähist. u. Arch. der Schweiz, (Zürich 1966), 183-208

HARMAND 1967: J. Harmand, L'armée et le soldat à Rome de 107 à 50 avant notre ère, (Paris 1967)

HERRMANN 1969: F.-R. Herrmann, 'Der Eisenhortfund aus dem Kastell Künzing', Saalburg Jahrbuch, 26, 1969, 129-41

HOGG 1969: A.H.A. Hogg, 'Pen Llystyn: a Roman fort and other remains', Archaeological Journal, 125, 1969, 101-92

HÜBENER 1973: W. Hübener, Die römischen Metallfunde von Augsburg-Oberhausen, Materialhefte zur bayerischen Vorgeschichte 28, (Kallmünz 1973)

HURST 1985: H.R. Hurst, Kingsholm. Excavations at Kingsholm Close and Other Sites with a Discussion of the Archaeology of the Area, Gloucester Archaeological Reports 1, (Cambridge 1985)

JACKSON & STONE forthcoming: R. Jackson & M.J. Stone (eds.), Roman Military Studies I, (forthcoming)

JAMES forthcoming: S.T. James, 'The fabricae, state arms factories of the later Roman Empire' in JACKSON & STONE, forthcoming

JARRETT & DOBSON, 1965: M.G. Jarrett & B. Dobson (eds.), Britain and Rome, (Kendal 1965)

JOHNSON 1983: A. Johnson, Roman Forts of the 1st and 2nd Centuries A.D. in Britain and the German Provinces, (London 1983)

KEIM & KLUMBACH 1951: J. Keim & H. Klumbach, Der römische Schatzfund von Straubing, Münchener Beiträge zur Vor- und Frühgeschichte Bd.3, (München 1951)

KEPPIE 1984: L.J.F. Keppie, The Making of the Roman Army from Republic to Empire, (London 1984)

KLUMBACH & BAATZ 1970: H. Klumbach & D. Baatz, 'Eine römische Parade-Gesichtsmaske aus dem Kastell Echzell Kr. Büdingen (Hessen)', Saalburg Jahrbuch, 27, 1970, 73-83

LEAHY 1980: K. Leahy, 'A first century military pendant from Normanby, South Humberside', Lincolnshire History and Archaeology, 15, 1980, 83-4

LEHNER 1928: H. Lehner, 'Ausgrabungen in Vetera 1927', Germania, 12, 1928, 20-5

MACMULLEN 1960: R. MacMullen, 'Inscriptions on armor and the supply of arms in the Roman Empire', American Journal of Archaeology, 64, 1960, 23-40

MACMULLEN 1984: R. MacMullen, 'The legion as a society', Historia, 33, 1984, 440-56

MANN 1963: J.C. Mann, 'The raising of new legions during the principate', Hermes, 91, 1963, 483-9

MANNING 1972: W.H. Manning, 'Ironwork hoards in Iron Age and Roman Britain', Britannia, 3, 1972, 224-50

MAXFIELD 1981: V.A. Maxfield, The Military Decorations of the Roman Army, (London 1981)

MEGAW & SIMPSON 1979: J.V.S. Megaw & D.D.A. Simpson, Introduction to British Prehistory from the Arrival of Homo Sapiens to the Claudian Invasion, (Leicester 1979)

MILLAR 1982: F. Millar, 'Emperors, frontiers and foreign relations, 31 B.C. to A.D.378', Britannia, 13, 1982, 1-23

NIBLETT 1985: R. Niblett, Sheepen: an Early Roman Industrial Site at Camulodunum, CBA Research Report 57, (London 1985)

OLDENSTEIN 1977a: J. Oldenstein, 'Zur Ausrüstung römischer Auxiliareinheiten. Studien zu Beschlägen und Zierat an der Ausrüstung der römischen Auxiliareinheiten des obergermanische-raetischen Limesgebietes aus dem zweiten und dritten Jahrhundert n. Chr.', Bericht der römisch-germanischen Kommission, 57, 1976 (1977), 49-366

OLDENSTEIN 1977b: J. Oldenstein, 'Zur Buntmetallverarbeitung in den Kastellen am obergermanischen und rätischen Limes', Bulletin des Musées Royaux d'Art et d'Histoire, 46, 1974 (1977), 185-96

PARKER 1932: H.M.D. Parker, 'The antiqua legio of Vegetius', Classical Quarterly, 26, 1932, 137-49

PEATMAN 1981: J. Peatman, Abbeydale Industrial Hamlet, (Sheffield 1981)

VON PETRIKOVITS 1970: H. von Petrikovits, 'Die Spezialgebäude römischer Legionslager', Legio VII Gemina, (Léon 1970)

VON PETRIKOVITS 1974a: H. von Petrikovits, 'Römisches Militärhandwerk. Archäologische Forschungen der letzten Jahre', Anzeiger der österreichischen Akademie der Wissenschaften, philosophisch-historische Klasse, 111, 1-23

VON PETRIKOVITS 1974b: H. von Petrikovits, 'Militärische Fabricae der Römer' in D.M. Pippidi (ed.), Actes du IXe congrès international d'etudes sur les frontières romaines, (Bucharest 1974), 399-407

VON PETRIKOVITS 1975: H. von Petrikovits, Die Innenbauten römischer Legionslager während der Prinzipatszeit, Abhandlungen der Rheinischer Westfälischen Akademie der Wissenschaften Bd.56, (Opladen 1975)

PLANCK & UNZ forthcoming: D. Planck & C. Unz (eds.), Kongress-Akten der Beiträge am Limes-Kongress in Aalen 1983, Sonderbd. Forschungen und Berichten zur Vor- u. Frühgeschicht. Baden-Württemburg, (Stuttgart forthcoming)

RACKHAM 1976: O. Rackham, Trees and Woodland in the British Landscape, (London 1976)

RAEPSAET-CHARLIER 1973: M.-T. Raepset-Charlier, 'Germania inferior et Germania superior', Latomus, 32, 1973, 158-61

RICHMOND 1933: I.A. Richmond, 'The four Roman camps at Cawthorn in the North Riding of Yorkshire', Archaeological Journal, 89, 1932 (1933), 17-78

RITTERLING 1904: E. Ritterling, 'Das frührömische Lager bei Hofheim i.T. Ausgrabungs- und Fundbericht', Annalen des Vereins für nassauische Altertumskunde und Geschichtsforschung, 34, 1904, 1-110, 397-423

ROBINSON 1975: H.R. Robinson, The Armour of Imperial Rome, (London 1975)

ROBINSON 1977: H.R. Robinson, 'The origins of some first century legionary helmets' in J.E. Bogaers (ed.), Studien zu den Militärgrenzen Roms II: Vorträge des 10. internationalen Limeskongresses in der Germania Inferior, (Köln 1977), 553-9

ROSS & FEACHEM 1976: A. Ross & R. Feachem, 'Ritual rubbish? - the Newstead pits' in J.V.S. Megaw (ed.), To Illustrate the Monuments. Essays Presented to Stuart Piggott, (London 1976), 230-7

SADDINGTON 1982: D.B. Saddington, The Development of the Roman Auxiliary Forces from Caesar to Vespasian (49 B.C.-A.D.79), (Harare 1982)

SANDER 1929: E. Sander, 'Frontin als Quelle für Vegetius', Philologische Wochenschrift, 49, 1929, 1230-31

SANDER 1932: E. Sander, 'Die Hauptquellen der Bücher I-III der epitoma rei militaris des Vegetius', Philologus, 87, 1932,

369-75

SANDER 1939: E. Sander, 'Die antiqua ordinatio legionis des Vegetius', Klio, 32, 1939, 382-91

SANDER 1962: E. Sander, 'Der praefectus fabrum und die Legionsfabriken', Bonner Jahrbücher, 162, 1962, 139-61

SANDYS 1927: J.E. Sandys, Latin Epigraphy. An Introduction to the Study of Latin Inscriptions, ed.2, revised by S.G. Campbell, (London 1927)

SCHENK 1930: D. Schenk, 'Flavius Vegetius Renatus: die Quellen der Epit. rei militaris', Klio, Beiheft 22, 1930

SCHLESINGER & GEER 1959: A.C. Schlesinger & R.M. Geer, Livy with an English Translation in Fourteen Volumes. XIV Summaries, Fragments, and Obsequens, Loeb Classical Library, (London 1959)

VON SCHNURBEIN 1974: S. von Schnurbein, 'Der römischen Militärlager bei Haltern, Bericht über die Forschungen seit 1899', Bodenaltertümer Westfalens, 14, 1974

SCHÖNBERGER 1978: H. Schönberger, Kastell Oberstimm: die Grabungen von 1968 bis 1971, Limesforschungen 18, (Berlin 1978)

SCHÖNBERGER 1979: H. Schönberger, 'Valkenburg Z.H.: Praetorium oder Fabrica', Germania, 57, 1979, 135-41

TAYLOR & WILSON 1961: M.V. Taylor & D.R. Wilson, 'Roman Britain in 1960. I Sites explored', Journal of Roman Studies, 51, 1961, 157-91

ULBERT 1969: G. Ulbert, Das frührömische Kastell Rheingönheim, Limesforschungen 9, (Berlin 1969)

WATSON 1969: G.R. Watson, The Roman Soldier, (London 1969)

WEBSTER 1979: G. Webster, The Roman Imperial Army of the First and Second Centuries A.D., ed.2, (London 1979)

WHITE 1984: K.D. White, Greek and Roman Technology, (London 1984)

WILSON 1974: D.R. Wilson, 'Roman Britain in 1973. I Sites explored', Britannia, 5, 1974, 397-460

ZADOKS-JOSEPHUS JITTA & WITTEVEEN 1977: A.N. Zadoks-Josephus Jitta & A.M. Witteveen, 'Roman bronze lunulae from the

Netherlands', *Oudheidkundige Mededelingen uit het Rijksmuseum van oudheden te Leiden*, 58, 1977, 167–95

THE PRODUCTION AND SUPPLY OF MILITARY LEATHERWORK IN THE FIRST AND SECOND CENTURIES A.D.: A REVIEW OF THE ARCHAEOLOGICAL EVIDENCE

C. van Driel-Murray

This paper will be principally concerned with Roman military leatherwork as represented in the archaeological record. Various aspects will be reviewed, but more in the hope of stimulating discussion and research than of offering any all-embracing answers to the problems in this particular field.

The place of leatherwork in the Roman army can be approached from two different points of view: there is the primary, archaeological level, with the identification of the pieces of equipment represented by the finds, and the direct evidence which this provides as to form, construction and development of individual pieces of equipment, and the secondary, interpretative level in which systems of manufacture and mechanisms of supply of both the raw material and of the items made from it are deduced from internal evidence. All aspects are, however, to a great extent governed by the nature and characteristics of leather as an organic product and the severe bias in the available evidence which this entails.

In the first place, leather is only conserved under special environmental conditions, which means in practice that the waterlogged deposits of wells, ditches and river courses form by far the richest source of Roman leatherwork. But these deposits, being in the main rubbish dumps, in themselves introduce considerable problems in the interpretation of the finds incorporated in them and their dating, quite apart from the northern and consequently military bias displayed in excavation of and research into this type of context. Furthermore, what leather survives in these contexts was almost invariably discarded as worn-out or useless: the off cuts of manufacture, damaged pieces removed and discarded on repair and the final jettisoning of worthless equipment. The evidence must be pieced together from what was consciously thrown away, and the problems of rubbish disposal and basic conservation are interwoven with any interpretation of the surviving material. Thus in an examination of Roman leatherwork we must be continually aware of the shortcomings of the available evidence: the lack of comparative material from Italy and Gaul especially, the paucity of leather other than footwear after the mid-second century, the difficulty in isolating military from civilian products and methods of manufacture in the rather mixed contexts of the <u>Limes</u>

forts and their *vici*, the absence of comparable find complexes from civilian sites and especially the fact that the conditions required for preservation virtually exclude the possibility of identifying the location of a workshop and its products. In addition, a whole category of leatherwork is missing from the record, for it is only the vegetable tanned leathers which are able to survive in waterlogged deposits. Alum-tawed[1] and oiled leathers - which may have been particularly important in the military context - fail to survive at all, save in dry and airless conditions.

The evidence for the manufacture of Roman military equipment is, therefore, severely restricted not only as a result of physical considerations, but also because of the lack of quantitative publication. Consequently, the discussion will be based principally on the evidence from six sites. This is not very much, although they do provide a fairly balanced picture, and future work will inevitably modify the conclusions reached here. For the first century we have the legionary fortress Vindonissa[2] (Switzerland), the auxiliary fort Valkenburg[3] and the outpost Velsen[4] (both in the Netherlands), while the second century is represented by the military *fabrica* on the Bonner Berg[5] (W. Germany) and also the auxiliary sites Zwammerdam[6] and Vechten[7] (Netherlands), although neither of these contexts is as well defined as one could wish.

Despite all the limitations, leather, because it is one of the few organic materials to be preserved in any quantity, is an unequalled source of information for many items of clothing and equipment otherwise known only from pictorial sources, if at all.[8] Problems of interpretation of function remain due to the disappearance of other organic components such as glue, wood or stuffing, while the disintegration of larger items into their constituent parts makes the accurate reconstruction of complex articles such as tents difficult. Still, the essential identification of a large number of items is clear enough, and emphasises the role of leather in an army which seems to have been unfamiliar with tarpaulin. Shoes, shield covers, tents, baggage covers and kit bags, saddles and horse trappings, briefcases, letter and tablet envelopes, cushion covers and purses,[9] are all certainly attested on military sites: the function of many other fragments can only be guessed at and many more purposes may be expected: awnings, smith's aprons, furnishings, wine and water skins,[10] flask covers,[11] sheaths and quivers as well as innumerable straps and fittings. The use of oiled or alum-tawed leather may be responsible for the otherwise unaccountable lack of horse harness, straps, belts and sheaths on Roman sites. Both kinds of leather were certainly widely used for these purposes in the 18th and 19th century cavalry and artillery regiments in the Netherlands and the use of oiled

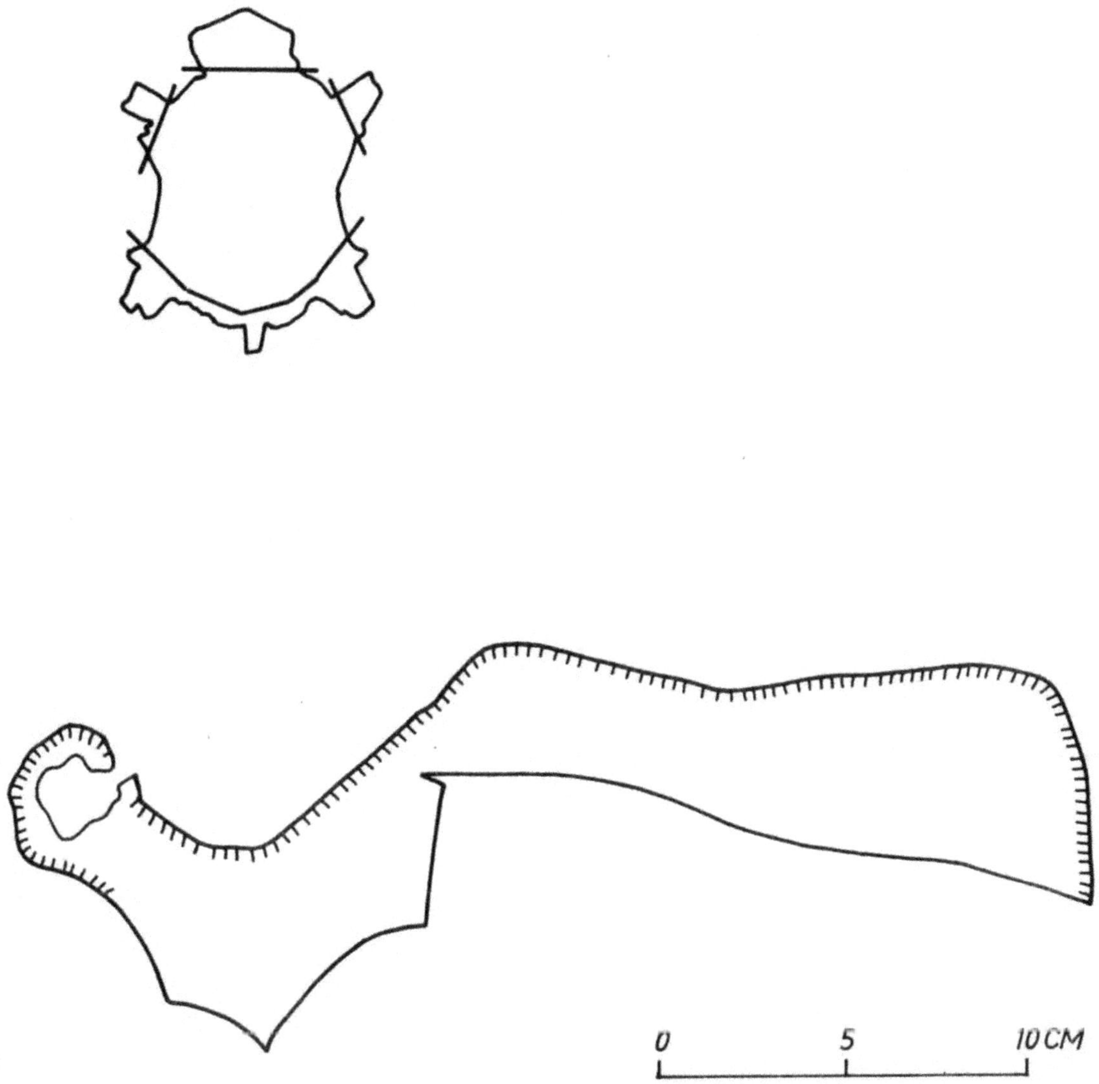

Fig. 1. a. Primary cutting lines marked on skin (not to scale).
b. Primary off cuts from Velsen with natural orifice (scale 1:2).

leather for horse harness was actually laid down in the current military specifications.[12] Any large-scale use of leather in clothing, as proposed by Gansser-Burckhardt must probably be discounted.[13] Not only are any positively identified examples extremely rare,[14] but it is also evident from the sources that the army looked primarily to textiles for its clothing.[15] Cuir bouilli might have been used for muscle cuirasses, but again, nothing has survived. Due to its low water resistance, alum tawed leather would have been unsuitable for out-door military clothing.

It will be clear that the Roman army in the first and second centuries must have required leather on a truly vast scale. Exact quantification is of course impossible, since we cannot assess replacement rates, but when it is realized that a tent for eight men requires about 70 complete goatskins, while the cover for each auxiliary shield demands a further 1½-2 skins, the amount is staggering. To fully equip a newly raised unit or to re-equip one at a time of emergency or following hostilities must have posed formidable problems of supply.[16] Furthermore, tanning is a relatively slow process and excessive slaughter of livestock for their skins alone will have severe repercussions on future supply.[17] By far the most satisfactory solution is to ensure the steady supply of raw materials, to hold adequate stocks and to keep equipment permanently in good repair - and this is exactly what the types of finds from military sites such as Bonn, Velsen and Valkenburg seem to show.

Before going into detail, we must first look at the evidence on which any hypotheses on the manufacture and maintenance of leather equipment must be founded. A basic misconception as far as the study of the manufacture of leather goods is concerned is that concentrations of finished products identify the presence of the craftsman making them.[18] Although this may sometimes be the case,[19] in general the converse will be true: the finished items are dispersed and ultimately end up, not back in the workshop but on the rubbish tip, where they reveal more about the processes of rubbish disposal than about manufacture. For questions of production we are dependent, not on the completed articles (although these can convey information on manufacture through details of technology and skin choice) but on the snippets of leather left after the completion of the item concerned. These off cuts are all too often ignored, yet it is these scraps which reveal whether a leather worker was present, whether he was making new items or only repairing old ones, what products were concerned, whether there was any craft specialization and what value was placed on the raw material used. Identification of the animal from which the skin[20] was taken allows further inferences as to the selection or availability of particular animals and the possibilities of import.

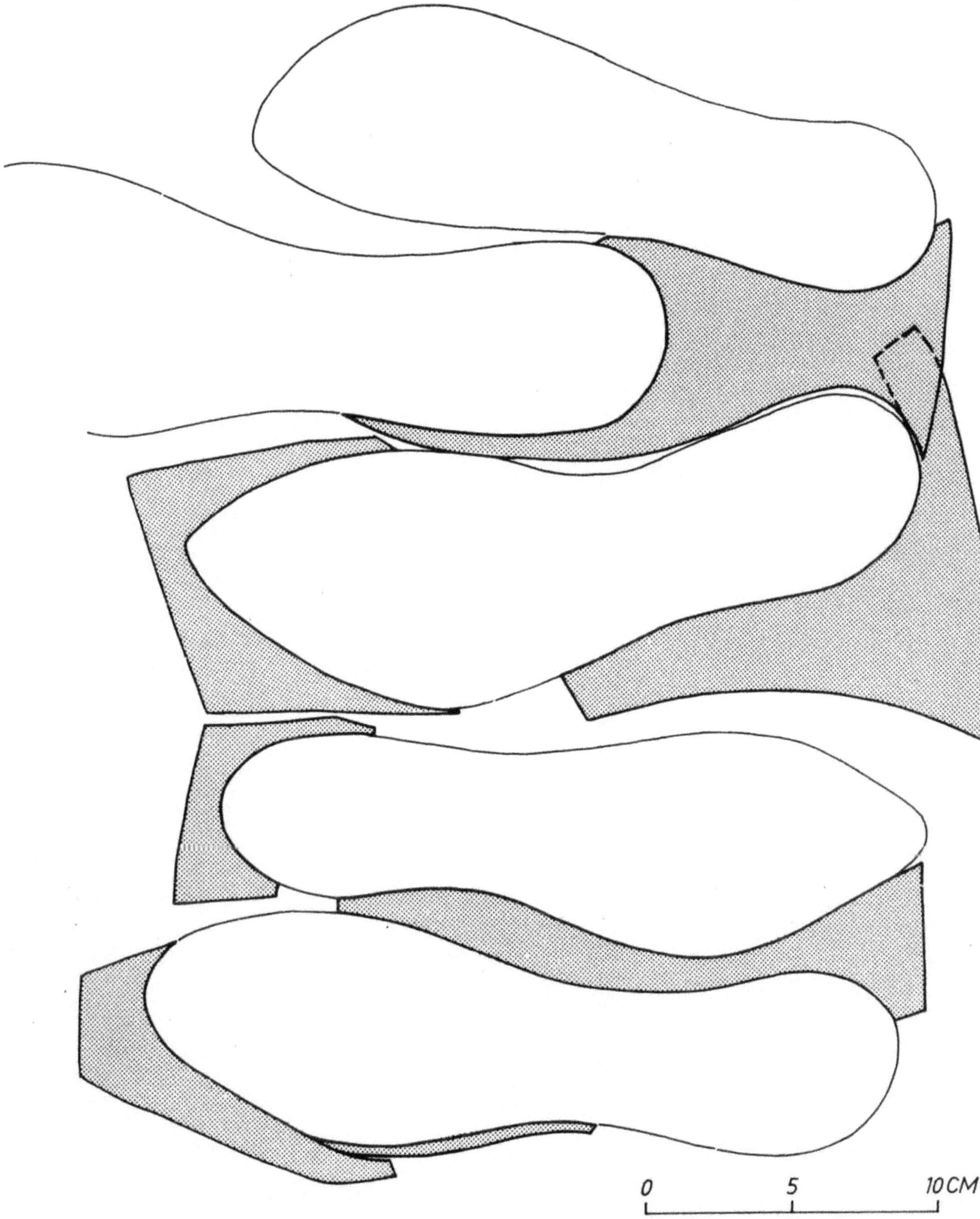

Fig. 2: Secondary off cuts from Velsen arranged in their relevant positions around sole shapes (also from Velsen). Note gaps where additional trimming would be required (scale 1:2).

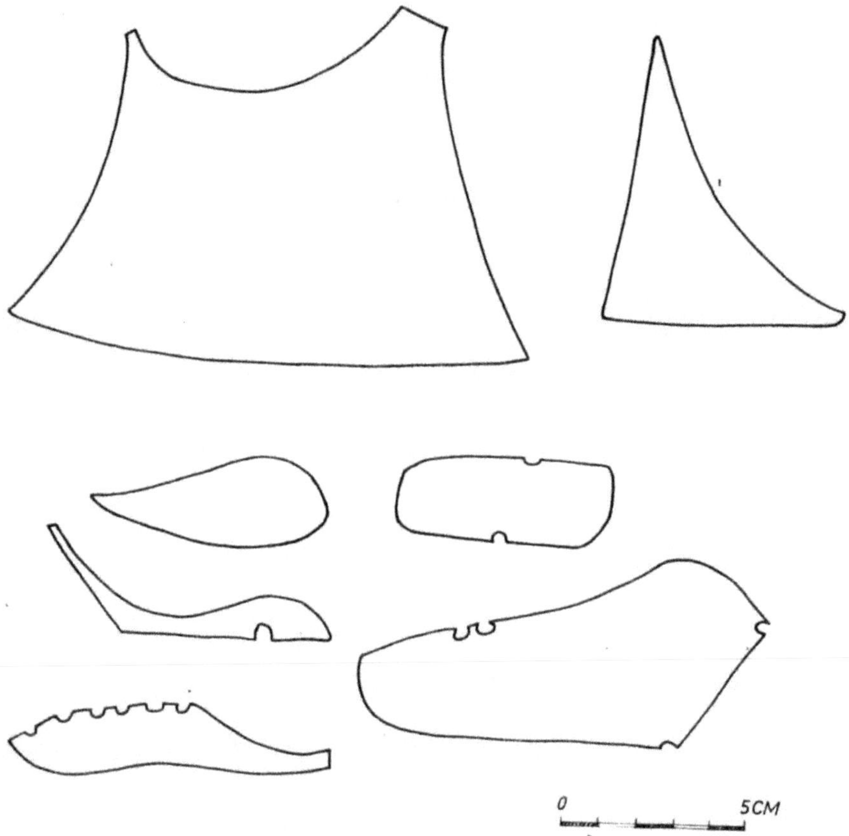

Fig. 3. Off cuts from Maastricht characteristic for the manufacture of single-piece footwear (scale 1:2).

It is perhaps worthwhile to look at these uninspiring fragments more closely. They may be divided into three categories, each reflecting a particular part of the manufacturing process:

1) Primary off cuts (Fig.1): useless or poor quality portions removed before cutting-out begins. This might also be done at the tannery to neaten the product, but Roman evidence from Xanten, Maastricht and Vindonissa[21] suggests that the skins were delivered entire to the leatherworker, with head, tails and even udders still attached. Characteristic are skin edges, cut away from the better parts in straight strips, and especially the heads and necks of goats, cut away from the shoulders in a straight line, as at Vindonissa and Bonn.[22]
2) Secondary off cuts (Fig.2): the pieces left between the cut out items, often as a silhouette. In fact, the distinction between primary and secondary off cuts is not always clear in Roman contexts, since articles tend to be cut directly from the unneatened skin, except where large sheets are required, as for tentage or shields. Smaller items were simply arranged around the less suitable or damaged areas which nowadays would have been removed beforehand. It is, however, the secondary off cuts which give most information on the type of items manufactured, and, because they often come from the better parts of the skin, these are also most suited for animal identification.
3) Trimmings: narrow slivers left after neatening a roughly shaped article, often done on completion of the item (e.g. removal of the excess from a made-up shoe sole, shown in Fig.2 as narrow white areas next to the soles).

By far the most distinctive are the off cuts from shoemaking (Fig.2) and these are especially useful in establishing the type of leatherwork being carried out in Velsen, Valkenburg and Bonn, which we will examine below. Although it occurs in a civilian context, a find consisting almost entirely of off cuts (and a single worn-out shoe) from Maastricht[23] provides a good illustration of the potential of a well-defined group (Fig.3). Here, the left over silhouettes were sufficiently characteristic - and the stratigraphy was sufficiently clear - to enable the identification of an individual workman, entirely specialized in the production of single-piece footwear (the so-called carbatinae[23a]) on the evidence of the off cuts alone.

Such restricted, individual, rubbish disposal as at Maastricht is rare. More usual is the situation at Zwammerdam and Vechten, with their extensive riverfronts, where the mixture of obviously military items with obviously civilian ones suggests that refuse from both the camp and its neighbouring vicus was thrown indiscriminately into the river. Apart from the intrinsic

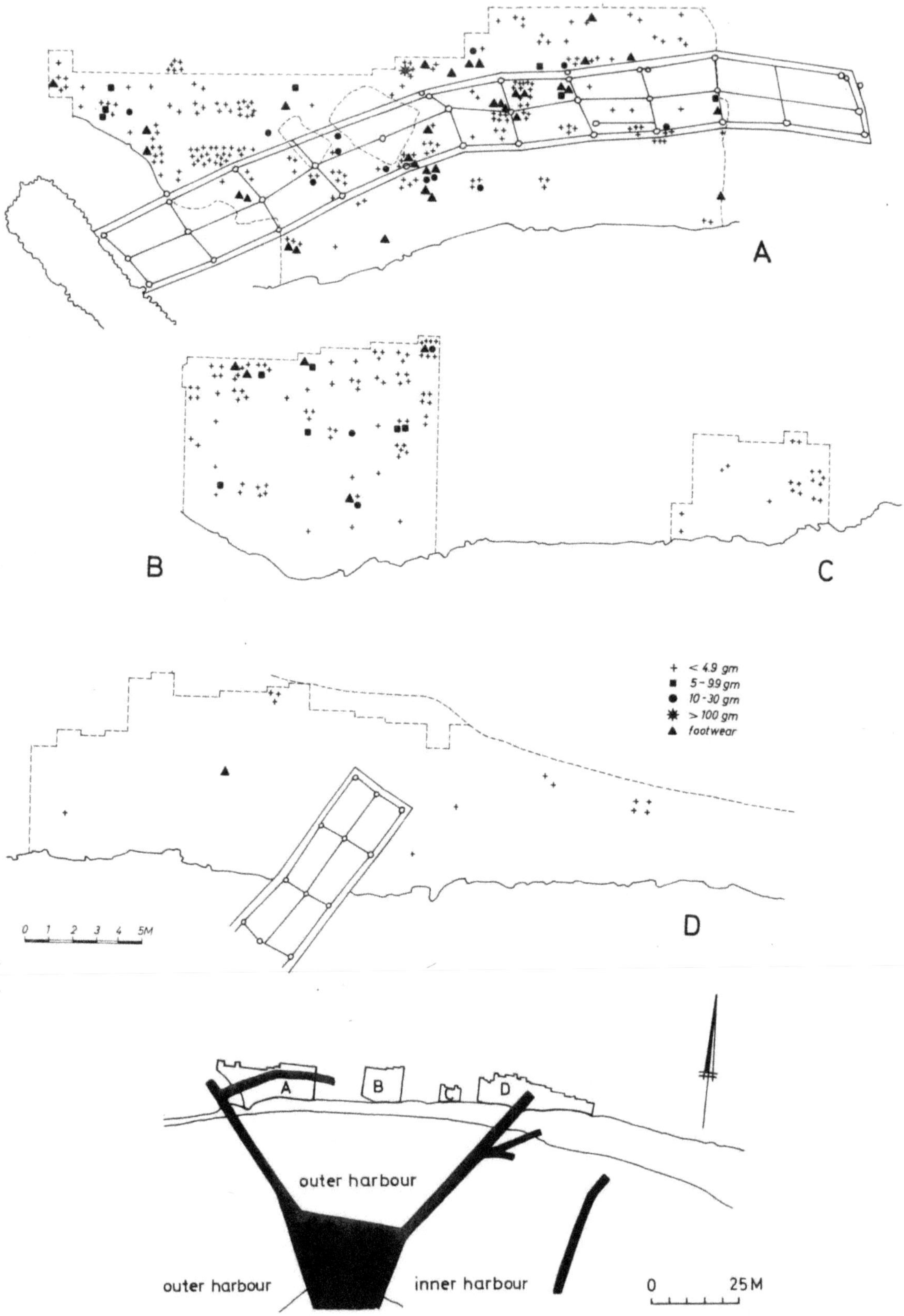

Fig. 4: Leather finds from Velsen excavated in one metre squares and plotted by weight.
Below: The harbour works at Velsen N.H. (plan, J.-M.A.W. Morel) with location of river-bed excavation trenches and the position of the jetties and moles.

interest of the objects recovered, there is little information on either production or supply to be gleaned from such mixed deposits. Similarly, along waterfronts, where the object is land reclamation, a concerted effort by authorities and/or individuals is likely to gather suitable material - including former rubbish dumps - from a wide area.[24] Plotting individual find concentrations might provide some differentiation in the 'primary' rubbish dumps (as opposed to re-deposited, 'secondary' rubbish) and could also lead to a better understanding of the way in which various categories of material were discarded.

Adverse conditions frequently hamper detailed work on waterlogged sites, but the distribution of leather finds from two of the first century sites in Holland, Velsen and Valkenburg, though not in themselves spectacular do illustrate two different processes of rubbish disposal which in turn clarify the system of manufacture and of the maintenance of leather equipment on these sites.

The actual site of the early fort at Velsen, occupied between c. A.D. 15-30 or a little later[25] is too shallow and too eroded for leather to be preserved within the defences. In the associated harbour, however, conditions are slightly better and due to the great care and precision in excavation, a quantity of poorly preserved leather was retrieved. When the finds are plotted by weight (Fig.4), the relationship to the westernmost pier and jetty is immediately apparent, with only occasional, much lighter items being washed further along by the current. A similar pattern emerges from the distribution of animal bone.[26] This implies that this particular pier was the chosen route for the disposal of camp refuse in open water, while the other piers were deliberately kept clear for shipping. The material found in this area is, therefore, likely to represent the range of activities undertaken in the camp, and might even reflect a concentration of rubbish-producing activities in the western half of the camp (see below, p.62 in connection with the evidence for an industrial quarter at Hofheim). The leather itself consists primarily of shoemaker's off cuts, sole trimmings and smallish fragments cut away from tents and shield covers, all of which suggests that manufacturing activities were restricted to the supply of the immediate necessities - footwear - while other equipment was merely kept in a state of good repair.

Although the leather from Valkenburg Z.H. was also clearly discarded deliberately, this is not a normal case of rubbish disposal. The preservation of so many virtually complete items of equipment can be attributed to the fact that a change in garrison coincided with the need to raise the site of the fort above the reach of sporadic flooding at some point in the occupation.[27] Presumably, damaged and even remotely

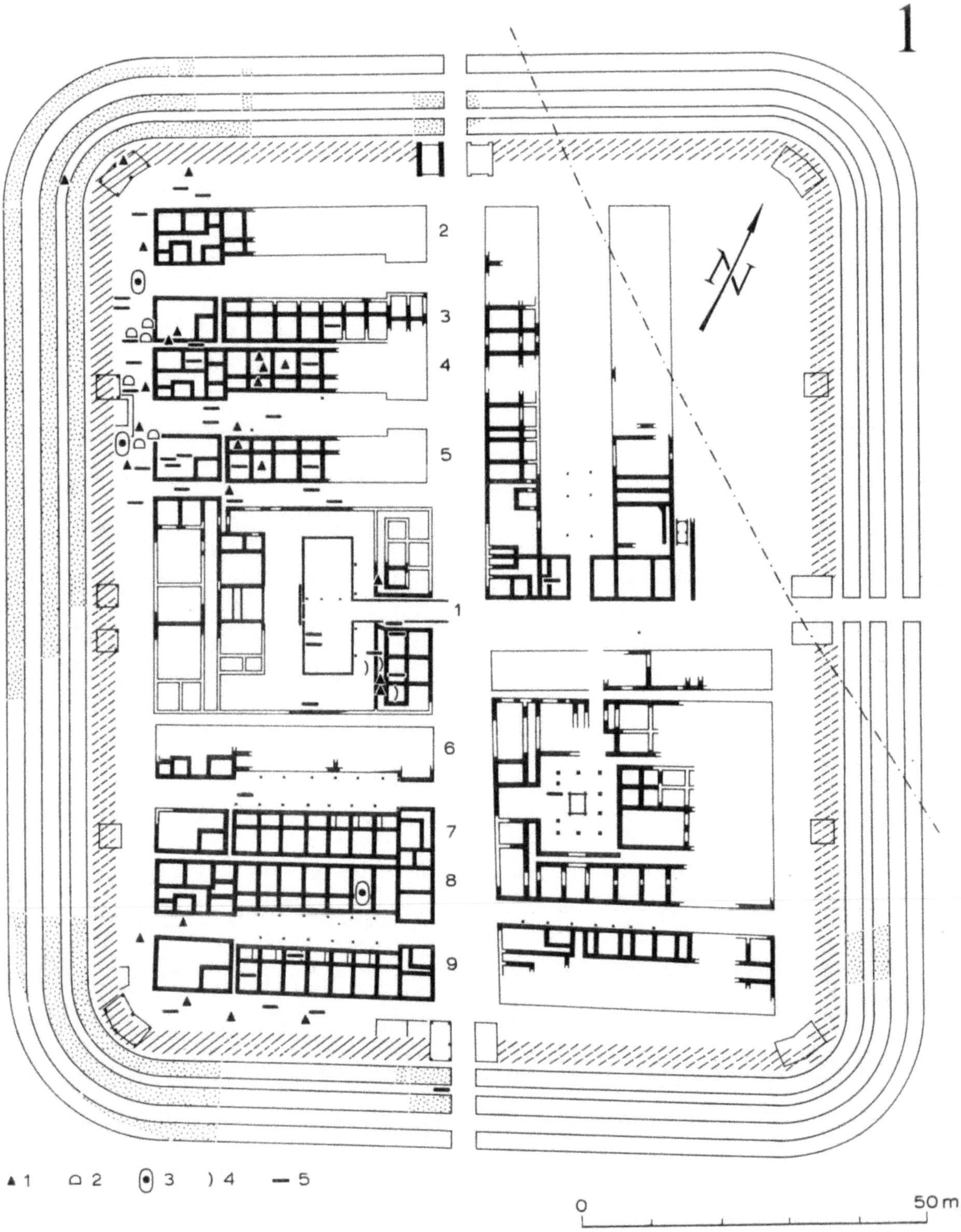

Fig. 5: Distribution of leather in Valkenburg 1 (location of find numbers based on GROENMAN-VAN WAATERINGE, 1967, fig.75a). 1. footwear, 2. tentage, 3. shield covers, 4. shoemaker's off cuts, 5. unclassified leather finds.

unserviceable equipment was abandoned to cut down on baggage. Here one might expect to find a fairly accurate reflection of the activities carried out within the camp, though not necessarily at the exact location (Fig.5). The shield cover in one of the arma of barrack block no.8 fits the context nicely, shoes scattered all over the site would reflect individuals sorting out their personal gear, while the concentration of tent and shield fragments at the western end of the barrack blocks 3 and 5 suggest the location of general stores for leather goods, from which new was issued and to which damaged items were returned for repair or rejection.[28] The fabrica-like function of the end buildings of barracks 3, 5, 7 and 9 became obvious when the structures in them were examined together with the finds,[29] but the distribution of leatherwork suggests a more specific function in terms of leather working and storage for the first two. Curiously, the concentration of large articles of equipment in these areas is maintained in the phase 2/3,[30] either because these two end buildings retained their function of leather stores, or because of the redistribution of rubbish from the earliest phase within this relatively short period of time (i.e. between c. A.D. 40-70). It is noteworthy that the find distribution suggests no other obvious candidate for a fabrica in the fort.

It should be pointed out, however, that although the find lists in the excavation records regularly mention the occurrence of 'footwear', other less obviously recognisable items are simply listed as 'leather'. Consequently, the distribution of footwear is likely to reflect the actual find pattern accurately, even though the other types of equipment, like shields, saddles and tents are under-represented on the plan.[31] The apparent concentration of shoemaker's off cuts in a wing of the principia presumably locates the camp shoemaker's shop[32] while a pack of leather from the same area, composed of pieces cut and discarded from items undergoing repair, suggests that general maintenance was also carried out in these wings. It looks as though these cubicles on either side of the principia entrance were in fact small workshops, and not actually part of the principia itself.[33]

The types of off cuts present in Valkenburg suggests a combination of shoemaking and general repairs, thus repeating the picture obtained for Velsen, but in marked contrast to the wide range of equipment produced at the legionary fortress Vindonissa. Here, the large and very varied collection of tools and off cuts attests not only to the manufacture of footwear, and the maintenance of equipment in general,[34] but also to the production of shield covers, tents, straps and other military paraphernalia. There is, of course, a considerable difference in size between the garrisons of Valkenburg and Vindonissa, which in itself might explain the relative scale of production, but in addition, the early garrisons of Velsen and Valkenburg were presumably initially entirely equipped from a base camp, and may have remained dependent on it until sources of supply became more settled and the camps more permanent. We may possess a record of the relationship between fortress and forward position in the earliest stages of conquest in a wooden writing tablet from Vindonissa itself.[35]

> "...*soleas clauatas fac mittas nobis ut abeamus. Cum veniemus*..."
>
> "send us the nailed shoes so that we may depart. When we arrive..."

The text, which is probably pre-Claudian in date, implies ready stores in the fortress of the shoes we know (from the presence of the off cuts) to have been made there. Campaigning forces may have sent in their requests for supplies when necessary. Both Velsen and Valkenburg would represent a slightly more developed stage, since these forward positions were already manufacturing their own footwear.

The boots with which the soldiers were fitted out have been found at a large number of first century sites, straddling the length of the Limes from Switzerland to northern Britain. All follow a similar cutting pattern, with an openwork, strap-like upper cut in one with a middle sole and sandwiched between an inner and outer sole, the whole unit being nailed together. There is no discernible regional variation, and uniformity even extends to identical nailing patterns on boots from sites as far apart as Vindonissa and Velsen.[36] It looks remarkably like a standard military issue, made to central specifications at the individual camps. A work roster from Vindolanda[37] shows the system of manufacture as deduced from the finds in working. No less than one third of the day's workforce is detailed to make shoes in the workshops. Presumably a skilled shoemaker (immunis, soldier, slave?[38]) cut out the patterns required the unskilled soldiers made up the footwear under supervision, rather as in the Netherlands prisoners and paupers made up military boots from ready-cut packs (with thread, eyelets and tacks all included) supplied by army shoemakers from the 1830s onwards.[39] If the slightly enigmatic "ad calc" of P. Gen. Lat. 1 really does mean 'to the shoes' as both von Premerstein and Fink attractively suppose,[40] this Egyptian work roster of A.D. 90-96 may reflect a similar situation in a legionary context. The products of the day's labour would be stored in the camp - in the case of Valkenburg 1, perhaps in the tabernae at the entrance of the principia, or in the end building of barrack 3 - awaiting distribution. In an account of two soldiers from Egypt, covering the year A.D. 81,[41] the standard deduction of 12 drachmas (i.e. 3 denarii) for boots and socks in each of the three pay periods is highly suggestive of a thrice yearly issue of these items.

The evidence for manufacture at the auxiliary camps extends only to the production of footwear, which implies that more specialized and longer lasting equipment probably continued to be made elsewhere, perhaps at the former base camps, at least until fully stable conditions pervaded. As with footwear, the leather goods from all sites are highly standardized in form, arrangement of individual components and use of specific seam and reinforcement types. Workmanship is highly professional with careful use and selection of leather, a minimum of wastage in cutting and complex, finely made seams. A system of regional workplaces provisioning groups of forts in the immediate neighbourhood is attractive, but is not supported by any positive evidence, neither is it the only way of achieving uniformity, as we have seen with the military boots. It would not be justified on present evidence to assume that it was the legionary fortresses which were supplying the auxiliary forts in their region with leather goods.[42] To judge for the types and quantities of off cuts, Vindonissa would have been quite capable of meeting all its own requirements, but with the exception of a single fragment of an oval shield cover,[43] there is no evidence that it was involved in any way with either the manufacture or maintenance of auxiliary equipment.

The leatherwork recovered from first century sites seems to suggest that auxiliary forts drew on supplies from more established centres (not necessarily legionary), at least in the earlier stages of conquest and occupation. Footwear was soon made in every camp, although auxiliary forces may have continued to rely on their former base camps for certain specialist pieces of equipment: there may also have been some form of local specialization between neighbouring auxiliary camps. The uniformity of the equipment - over a wide geographical area and for such a long period in time implies a well developed central control over manufacture, which invites comparison with the standard models which were issued regularly to Dutch garrisons in the early part of the 19th century, and to which all equipment, whether made privately, on contract or in army workplaces, had to conform as regards style, manufacture, quality and colour.[44] Such a system of central specification would explain the remarkably close correspondence between equipment found at widely separated sites quite adequately, without the necessity of imposing central manufacture as such.

Perceptible changes in the composition of later leather complexes may be linked in part to what seems to be an overhaul of military affairs in the first half of the second century. The effects of the reorganization are seen not only in the introduction of new equipment,[45] but also in the acceptance and perhaps also the extension of trends in supply which were already apparent in the later first century.

In leatherwork, the changes seem to be reflected in a loosening of uniformity which is especially noticeable in the footwear. The standard, three-piece military boot disappears from the archaeological record early in the second century and soldiers turn to the normal closed footwear of the types used by civilians.[46] None of the typical military boots occur at the _fabrica_ on the Bonner Berg. The shoes are all of ordinary civilian types, though concentrated in the larger sizes (sizes 38-42) and thus indicative of a predominantly adult male population.[47] Sites in northern Britain show the same variety and again, no standard military boot can be identified.

The type spectrum of leather finds from the second century legionary _fabrica_ on the Bonner Berg contrasts markedly to what we have seen at first century sites. Although little documentation of the 1906 excavation remains, the leather seems to have been collected with considerably more thoroughness than the pottery, and so probably presents a fairly accurate picture of the activities on the site even though no exact find spots are recorded.[48] In contrast to Velsen, Valkenburg and, more especially, Vindonissa, there is little evidence for the manufacture of footwear (possibly only repair soles) and the _fabrica_ seems to have specialized in the making and the repair of military equipment such as shields, tents and baggage covers. This site was in use for too long, and the place of the leather finds within the time range is too uncertain for the activities to be directly associated with the re-equipment of the _Legio I Minervia_ after its return from the Dacian wars in 107, even though the _fabrica_ was probably founded around this date. The leather finds therefore most likely represent the normal work of a _fabrica_ in ensuring that equipment was at all times in top condition - with the exception of footwear. Since, in many respects, the refuse from the Bonner Berg is closely comparable to that from Vindonissa in its context, the absence of not only the standard military boot, but also of evidence for the manufacture of the footwear itself does suggest a modification of existing practices. The suspicion arises that for footwear at least, a different system of supply was developing, one in which civilian shoemakers, working within their established traditions and to their normal models were incorporated. The activities on the Bonner Berg, with its evidence for metal smelting and general repairs is closely comparable to the situation described in the work roster P. Berlin inv. 6765[49] (discussed by M. Bishop elsewhere in this volume, and to whom I am grateful for bringing this text to my notice), which reveals a combination of soldiers, _immunes_, civilians and slaves working on a variety of equipment, including shields, in an unspecified military _fabrica_. This incorporation of civilians in the workforce,[50] together with the tendency for manufacturing to move outside the confines of the camp[51] already in the first century, prepares the ground from increasing civilian involvement with military supply. Furthermore, the growth of the vici may have made individual purchase, or the placing of contracts (or imposition of levies) more attractive, especially in smaller forts. From here, of course, it is but a short step to requisitioning.

The presents of shoes and socks received by an early second century auxiliary soldier in Vindolanda[52] may have been luxuries over and above the military issue, but the gift may also be a reflection of the increasing personal responsibility (and choice) which seems to be implied by the persistent requests from Egyptian soldiers to their family for food, clothing and money.[53] A 'privatization' of personal equipment might explain the variety in footwear as well as the more exuberant decorative schemes on armaments.[54] On the other hand, the military issue - of clothing especially - may have been somewhat on the short side to encourage thrift. The problems of keeping equipment in top condition whilst guarding against unnecessary expenditure exercised the ingenuity of Dutch military planners throughout the 17th and 18th centuries, eventually culminating in 1796 in a resolution stipulating a twice yearly issue of clothing, followed in 1807/08 by a detailed specification of the length of time each individual item of clothing and equipment was expected to be serviceable. This varied from six months for infantry shoes to 16 years for a saddle.[55] The material was military issue, and, as in the first century, a standard sum was deducted from pay to cover the cost of replacement at the fixed intervals. In the first century account of two soldiers from Egypt mentioned above (note 41) clothing seems to have been issued once a year, shoes and socks three times, although one of the soldiers is perhaps charged extra in the first pay instalment for additional gear or for replacements. In the Dutch system, clothing could be retained by the soldier on discharge if less than one quarter of its nominal life remained.[56] An additional charge was made for replacements betweentimes, however necessary, and it is perhaps in a like system that the requests for family aid should be interpreted.

Whether equipment such as shield covers, tents etc. was ever produced widely in the second century auxiliary camps is, in the absence of unambiguous contexts, difficult to say. Conditions for the preservation of leather are poor in the higher levels of Valkenburg (periods 4-6), while in Vechten and Zwammerdam the difficulties in both dating and disentangling military from possibly civilian production hamper the assessment of the activities represented by the finds. The possibilities open for variations in the methods of supply to individual camps will no doubt have depended to a great deal on the availability of local resources and skills. That some sort of control continued to be exercised over the production of strictly military equipment made of leather is likely, since the variety of second century footwear contrasts markedly with the continuing first century standards of workmanship and uniformity of pattern and stitching on both tentage and shield covers. Footwear apart, there is still a very close resemblance between the leather from Velsen (early first century), Valkenburg (mid first century) and Vindonissa (first century) on the one hand, and the Bonner Berg (early second century), Newstead,[57] Birdoswald[58] and Bar Hill[59] (all early-mid second century) on the other. The manufacture of most of the items of leather required by the army lies well within the capabilities of the average leatherworker, especially if, as seems probable, standard patterns were available. Smaller camps, of course, may not have had a separate specialist for leatherworking at all: rather slovenly stitching and strange seam types at Zwammerdam might betray soldiers carrying out the necessary repairs themselves[60] on equipment originally made according to standard procedure by a specialist working at another camp: more a case of inter-camp cooperation than of 'central supply'. It is this rather than distribution from a legionary centre which might characterize the organization of supplies to the small, closely spaced camps along the Rhine Limes. It may be that the pressure of keeping up with running repairs compelled the shedding of one of the fabrica's tasks: the provision of footwear was perhaps one of the easiest branches for civilians to take over, especially in view of the growing markets for such products outside the camps.[61]

Thus in the first half of the second century we seem to have a dual system of provisioning. Supply of footwear seems to have moved into the civilian sphere, whilst some sort of military control still extended over the other items of leather equipment. Whether these were still manufactured by soldiers - in or outside the camps - or by independent civilians is, in the absence of suitable find concentrations unknown. Lack of well defined complexes for the later second century and beyond complicates the assessment of any trends which might otherwise be discernible in the earlier material. Did the military establishment continue to provide its own tents, shields and baggage covers or were these also handed over to civilians? Or is civilian involvement in leatherworking confined to the provision of footwear? It is perhaps noteworthy that there are no shoe factories listed in the Notitia Dignitatum, and in 311-312 it was necessary to requisition shoes,[62] even though (admittedly at an earlier date), hides were bought or levied for the purpose of making arms (by whom is not specified: at this time the military authorities could well have provided civilians with the raw materials for making up under contract or as an imposition).[63]

It is also significant that no obviously military leatherwork was recovered from the wells at the Saalburg, despite extremely good conservation and the very great interest of the excavator, L. Jacobi, in such perishable materials.[64] If the vicus was actively concerned with the provision of military leatherwork, we might have expected more evidence of this amongst the clearance debris dumped into disused wells after the destruction of the settlement in the 230s. The absence of military leatherwork here contrasts with the situation at Zwammerdam and Vechten (both sites extend into the third century), but it might also be the result of a general running down of the Saalburg garrison at that time and the fact that the wells for the most part represent a single clearance horizon c. 230,[65] while the dumps at the other two sites are an accumulation over a longer period and perhaps from a less restricted area of the settlement. Here again, the character of the leather-containing contexts at the Saalburg strongly influence the type of material recovered, and a direct comparison between destruction debris and normal refuse accumulation is impossible, unless far more is known about the distribution of activities within the settlements.

An additional aspect to be considered is the supply of the raw material itself. As leather is a by-product of meat consumption, the osteological evidence from military sites provides indirect evidence for the type, and possibly the quantity of leather available locally. Examination of leather from military sites along the Rhine Limes shows that cattle hides were used chiefly for footwear and the large rectangular shield covers[66], while goatskin was utilized for virtually all other equipment. Gansser-Burckhardt, who examined the leather from Vindonissa and who was the first to discuss the entire subject in any detail, considered that the goatskins were imported from Italy and even found some improbable 'Old Italic' inscription to prove it.[67] However, in view of the restricted use of land transport for bulk goods, mule trains carrying skins over the Alps seem an unlikely prospect, even in military contexts. In fact, local sources must have been readily available, for Switzerland to this day maintains large herds of goats. Further north, the situation is more complex, for the numbers of goats implied by the leather identifications fail to materialize in the osteological record.[68] It is possible that goat meat was not consumed in the urban and military settlements, and if the animals were raised primarily for milk (cheese) and skins on the least hospitable soils, they might leave little trace archaeologically. Goat bones continue to be infrequent in medieval domestic sites. They are, however, present in considerable numbers amongst the refuse of a medieval tannery complex at 's Hertogenbosch[69] and the discrepancy between the animal bones and the skin identifications at Roman sites may, therefore, be a consequence of the lack of excavation on suitable industrial locations. The difficulty may also in part lie with the actual identification of the skins, which is based on the microscopic examination of the arrangement of the hair follicles on the grain side of the leather. Quite apart from the possibility of any structural change in the skins resulting from an additional 2000 years of control and breeding, it is difficult to distinguish between even modern sheep and goat skins.[70] The more primitive hairy breeds of (modern) sheep display a follicle pattern which approaches that of the modern goat, so the improved Roman wool bearing sheep would presumably produce leather closely comparable to that of modern average wool breeds.[71] So in a situation where goats, native sheep and improved breeds can be expected, the choice lies between a continuum of goat/hair sheep/wool sheep: difficult at the best of times and even more so with archaeological samples, which generally come from worn-out and discarded items which have been subjected to all kinds of secondary stresses. In addition, the structure of the follicles is not the same over the whole skin: neck, leg, axillia and belly portions may differ considerably from the best parts of the back. Some of the 'goats' may, then, turn out to be 'hairy sheep'. On the other hand, the leather from even hairy sheep is flabbier than that from goat, which would make it unsuitable for purposes such as tentage, while the tendency for complete dermal/epidermal separation which is a characteristic for all tanned sheepskins is observed only very rarely on Roman skins which would on other grounds be classed as 'goat'. Furthermore, the Romans themselves were well aware of the differences in quality between goat and sheepskins[72] and the military at any rate are unlikely to have been fooled by the substitution of sheep, however hairy. The necessity of identification is one reason why horns and hoofs are left on the skin.[73] Vast herds of goats were certainly kept in Italy[74] and the skins could have reached the Rhine forts by water, which would have been less expensive than overland. But even so, it is difficult to see the army as being entirely dependent upon long-distance and slow transport for such essential material, especially in view of the attempts to secure local sources for virtually all its basic requirements.[75] In sum, I would regard the incompatibility as an effect of the distortion introduced by policies of excavation: as in the medieval example quoted, the excavation of a Roman tannery would probably redress the balance.[76]

Meat consumption by the soldiers themselves would release a steady supply of cattle hides, but this may have been insufficient to meet requirements. To explain the observed wealth of Roman imports to Öland, Ostergötland and Västergötland, Hagberg[77] makes out a strong case for a flourishing trade based on the export of cattle hides to the Roman Empire. Although modern leather buyers are sometimes able to tell the country of origin of particular hides, this is impossible with ancient material since there are so many variables (climate, breed, size, age, nutrition etc.) quite apart from the present condition of the surface of the leather. There are, however, two features which occur on the leather itself and which provide a more direct hint of some form of trade in hides and skins. On a very large number of off cuts from Velsen and Xanten swarms of shallow depressions occur on the grain side of the leather. These closely resemble the depressions on modern hides from Africa, which are caused by strewing raw salt or so-called 'salt earths' over the freshly flayed skins to preserve them until transportation to the tanneries in the towns.[78] These 'salt pocks' occur chiefly on cow hide and it would be tempting to associate them with evidence from ancient sources that hides not only formed an important part of the levies imposed on newly conquered areas, but were also a major item of export from unconquered barbarian lands.[79] Of course, salt preservation could have been practised within the empire as well, nor is it the only method of limiting putrefaction,[80] but it is interesting to note that Xanten is a frontier post, presumably active in trade across the Rhine, while Velsen is an outpost, dependent for materials either on support posts situated in the secured hinterland or on local tribute/exaction. At this particular site, the oxhides demanded by Olennius, which were the immediate cause for the rebellion of the Frisians in A.D. 28, take on a new significance.[81]

Another feature which may be associated with trading in hides and leather are the stamped inscriptions, often initials or names, occurring on off cuts from a number of sites, both military and civilian.[82] Certain marks were applied to the skins prior to tanning (hence often illegible: some of Gansser-Burckhardt's 'Old Italic' inscriptions are of this type). At Vindonissa the word TOTA occurs several times and it also appears on an off cut from Zwammerdam.[83] as skins were - and are - usually handled in bundles, TOTA might refer to a whole batch being claimed for official use.[84] Other markings may have something to do with toll payments and import duties. It is perhaps significant that letters are often well formed seriffed capitals, with names frequently in the tria nomina form, a status one could hardly expect for first or second century vicus tannery owners or stock farmers. These stamps do seem to place the supply of some hides and skins in a rather more official sphere of activity. Again, we may speculate on the import of both tanned and salt-cured or dried hides from Free Germany, with official stamps recording either the payment of tolls, sanction to trading activity or actual purchase for official requirements. Other marks applied after tanning probably identify the owner of the tannery and thus serve as quality controls in much the same way that medieval and modern tanners have individual marks so that substandard products can be traced back to source.[85] The absence of any military markings and the fact that the same mark rarely occurs twice on the same site rather suggest that tanneries were small scale, private enterprises. Tanners, like potters and some tile makers may have worked entirely for the army, though not part of the force themselves.[86] In legionary camps at least there would have been sufficient demand to support such specialist civilians, and since the army itself was a major supplier of hides as well as a major consumer of the tanned product, a fairly close interdependence between army and local tanners is to be expected.

Whether the army itself was also involved in the actual tanning is unclear. Tanneries are particularly elusive in the archaeological record, in part because of the extreme simplicity of the equipment used, but also because they tend to be situated outside the settlements (downwind) in areas where they are less likely to be located by excavation. A large, industrial tannery at Pompeii had fixed vats, but wooden tubs and even pottery vessels provide suitable receptacles for the tanning liquor. A simple method of tanning, known as 'bag tanning' would require no containers at all, since the tanning liquor is poured into the sewn-up skin, which is suspended to allow the liquid to pass through it, thus simultaneously tanning the skin.[87] Unhairing and fleshing can be carried out with rib bones or wooden scrapers:[88] large, two handled fleshing knives are rare in the archaeological record and may be associated only with particularly large scale tanneries, or with the preparation of special thin leathers.[89] In the absence of the sort of environmental evidence from Vindolanda,[90] the surest evidence for the location of industries dealing with skins is the over-representation of horn cores and metapodials in the osteological record, since these are often taken off with the skin when the beast is flayed.[91] Once again, the industry is recognised, not by its products but by its characteristic refuse. However, when a tannery deals with salted or dried skins, even these signs may be absent, since horns and hoofs are frequently removed in the preliminary cleaning to avoid damage in stacking and bundling.

The published evidence for the location of a tannery at Hofheim[92] may be interpreted rather in terms of the preliminary treatment of skins. The whole north-west corner of the earthen fort was occupied by storage facilities and workshops, and that some activity concerning skins was carried out there is likely not only from the presence of water tanks but especially from the large numbers of cattle horn cores which were found in the rubbish dumps between the buildings R and T (Fig.6). Judging from the pattern of refuse disposal, an analogous industrial quarter, centred on two boat-houses, might be present at Velsen

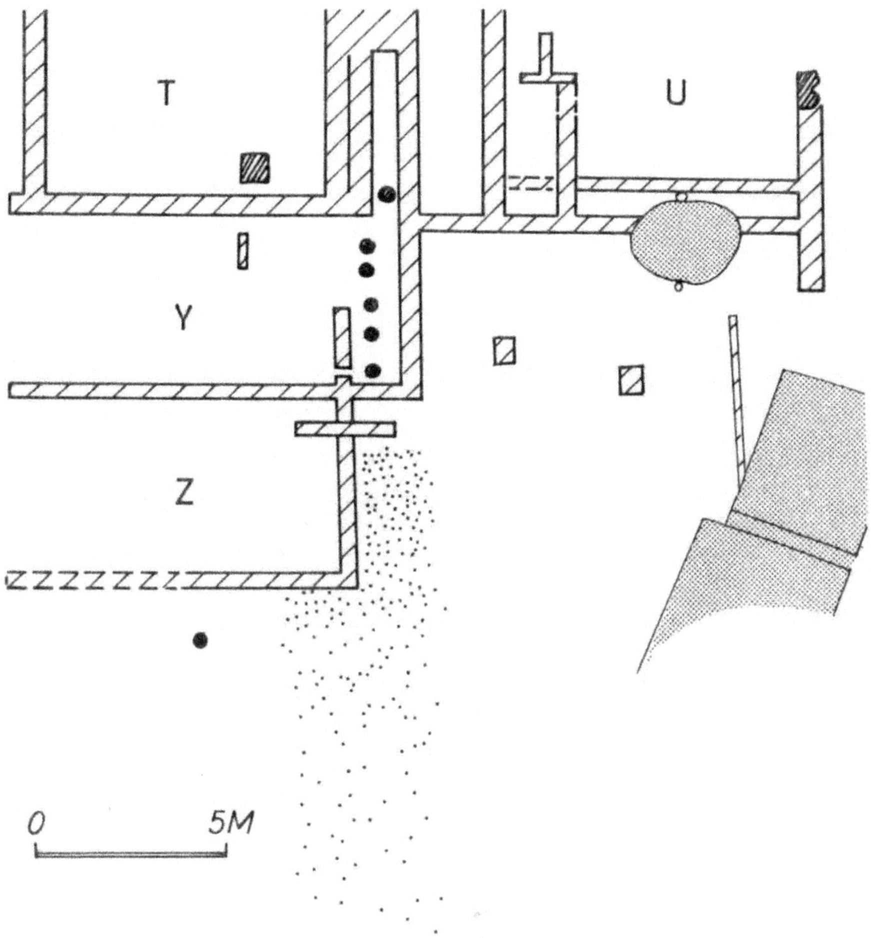

Fig. 6. Enlargement of Ritterling's plan of the industrial quarter at Hofheim with post holes of drying frames. Shaded areas represent water tanks (after RITTERLING, 1912, Taf.III).

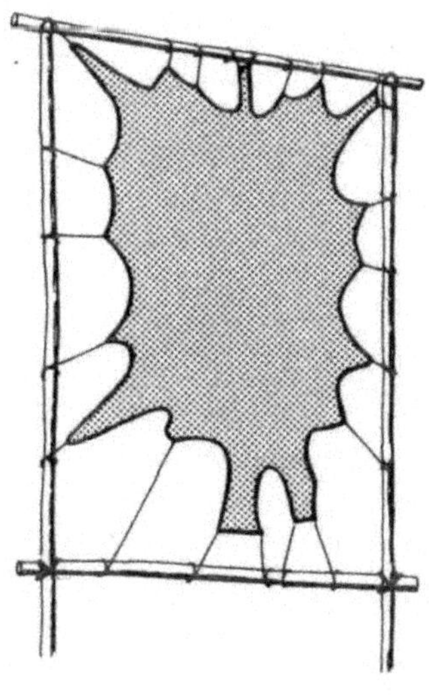

Fig. 7. Hide drying frame (from ATEN et al., 1955, fig.42).

(above p.51). The over-representation of cattle horn cores[93] from the riverine refuse dumps suggests that activities associated with hide processing were also a feature of this early fort. However, the forward position of both Velsen and Hofheim perhaps makes it unlikely that so lengthy a process as tanning was carried out there, but hides from cattle slaughtered for consumption were probably cleaned, prepared and dried or salted, ready for bundling and transport to the more secure hinterland. Later, with more settled conditions and the growth of the vici, the freshly flayed hides could be passed directly to local, civilian tanners. Ritterling remarks[94] on the formless swarm of post holes to the south of building Z and skirting the rubbish tips, tentatively suggesting that these posts may have had something to do with stretching and drying hides. The dimensions and depth of the posts seem excessive for simple flat pegging, and if the plan is to be believed, the posts appear to be set in pairs, about one metre apart. This suggests that the Romans used a system of post frames to which the fresh, cleaned hides were laced (Fig.7). Today, this method is advocated by the Food and Agriculture Organization of the United Nations, since it achieves more thorough and even drying, avoiding the very considerable loss by putrefaction which flat pegging brings to traditional tanning industries in the developing countries.[95] In addition, a roof can be thrown over a series of frames for protection, though the scatter of posts at Hofheim suggests rather the treatment of hides on an individual and irregular basis.

Environmental evidence from Vindolanda confirms that tanning was being carried out at this particular auxiliary fort[96] but the absence of similar exceptionally well preserved, in situ, contexts at other sites means that it is difficult to assess how common this activity was in the other permanent forts. In view of the general need for water and the tendency to site noisome industries away from the settlements, the lack of any specific osteological, environmental or structural evidence for tanning at Vechten, Valkenburg and Zwammerdam may only mean that it has so far eluded the excavators. At Bonn, Lehner may not have recognized the significance of bones and horn cores (though Ritterling did at Hofheim) but he would surely have mentioned them if they had been present in any quantity. For Gansser-Burckhardt it was obvious that if there had been tanneries, these would have been situated in the (unexcavated) valley below Vindonissa.[97]

Documents from Egypt seem to indicate that even quite small numbers of skins could be bought (or requisitioned) for military purposes,[98] and that supplies of leather might be levied from a town defaulting on its municipal dues.[99] In each case, the actual tanning was left to the civilians, which is what the absence of military marking on leather would already lead us to expect. In detail, the sources of supply will have varied according to the local situation, but it is clear that the army exploited every possible method to meet its requirements.

SUMMARY

It is only fair to emphasise the shortcomings of leatherwork as evidence for the mechanics behind the production of Roman military equipment. The mere fact of preservation already introduces bias for there is little chance of finding leather at all outside waterlogged deposits, and these are perhaps the most difficult of all deposits to interpret or date. We are entirely dependent on refuse, so a clear definition of material leading to the location of workshops and identification of the specific products made there is simply not feasible. Even if it can be shown that leather equipment for military use was being made outside the camps, in the *vici*, this is not to say that production had passed out of military control, or that military personnel were not involved. Once camps were established on permanent sites there would probably be a tendency for the workshops to move outside the walls in any case, as seems to have occurred at the Saalburg.

In many respects, leather is an atypical product and the organization of supplies need not follow the trends established for other pieces of equipment. To start with, it is a perishable by-product of stock raising for consumption. The scale of the army's requirements was probably greater than for any other military essential apart from foodstuffs and textiles, which would tend to put the entire organization of manufacture on a different footing to that of, for instance, small bronzes. That leather is used in large quantities in several very basic and essential pieces of equipment which had to be uniform - shields and tents - would also favour central control or centralized manufacture.

All I can say at the moment is that the army probably obtained its leather via imports from outside the Empire or from small private manufacturers. In the first century the army was entirely self-sufficient in the manufacture of leather equipment, with workshops presumably concentrated around certain base camps. Only essential maintenance and - equally essential - the production of footwear was carried out at the auxiliary camps themselves, although the hides obtained via the food supply were processed and were presumably dispatched to the base camps for tanning and use in the production of new equipment (Fig.8). All equipment was highly standardized and was obviously made by professional craftsmen. During the early part of the second century, the standard issue footwear ceases to be used and production of shoes seems to move into the hands of civilian shoemakers, though the exact form which the arrangement took is open to several interpretations. The continuing uniformity of strictly military equipment such as shields and tents suggests that the production of these remained under military control (Fig.9). Though central workshops could have been used, a closely supervised system of contracting out to civilian craftsmen cannot be entirely excluded. This might indeed form a better framework in which to fit the apparent extension of private enterprise and the increasing laxness in matters of uniformity and quality which is suggested by the less meticulous workmanship at Zwammerdam.

Since Roman leather complexes in the Netherlands cease entirely after 250, and large, datable groups are not available after even the mid second century, this review is of necessity incomplete. Future regional studies and an increasing density of finds spots are certain to clarify the processes of military manufacture and supply especially in the second century, as well as providing additional evidence for the range and development of leather equipment.

ACKNOWLEDGEMENTS

I would like to thank all those who read and commented on this article, and especially Drs J.G. Kerkhoven, curator at the Koninklijk Nederlands Leger- en Wapenmuseum 'Generaal Hoefer', Delft, whose assistance with the details of 18th and 19th century military provisioning was invaluable, and A. Visser, IPP, Amsterdam, who produced the drawings.

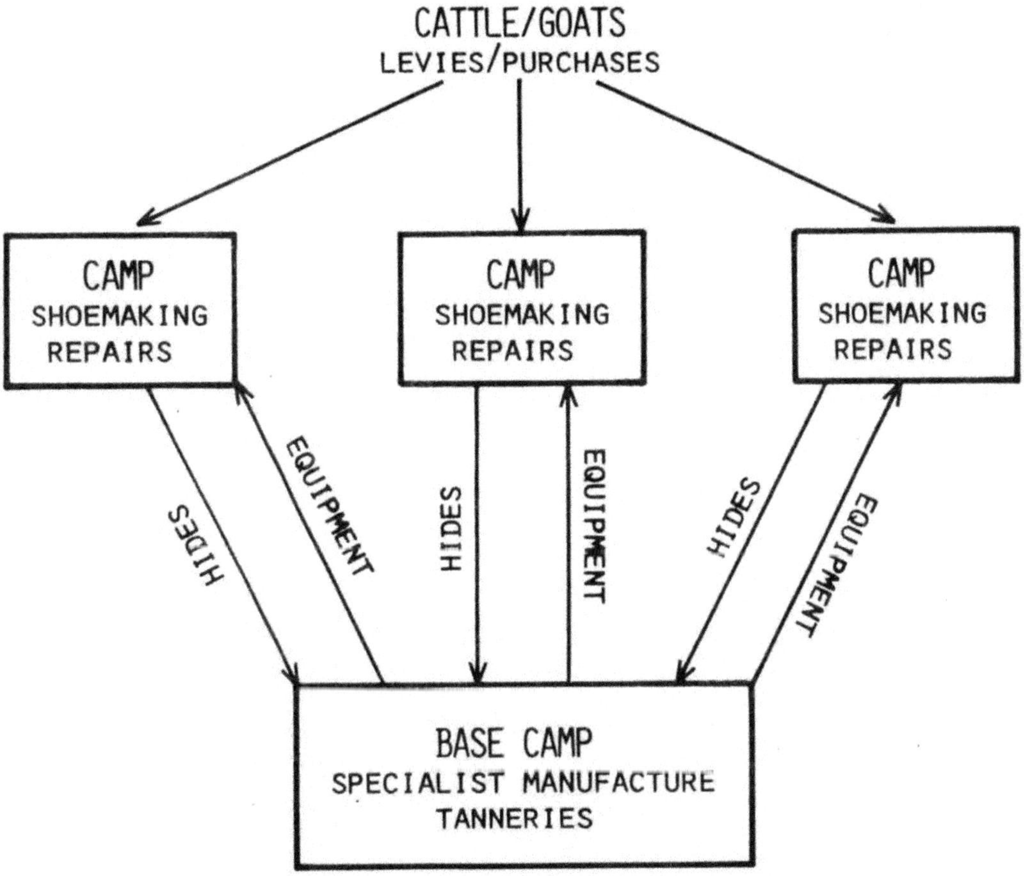

Fig. 8. Leather supply in periods of forward movement.

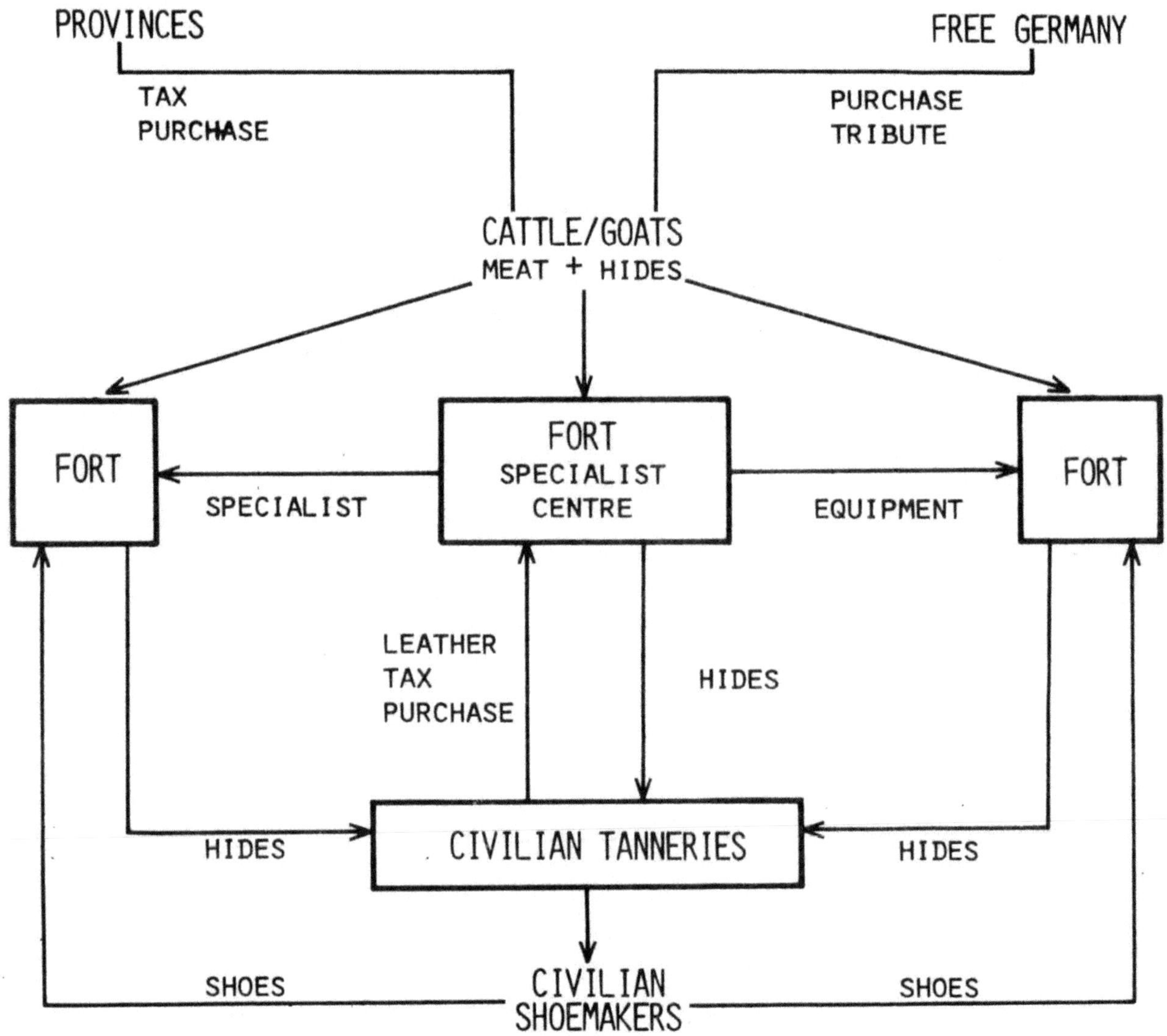

Fig. 9. Leather supply in periods of consolidation.

NOTES

1. Alum tawed leather, aluta (A. Mau in R.E. I (1894) under aluta, Pliny Nat. Hist. XXXV, 190), was known in Roman times. This mineral tannage produces a light, soft leather, white in colour (hence the medieval Whitetawyers) and suited to staining, though with a low water resistance (REED, 1972, 61-4). Alum production in Egypt was under a state monopoly, cf. JOHNSON, 1936, 387 no.242 (P.Oxy 2116).

2. GANSSER-BURCKHARDT, 1942, henceforth V.L.

3. GROENMAN-VAN WAATERINGE, 1967, henceforth Valkenburg.

4. Unpublished.

5. VAN DRIEL-MURRAY & GECHTER, 1984, henceforth B.B.

6. VAN DRIEL-MURRAY, 1977b, later material unpublished.

7. Unpublished.

8. For the problems of distinguishing leather from cloth on sculpture, cf. ROBINSON, 1974.

9. Most of these items are represented in the extensive leather complexes from Vindonissa (note 2), Valkenburg (note 3) and the Bonner Berg (note 5), all with further references. In addition, for writing tablet envelopes see BAATZ, 1984 and Valkenburg 126-9 and for saddles, LAWSON, 1978, 143-57.

10. V.L. 56-8, and reliefs such as that illustrated in GOLD DER THRAKER, cat. no. 493, 229-30.

11. DOPPELFELD, 1980, 299; CLOSE-BROOKS, 1977/8.

12. TEUPKEN, 1823 Sect. 4, 69-72, Sect. 5, 73-4.

13. ROBINSON, 1974; GROENMAN-VAN WAATERINGE, 1974, 80-1.

14. B.B. 24-5 for a summary.

15. WILD, 1970, 9.

16. Tents would be especially susceptible to long term storage and it is here that one might expect a crisis when it came to sudden unexpected campaigns after a long period in permanent forts.

17. The difficulties experienced by the Dutch authorities in

1914-1918 in obtaining sufficient leather of adequate quality, mainly for the manufacture of boots and harnesses, graphically illustrate a crisis situation. Even so, it was the rumour of inadequate footwear which virtually brought the cabinet to collapse, not the abysmal conditions in which recruits were housed. BOSBOOM, 1933, 105ff. and 164ff.

18. As for example in the earlier reports concerning the fabrica on the Bonner Berg. Since 1907 (LEHNER, 1907, 239 to VON PETRIKOVITS, 1974, 104) the site has been referred to as a shoemaker's workplace purely on account of the shoes and soles found there. All the off cuts and seamed fragments - some large enough to make their purpose as shield covers or tent panels obvious - were still interpreted in relation to the easily recognizable shoes and the true function of the fabrica in making and repairing items of military equipment was totally obscured. B.B. 3.

19. As for example where old leather is stripped off a wooden base which is to be re-covered, as is the case with the shields at the Bonner Berg, and, in a medieval context, sword sheaths from Leiden, cf. VAN DRIEL-MURRAY, 1980, 39.

20. Technically, 'skin' refers to small animals such as sheep and goats while 'hide' refers to large ones such as cattle, but here skin is also used for the untanned pelts of both.

21. V.L. 7-9, others unpublished.

22. V.L. 73, 74, Abb.2. B.B. 1-5, Pl.13.

23. Valkenburg 143; PANHUYSEN, 1984, 49-50 and VAN DRIEL-MURRAY, forthcoming.

23a. In fact a Greek word, but it is widely accepted to describe the single piece footwear of the Roman period: BUSCH, 1966, 166; CURLE, 1911, 152.

24. There are numerous medieval examples in the Netherlands, such as at the Boommarkt, Leiden, where individual, sporadic 'bucket fulls' of household and industrial rubbish were clearly visible amongst the centrally organized clay and sand fill of the reclamation. In another context, but illustrating the consequences for dating of the material in such levelling operations, see HARTLEY, 1972, 46 and M. Rhodes in MILNE & MILNE, 1982, 87-8.

25. Most recent survey, MOREL, in press; earlier summary, MOREL & DE WEERD, 1980.

26. MOREL et al., forthcoming.

27. Valkenburg, 195-7; GLASBERGEN & GROENMAN-VAN WAATERINGE, 1974, 20; DE WEERD, 1977, 271.

28. Before the fabrica-like function of the end buildings became clear, the distribution was seen rather as the result of the construction activities associated with the new lay-out of the camp, Valkenburg 212.

29. GLASBERGEN & GROENMAN-VAN WAATERINGE, 1974, 11, 25.

30. Valkenburg Fig.75b.

31. The composition of 42 of the 79 find numbers stated in the field notes to contain leather, and attributable to Valkenburg 1, can no longer be established since the numbers assigned became separated from the objects during storage. The leather survives as 'stray finds' and is described as such in Valkenburg 35-42. The same fate has overtaken many of the shoe soles, but these were readily identified in the field, so the distribution can be plotted from contemporary records, even if individual finds can no longer be correlated with their original find number. The majority of stray finds (nos. 1-78 and possibly 79-84) come from the 1941 and 1942 campaigns of excavation, when work concentrated almost exclusively on the north-western sector of the fort and on the principia. This suggests that at least 5, and perhaps 8, more fragments of shield covers could be added to the concentrations around the end buildings 3 and 5 and their barracks, as well as more tentage, purses and tablet envelopes (Valkenburg 35).

32. Valkenburg 196.

33. VON PETRIKOVITS, 1975, 49 for the varied uses to which the tabernae could be put.

34. V.L. 7-22, 98-109; GROENMAN-VAN WAATERINGE, 1974, 82.

35. BOHN, 1925, 14.

36. On the aspects of dating this and other types of footwear, as well as parallels and references, VAN DRIEL-MURRAY, in press. This true military boot (coinciding with the literary references to caligae as the typical soldier's boot) is illustrated in Valkenburg, Fig. 46 and LINDENSCHMIT & LINDENSCHMIT, vol. IV, Taf.37.

37. Tab. Vind. 1, BOWMAN & THOMAS, 1983, 77-9.

38. The slave Peregrinus, a sutor caligarius from Carnuntum (SWOBODA, 1964, 179) may have worked in such a mixed force, possibly in a skilled capacity.

39. KLAPPER, 1836, Sect. 977.

40. P. Gen. Lat. 1, verso, part V. Cf. VON PREMERSTEIN, 1903, 1-46 and FINK, 1971, 106-14. It seems hardly credible in this case that the P. Clodius Secundus mentioned in this roster spent two whole days polishing up his centurion's shoes: it is more likely that he was detailed to the manufacture of shoes for the use of Helius' century, in the manner described in the Vindolanda document, note 37. NICOLE & MOREL, 1900, prefer activities in the building trade.

41. P. Gen. Lat. 1, recto, part I, FINK, 1971, 243-9.

42. For a different view, VON PETRIKOVITS, 1975, 138.

43. V.L. Abb.54b.

44. TEUPKEN, 1823, 94-7.

45. OLDENSTEIN, 1976, 106-9; B.B. 4. Adoption of oval shields by legionaries, B.B. 4, 36 and VAN DRIEL-MURRAY, in press. The oval shields had been in use for long enough to require repairing at the fabrica of Leg. I Minervia at Bonn, which was in use from about A.D. 110-40.

46. VAN DRIEL-MURRAY, in press.

47. In view of the very mixed nature of the work force listed on P. Berlin inv. 6765 (ChLA 409), the small sizes could well have belonged to children or apprentices working in the fabrica, rather than to exceptionally young (or small) soldiers, as I previously suggested in B.B. 23-4.

48. LEHNER (1907, 239) clearly considered the leather as of exceptional interest and collected it with obvious care, even down to small snippets. He was much more selective with the pottery.

49. ChLA 409.

50. VON PETRIKOVITS, 1974, 17.

51. VON PETRIKOVITS, 1974, 7.

52. Tab. Vind. 38. BOWMAN & THOMAS, 1983, 132-5.

53. E.g. BGU 814, translation JOHNSON, 1936, 478 and P. Mich. VIII, 468, (CPL 251), where amongst other things, shoes are demanded.

54. MACMULLEN, 1960.

55. TEUPKEN, 1826, 51; HARDENBERG, 1858, 214-20.

56. The 20 denarii for his share in a tent credited to the account of a deceased soldier in P. Columbia 325 (GILLIAM, 1967, 238) might indicate that a similar system, with nominal scales of value related to the length of service left in major items of equipment, was also current in the Roman period.

57. CURLE, 1911, 150-3.

58. MACINTYRE & RICHMOND, 1934.

59. ROBERTSON, SCOTT & KEPPIE, 1975, 59-91.

60. Cf. WILD, 1977, 30 for a similar situation with textiles.

61. In the Dutch mobilization of 1914-15, army bootmakers were fully occupied with repairs, thereby compelling the enlistment of private workshops and even retailers for the provision of boots for the first time. BOSBOOM, 1933, 164ff.

62. All 700 pairs of them. How the problem of sizes was tackled is not recorded, PSI 886 quoted by JONES, 1964, 625 and note 35.

63. PSI 465 (A.D. 265?) translation, JOHNSON, 1936, 633, no.387, 106 hides 'for the manufacture of arms' levied on the town of Oxyrhynchus. Hides for military purposes are also mentioned in Tacitus Ann. IV,72.

64. JACOBI, 1897; BUSCH, 1965.

65. VAN DRIEL-MURRAY, manuscript in circulation, The Saalburg Wells. A context reconsidered.

66. V.L. 64-75, confirmed by later work on Rhine Limes sites.

67. V.L. 102, 106-8.

68. E.g. VAN WIJNGAARDEN-BAKKER, 1970; MAUSER, 1975; VERHAGEN, 1982, table 4.

69. PRUMMEL, 1978.

70. REED, 1971, 44.

71. Ryder in WILD, 1977, 34-7 and table 3; RYDER, 1983, 178, fig. 2.6; WILD, 1970, 5.

72. Reflected in price differences, Diocletian's Edict, VIII, 11-14.

73. SCHMID, 1973.

74. SHA Aur. 10.2 record 15,000 goats against 10,000 sheep being brought as booty to the estates of Valerian.

75. WILD, 1977, 30.

76. A line of research which I am attempting to follow up with Mrs. S.Y. Vons-Comis, a textile specialist, is the comparison of the kemp/wool ratio visible on the skins with that from textiles coming from the same site. This might give some idea of the range of possibilities which could be covered by the term 'sheepskin' in the Roman period.

77. HAGBERG, 1967, 122-5.

78. ATEN et al., 1955, 51ff.

79. Tacitus Ann. IV,72; CIL VIII,4508, tolls on hides from Numidia, both dressed and fresh.

80. Application of earth containing mineral salts and drying are also possibilities. For salted hides in Italy, cf. Cato De Re Rustica CXXXV,3-4.

81. Tacitus Ann. IV,72.

82. VAN DRIEL-MURRAY, 1977a.

83. VAN DRIEL-MURRAY, 1977a, and V.L. 99-102.

84. SHA Claudius 14.3 refers to hides in bundles of 10, for use in tent making. If only one in ten hides or skins were marked, this would account for the relative scarcity of inscriptions on leather.

85. EGAN, 1979.

86. VON PETRIKOVITS, 1974, 17.

87. MANN, 1960, 97ff. An off cut showing the edge stitching

appropriate to this method is perhaps visible on V.L. Abb.15, b & c.

88. Wooden leather scrapers, Diocletian's Edict, XIII.10; wooden scrapers and slickers from Vindolanda, BIRLEY, 1977, 123-4; possible medieval bone scrapers, PRUMMEL, 1978, 410, and bone slickers from Vindonissa, V.L. 21.

89. GAITZSCH, 1980, 63ff.

90. BIRLEY, 1977, 124.

91. GROENMAN-VAN WAATERINGE, 1970; PRUMMEL, 1978.

92. RITTERLING, 1912, 54-5, 61-62 and Taf.III.

93. Dr. L.H. van Wijngaarden-Bakker, pers. comm.

94. RITTERLING, 1912, 61.

95. ATEN et al, 1955, 43-51.

96. BIRLEY, 1977, 124.

97. V.L. 18.

98. P. Grenf. II,51 (translation JOHNSON, 1936, 628 no.375), recording the purchase of 4 goat skins.

99. PSI 465 (see note 63).

ABBREVIATIONS

B.B.	VAN DRIEL-MURRAY & GECHTER, 1984.
BGU	Berliner griechische Urkunden (Aegyptische Urkunden aus den königlichen Museen zu Berlin) 9 vols, Berlin, 1892-1937.
ChLA	Bruckner, A. & R., Marichal, Chartae Latinae Antiquiores, Olten and Lausanne, 1954-
CIL	Corpus inscriptionum Latinarum, Berolini 1863-
CPL	Cavenaile, R. Corpus Papyrorum Latinarum, Wiesbaden, 1958.
Valkenburg	GROENMAN-VAN WAATERINGE, 1967.
V.L.	GANSSER-BURCKHARDT, 1942.
P. Oxy.	The Oxyrhynchus Papyri, ed. B.P. Grenfell, A.S. Hunt et al., London, 1898-
PSI	Papiri greci e latini. Florence, 1912-
R.E.	Pauly-Wissowa Realencyclopädie der classischen Altertumswissenschaft, Stuttgart, 1894-
SHA	The Scriptores Historiae Augustae, translation D. Magie, London, 1921.

BIBLIOGRAPHY

ATEN et al., 1955: A. Aten, R. Faraday-Innes, & E. Knew, Flaying and Curing of Hides and Skins as a Rural Industry, F.A.O. Agricultural Development Paper No.49 (Rome 1955)

BAATZ, 1983: D. Baatz, Lederne Gürteltaschen römischer Soldaten? Archäologisches Korrespondenzblatt 13, 1983, 359-61.

BIRLEY, 1977: R. Birley, Vindolanda. A Roman Frontier Post on Hadrian's Wall (London 1977)

BOHN, 1925: O. Bohn, Hölzerne Schrifttäfelchen aus Vindonissa, Anzeiger für schweizerische Altertumskunde n.s. 27, 1925, 8-15

BOSBOOM, 1933: N. Bosboom, In moeilijke omstandigheden (Augustus 1914-Mei 1917) (Gorinchem 1933)

BOWMAN & THOMAS, 1983: A.K. Bowman & J.D. Thomas, Vindolanda: The Latin Writing Tablets Britannia Monographs no.4, 1983.

BUSCH, 1965: A.L. Busch, Die römerzeitlichen Schuh- und Lederfunde der Kastelle Saalburg, Zugmantel und Kleiner Feldberg, Saalburg Jahrbuch 22, 1965, 158-210

CLOSE-BROOKS, 1977/8: J. Close-Brooks, A Roman iron flask from Newstead, Proceedings of the Society of Antiquaries of Scotland 109, 1977/8, 372-4

CURLE, 1911: J. Curle, A Roman Frontier Post and its People. The Fort of Newstead in the Parish of Melrose (Glasgow 1911)

DOPPELFELD, 1980: O. Doppelfeld, Das fränkische Frauengrab unter dem Chor des Kölner Domes, in O. Doppelfeld & W. Weyres, Die Ausgrabungen im Dom zu Köln (Mainz 1980), 264-308

VAN DRIEL-MURRAY, 1977a: C. van Driel-Murray, Stamped leatherwork from Zwammerdam, in B.L. van Beek, R.W. Brandt & W. Groenman-van Waateringe, Ex Horreo, Cingula IV (Amsterdam 1977), 151-64

VAN DRIEL-MURRAY, 1977b: C. van Driel-Murray, The leatherwork, in J.K. Haalebos, Zwammerdam-Nigrum Pullum, Cingula III (Amsterdam 1977), 249-81

VAN DRIEL-MURRAY, 1980: C. van Driel-Murray, Veertiende-eeuwse zwaardscheden uit Leiden, Bodemonderzoek in Leiden, Jaarverslag 1979 (1980), 35-44

VAN DRIEL-MURRAY, in press: C. van Driel-Murray, Shoes in perspective, in Papers of the XIII Congress of Roman Frontier Studies, 1983 Aalen

VAN DRIEL-MURRAY, forthcoming: C. van Driel-Murray, Een romeinse schoenmaker aan de Houtmaas Maastricht, Publications de la Société Historique et Archéologique dans le Limbourg

VAN DRIEL-MURRAY & GECHTER, 1984: C. van Driel-Murray & M. Gechter, Funde aus der fabrica der legio I Minervia am Bonner Berg, Rheinisches Ausgrabungen 23 (Beiträge zur Archäologie des römischen Rheinlandes 4), 1984, 1-84

EGAN, 1979: G. Egan, Two leather offcuts with quality-control marks from the moat of Oxford Castle, Oxoniensia 44, 1979, 101-3

FINK, 1971: R.O. Fink, Roman Military Records on Papyrus, Philological Monographs of the American Philological Association 26 (1971)

GAITZSCH, 1980: W. Gaitzsch, Eiserne römische Werkzeuge, B.A.R. International Series 78 (1980)

GANSSER-BURCKHARDT, 1942: A. Gansser-Burckhardt, Das Leder und seine Verarbeitung im römischen Legionslager Vindonissa (Basel 1942)

GILLIAM, 1967: J.F. Gilliam, The deposita of an auxiliary soldier, Bonner Jahrbücher 167, 1967, 233-43

GLASBERGEN & GROENMAN-VAN WAATERINGE, 1974: W. Glasbergen & W. Groenman-van Waateringe, The Pre-Flavian Garrisons of Valkenburg Z.H., Cingula II (Amsterdam 1974)

GOLD DER THRAKER, 1979: Gold der Thraker: archäologische Schätze aus Bulgarien (Mainz 1979)

GROENMAN-VAN WAATERINGE, 1967: W. Groenman-van Waateringe, Romeins lederwerk uit Valkenburg Z.H. (Groningen 1967)

GROENMAN-VAN WAATERINGE, 1970: W. Groenman-van Waateringe, Ossa Absentia, Inaugural Lecture, University of Amsterdam (Groningen 1970)

GROENMAN-VAN WAATERINGE, 1974: W. Groenman-van Waateringe, Römische Lederfunde aus Vindonissa und Valkenburg Z.H.: ein Vergleich, Jahresbericht Gesellschaft Pro Vindonissa 1974, 62-84

HAGBERG, 1967: U.E. Hagberg, The Archaeology of Skedemosse II (Stockholm 1967)

HARDENBERG, 1858: H. Hardenberg, Overzigt der voornaamste bepalingen betreffende... het Nederlandsche Leger ('s Gravenhage 1858)

HARTLEY, 1972: B.R. Hartley, The Roman occupation of Scotland: the evidence of samian ware, Britannia III, 1972, 1-55

JACOBI, 1897: L. Jacobi, Das Römerkastell Saalburg bei Homburg v.d. Höhe (Homburg v.d. Höhe 1897)

JOHNSON, 1936: A.C. Johnson, Roman Egypt, vol. II of Tenny Frank (ed.) An Economic Survey of Ancient Rome (Baltimore 1933-40)

JONES, 1964: A.H.M. Jones, The Later Roman Empire, vol. II (Oxford 1964)

KLAPPER: Klapper op de bestaande krijgsregeling der Landmacht, samenst. J.H. Kesman ('s Gravenhage 1836)

LAWSON, 1978: A.K. Lawson, Studien zum römischen Pferdegeschirr, Jahrb. des Römisch-Germanischen Zentralmuseums Mainz 25, 1978, 131-72

LEHNER, 1907: H. Lehner, Bonner Jahrbücher 116, 1907, 293

LINDENSCHMIT & LINDENSCHMIT, 1864-1900: L. Lindenschmit & L. Lindenschmit Sohn, Die Alterthümer unserer heidnischen Vorzeit, vol. I-IV (Mainz 1864-1900)

MACINTYRE & RICHMOND, 1934: J. MacIntyre & I.A. Richmond, Tents of the Roman army and leather from Birdoswald, Trans. Cumberland & Westmorland Antiquarian Soc. n.s. XXXIV, 1934, 62-90

MACMULLEN, 1960: R. MacMullen, Inscriptions on armor and the supply of arms in the Roman Empire, American Journal of Archaeology 64, 1960, 23-40

MANN, 1960: I. Mann, Rural Tanning Techniques, F.A.O. Agricultural Development Paper no. 68 (Rome 1960)

MAUSER, 1975: H.J. Mauser, Untersuchungen an Tierknochenfunden aus dem römischen Xanten (Kiel 1975)

MILNE & MILNE, 1982: G. Milne & C. Milne, Medieval Waterfront Development at Trig Lane, London, London & Middlesex Archaeological Society, special paper 5 (1982)

MOREL & DE WEERD, 1980: J.-M.A.W. Morel & M.D. de Weerd, Early Roman harbours in Velsen, in W.S. Hanson & L.J.F. Keppie (eds.) Papers of the XII Congress of Roman Frontier Studies, Stirling 1979 B.A.R. International Series 71 (1980)

MOREL, in press: J.-M.A.W. Morel, Velsen. An early Roman harbour and fort in North Holland, in Papers of the XIII Congress of Roman Frontier Studies, Aalen 1983

MOREL et al., 1984: J.-M.A.W. Morel, H. Kamermans, A. Visser, Spatial analysis of bone refuse in a Roman harbour, in Proceedings Workshop on Data Management of Archaeozoological Assemblages (Amsterdam 1984 - in press)

NICOLE & MOREL, 1900: J. Nicole & Ch. Morel, Archives militaires du ler siècle (Genève 1900)

OLDENSTEIN, 1976: J. Oldenstein, Ausrüstung römischer Auxiliareinheiten, Bericht der Römisch-Germanischen Kommission 57, 1976, 49-128

PANHUYSEN, 1984: T.A.S.M. Panhuysen, Maastricht staat op zijn verleden (Maastricht 1984)

VON PETRIKOVITS, 1974: H. von Petrikovits, Römisches Militärhandwerk: archäologisches Forschungen der letzten Jahre, Anzeiger der Österreichischen Akademie der Wissenschaften, Phil. Hist. Klasse 111, 1974, 1-21

VON PETRIKOVITS, 1975: H. von Petrikovits, Die Innenbauten römischer Legionslager während der Prinzipatszeit, Abhandlungen der Rheinisch-Westfälischen Akademie der Wissenschaften 56 (1975)

VON PREMERSTEIN, 1903: A. von Premerstein, Die Buchführung einer ägyptischen Legionsabteilung, Klio 3, 1903, 1-46

PRUMMEL, 1978: W. Prummel, Animal bones from tannery pits of 's-Hertogenbosch, Berichten Rijksdienst voor het Oudheidkundig Bodemonderzoek 28, 1978, 399-422

REED, 1972: R. Reed, Ancient Skins, Parchments and Leathers (London 1972)

RITTERLING, 1912: E. Ritterling, Das frührömische Lager bei Hofheim i.T., Annalen des Vereins für Nassauische Altertumskunde und Geschichtsforschung 40, 1912

ROBERTSON, SCOTT, & KEPPIE: A.S. Robertson, M.E. Scott, & L.J.F. Keppie, Bar Hill: a Roman Fort and its Finds, B.A.R. 16

(1975)

ROBINSON, 1974: H. Russell Robinson, Roman body armour in the first century A.D., in E. Birley, B. Dobson & M. Jarrett, Roman Frontier Studies, Eighth International Congress of Limesforschung, Cardiff 1969 (Cardiff 1974), 5-12

RYDER, 1983: M.L. Ryder, Sheep and Man (London 1983)

SCHMID, 1973: E. Schmid, Ziegenhörner als Gerberei-Abfall, Schweizerische Volkskunde 63, H. 5/6, 65-6

SWOBODA, 1964: E. Swoboda, Carnuntum: seine Geschichte und seine Denkmäler (Graz 1964)

TEUPKEN, 1823: J.F. Teupken, Beschrijving hoedanig de koninklijke Nederlandsche troepen en alle in militaire betrekking staande personen gekleed, geëquipeerd en gewapend zijn ('s Gravenhage 1823)

TEUPKEN, 1826: J.F. Teupken, Vervolg ('s Gravenhage 1826)

VERHAGEN, 1982: M. Verhagen, Een archeo-zoologische onderzoek naar de dierlijke resten uit de opgraving Valkenburg 1980, Unpublished undergraduate thesis (Amsterdam 1980)

DE WEERD, 1977: M.D. de Weerd, The date of Valkenburg 1 reconsidered, in B.L. van Beek, R.W. Brandt & W. Groenman-van Waateringe, Ex Horreo, Cingula IV (Amsterdam 1977), 255-89

WILD, 1970: J.P. Wild, Textile Manufacture in the Northern Roman Provinces (Cambridge 1970)

WILD, 1977: J.P. Wild, The Textiles from Vindolanda, 1973-1975 (Hexham 1977)

WILD, 1982: J.P. Wild, Wool production in Roman Britain, in D. Miles (ed.), The Romano-British Countryside, B.A.R. 103 (1982), 109-22

VAN WIJNGAARDEN-BAKKER, 1970: L.H. van Wijngaarden-Bakker, Dierresten uit het castellum te Zwammerdam, Helinium 10, 1970, 274-8

MANUFACTURE AND SUPPLY OF THE ROMAN ARMY WITH BRONZE FITTINGS[1]

J. Oldenstein

In this paper, I should like to put forward my ideas concerning the manufacture and supply of bronze fittings to the Roman army.[2]

First, I wish to explain my theories relating to the production of material with the aid of some examples; I am well aware that the material base for them is not very strong. Principally I want to present a model for the manufacture and supply of Roman military equipment.

During my research into this subject on the *limes* in Germany in the second and third centuries, I had the feeling that the organisation of manufacture and distribution of military equipment during the first century was different to that of the second and third centuries. However, I did not then have sufficient time to work on first century equipment, but now that it is possible, I can see the problems more clearly than before.

I want to begin by outlining my theories. I think the situation in the fourth century, with those central Roman weapon factories which appear in the 'Notitia Dignitatum', is not the same as that found in the first three centuries. From the time of Diocletian, perhaps as early as Gallienus, we have to reckon with weapon and equipment production by the state authorities. To work in state weapon factories was a military duty, as we learn from various inscriptions and from the 'Notitia Dignitatum'.[3] It was long thought that the situation described in literary and epigraphic sources for the fourth century would also apply for the first three centuries: namely weapon manufacture by the state authorities.

The large number of finds of very similar equipment, from Britain to Syria and Africa, seemed to justify these ideas.[4] Especially the finds from Dura Europos, which are sometimes identical to those from the Danube and Rhine *limes*, and from Britain; these seemed to prove that there were large central factories which produced the equipment and sent it to different units throughout the empire. It was thought that those factories were stationed in Gaul or in Italy; even now, however, no factory has been found in these areas.

However, there are many finds from the forts and fortresses

along the Rhine- and Danubian frontier, from Britain and from Africa, which contradict this argument. In well-excavated forts, and in their *vici*, substantial evidence for metalworking had come to light.[5] We know of crucibles from various forts. These clearly show that, at those places, we can reckon with a smaller or larger bronze industry. In addition, collected waste bronze material - broken and cut pieces, failed castings and moulds for example - found in the forts and *vici* show that the *fabricae* of the forts and fortresses produced or repaired bronze items to a greater or lesser extent.[6] The discovery of moulds plainly shows that new articles were also produced.

We can infer a similar picture from literary sources and inscriptions. They give no hint of weapon factories under state authority during the first three centuries. On the other hand, there is evidence for a lot of private craftsmen and traders, producing and dealing with weapons and military equipment. The army was a customer of these craftsmen and traders: in all probability, larger orders were given to the bronze manufactories in the provinces only when a lot of equipment was needed rather quickly. I think the reason for giving large orders to such bronze-working factories was the existence there of highly specialized craftsmen and the equipment for producing military equipment. I do not believe that these factories only produced military equipment, but they had the capacity to make it if desired.[7]

Thus far on general suppositions. It is hardly possible, without intensive special studies in this subject in the different provinces, to give any more than a general approach to the question of the production and distribution of Roman military equipment at the moment.[8] Still more difficult is a further problem - that of the structure of the workshops and the manufactories and how they were organized. Were there large factories or only small workshops? How were the products bought or sold? Were there markets, where military personnel could buy larger amounts of equipment, or have we to think in terms of traders, going from fort to fort, offering their goods? What proportion of equipment was produced in the *fabricae* of the legions and the auxiliary units?

For the moment I do not think that any one of these various possibilities is the whole answer. We have to count on a whole range of production and trading methods existing side by side. As already mentioned, some evidence of bronze working has been found in forts and fortresses.[9] In my opinion, those *fabricae* were not able to equip a legion or an auxiliary unit completely. Nor was this at all necessary: when a new unit occupied an old fort, or built a new one, the soldiers normally arrived fully equipped. The work of the *fabricae* is mainly to be seen in the repair of weapons. It is also conceivable that these fabricae were able to complete parts of belt mountings which had been broken, for example. To make a mould, the craftsmen could use a similar belt fitting. Master inscriptions on swords and sword scabbards show that weapons and equipment were produced away from the forts and their vici.[10] In the case of the sword scabbard from Straßburg, the master Quintus Nonienus Pudes calls the place where he made the piece "ad ara", at Cologne.[11] Just how large the workshop of Quintus Nonienus Pudes was, we do not know. Whether he produced only sword scabbards and no other equipment unfortunately also remains unknown. Furthermore, there are inscriptions from Italy of private sword and shieldmakers,[12] and in Mainz the inscription of a sword trader was found.[13] This evidence makes it clear that the weapons were produced and sold by private persons.

Unfortunately, the inscriptions do not give any criteria by which the extent of the craftsman's work, or that of his workshop, can be judged. But the mere question of the capacity of such a workshop is very important, if we want to get on the track of the production and distribution of Roman military equipment.

On what occasions was it necessary to produce a large amount of weapons and equipment suddenly? There are various answers. First of all, when a legion was newly founded: this was, however, a rare occurrence from the first to the third century. There were (for example) the XXII, raised by Caligula,[14] VII by Galba, I and II Adiutrix, IV and XVI by Vespasian. During the Marcomannic wars the legions I, II, and III Italica were raised and Septimius Severus recruited three new legions, the I, II, III Parthica. Most of the auxiliary units were founded during the first century, while the second saw the foundation of many numeri.[15] On average, then, it seems likely that only every ten years was a unit equipped.

Another situation calling for the production of a great deal of equipment was in advance of a campaign of conquest. Normally such campaigns were not hastily planned and there was enough time to collect weapons and equipment, which were then stored in the armamentaria, or to produce new weapons. After the middle of the first century the pace of conquest slackened. Britannia and Upper Germany came into the empire. Dacia, Mesopotamia and Arabia were conquered by Trajan. In cases of disputed succession to the throne and of civil war, quick decisions were necessary. Rapidity of action was important for the victory. We have passages in literary and epigraphic sources showing what happened in two such cases. When Vespasian came from the east to fight against Vitellius, he had the weapons for his army produced in large, rich cities.[16] We know furthermore from an inscription, found in Mainz, that a high-ranking decorated officer, Annianus, was sent to Northern Italy to recruit soldiers and to procure weapons in Milan.[17] This happened at the time when Maximinus Thrax came from the east to fight against Balbinus, who was chosen by the senate.

Both situations have a similar background. Although there are nearly one hundred and seventy years between them, both are civil war situations. It seems, furthermore, that in both cases Vespasian and Balbinus were not able to call on a reserve of weapons and equipment in their own provinces, and in both cases larger, rich cities were chosen for the manufacture of the missing weapons. One further situation where a lot of weapons would be needed is, in my opinion, after bloody campaigns and defensive battles: the loss of weapons and equipment had to be made good. It must be considered whether the capacity of military fabricae was sufficient for this or whether the resources of private workshops were also needed. All the arguments I have so far adduced show that there is no necessity to think that there was a state monopoly of weapon production in the first three centuries. The examples of both Vespasian and Balbinus demonstrate this. On all those occasions where a large number of weapons and equipment was needed, no workshop was seemingly able to deliver the required weapons, except for a lot of workshops specializing in iron, bronze, bone and leather working. Those workshops had the technical "know how" to produce the necessary equipment extremely rapidly.[18]

However, original finds from the forts and fortresses also support this thesis. During the second and third centuries, we find a lot of mountings and belt fittings which are typical of only one province.[19] On the other hand, there are still pieces common to all parts of the Roman empire.[20] These pieces are only similar and not from the same mould.[21] Pieces made from the same mould are seldom widely distributed; such pieces are even found in neighbouring forts but never over any distance.[22] I have tried to interpret this different distribution as follows. We have to count, on the one hand, with large factories in one province supplying many forts and fortresses; on the other with a lot of smaller workshops producing equipment at the same time. Those workshops supplied some in the closer or more distant surrounding area. The fabricae of the forts and fortresses repaired and produced for the needs of their own unit. Sometimes I think it possible that this or that piece was exchanged with a soldier from the next fort. It is not necessary to think that all those pieces, being similar and found in a number of provinces of the empire, were produced in large equipment and weapon manufactories. The fittings were normally of such a form that it was possible to make clay impressions from them. A specialist is not necessary for this work. If the patterns came from only one source, it is no longer necessary to think that all similar pieces were produced in central workshops or factories. Centurions, for example, very often came from the praetorian guard at Rome. From this point, they moved over the whole empire and to different units. If they bought parts of their equipment in Rome, before they went to their new units, very similar material would reach the furthest corners of the empire. The craftsmen in the fabricae of Dura Europos, Alexandria, Cologne and York would now have patterns for producing new pieces, pressing these new fittings in clay, to get their moulds. They could now produce belt fittings in York or Dura Europos of the form currently "en vogue" at Rome. High ranking officers of the legions and auxiliary units likewise changed province.[23] With them, or the companions of such officers, patterns for equipment could be spread from province to province very quickly. This kind of distribution and production assumes a special kind of economic structure in a province. It is hard to believe that we would find a lot of craftsmen and workshops during a campaign. A Roman province had to reach a special grade of Romanization before these things would come about. This implies that there had to be a certain level of prosperity and public safety in such a province, before traders and craftsmen would come to earn their money.

If we follow this model, the supply of equipment to Roman troops would take different forms at different times. Take, for example, the Roman province Britannia. When the conquest of the British Isles was planned by Claudius, the legionaries and auxiliaries brought their equipment with them from the continent. Now Britain was not conquered in one day: the Roman troops needed nearly eighty years, until Hadrian's Wall was built as the northern frontier. During the expedition of Claudius, the large expeditionary army could not be supplied from the land that it was conquering. There were no Roman towns and hardly any Roman or romanized traders and craftsmen. Supplies, especially of equipment and weapons, had to come from the continent; large issues came from Italy or Gaul. The situation in the two Germanic provinces would be similar, I think. In a second phase, after a certain consolidation in the province, the economic branches in a province would develop in such a way as to be able to take over the largest of the supply of the units stationed in the new province. In a third phase, the provinces would become self-sufficient, so that they only imported those things which could not be produced in the province itself.because of the lack of particular resources and the craftsmen who worked with them.

If we try to verify this model with the material found in the forts and fortresses, I believe that it is possible to come to similar conclusions. There is one example which is especially qualified to demonstrate it. I want to show it with the aid of those silvered, nielloed and cast pendants which are heart-shaped and belong to horse harness.[24] Pieces of this kind are seldom found in Augustan contexts. From Claudian times, they are very often found in Northern Africa, Germany, Britain and Syria.[25] In the later Flavian period we do not find such fittings any more; from this time, we have only simple bronze-sheet pendants without any decoration. These pieces imitate the earlier ones. It is noticeable that the pieces of the Claudian/early Flavian period are finished to an extremely high technical standard. They must have been produced in a two-part mould or in a mould with a wax impression which could be used only once: I think these pieces were cast by the lost-wax process. Silvering and niello are also indications of highly trained specialists. This kind of equipment can not be seen as the work of small workshops. I think this fact points to production by large specialized manufactories in Gaul or Italy: in those factories the equipment of an expeditionary army was produced. The supply of the equipment during the first phase of conquest also came from here. In the new provinces, there was no industry and no trading organization yet capable of taking over the supply of an army.

It is important that we only find the fine silvered pendants during a relatively short period and not in all parts of the empire. After the Claudian period, we only find them in those provinces which were parts of the empire for a short time, or where conquest was still in progress. Britain is a good example for clarifying what I mean. Heart-shaped pendants from horse harness, which we can date in the Claudian/Neronian period, are distributed up to the Longthorpe-Wroxeter line, and in the West up to the beginning of the Welsh mountains.[26] This area is that part of Britain which was conquered by the Roman army up to the rebellion of Boudicca. We find typical pendants of the Neronian/Flavian period in the south too, but more so in those parts of the province which were conquered after A.D. 60 - in the Welsh mountains and in the area north of the Longthorpe/Wroxeter line up as far as Newstead.[27]

A similar picture, albeit not so clear, is to be seen in those parts of Upper Germany lying to the right of the Rhine valley, which came to the province during the period of Vespasian.[28] We can find similar pendants in Pannonia, Syria and Africa.[29] It is not impossible that these pieces are connected with the civil war of 68/69 in these provinces.

Most striking is the fact not only that these fittings are mostly found in those areas where conquest was taking place or where we can consider military action likely, but also that as yet no heart-shaped silvered pendant is known from any military context after the period of Domitian. In the Domitianic period, the conquest of Britain and Upper Germany was completed and the position was consolidated in these provinces; the occupation was no longer of the same nature.

In the forts of the Upper German/Raetian limes, which were built in A.D. 83/85, we do not find these silvered heart-shaped pendants; instead simple bronze-sheet pendants without any decoration suddenly appear.[30] It seems to me that these pendants are products of workshops in the new province, which now slowly begin to take over the supply of the provincial units.

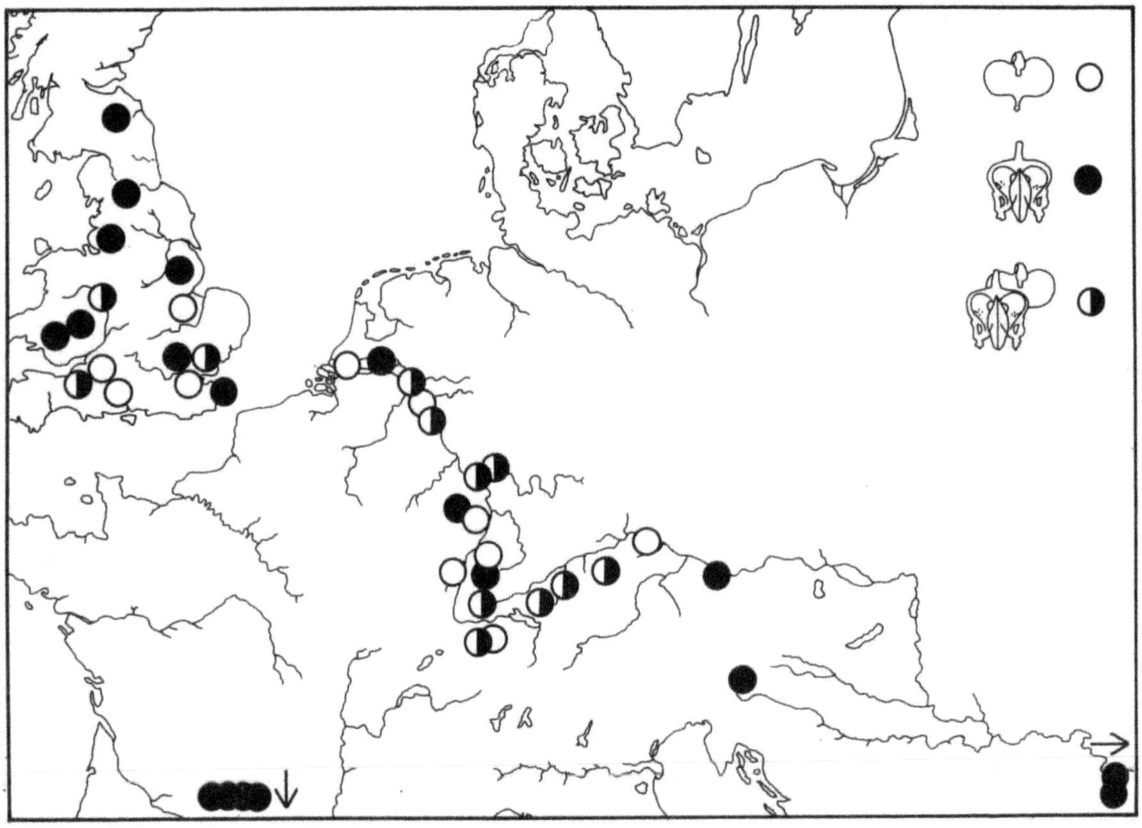

Fig. 1: Distribution of Claudian/Neronian (white) and Neronian/Flavian (black) pendants (after A. Böhme in SCHÖNBERGER, 1978, Abb.78 & 79)

From the beginning of the second century, we find more and more hints in the forts and in civil centres of the various provinces that repair in the main, but also production, has begun. I have collated the different evidence in my thesis. This position began to change from the middle of the third century, with the rapid deterioration in the economic condition of the provinces, so that private industry was no longer able to supply the Roman army. Under Diocletian, the equipment and weapon industry became a state monopoly. The inscriptions and the literary sources of that time also clearly reflect the changed situation.[31]

It seems sensible to me look at another branch of industry, samian production, in the same way. We can recognize similar phenomena. The highly specialized potters begun to produce in Italy and distributed their goods up to the Rhine and to the Danube. With increasing Romanization, the pottery centres came over from South and Central Gaul, nearer and nearer to their customers at the Rhine and Danube frontier. In the middle of the second century, the large potteries of Trier and Rheinzabern came into production. They now had more or less direct contact with their customers.[32]

To summarize my ideas on the supply of Roman weapons and equipment, it must be said that such a summary can only be in the form of a model, but I think this model is more than probable.

In a phase of occupation, most of the equipment and weapons needed by the expeditionary army were produced in Italy or Gaul. These well-romanized parts of the empire had the economic base for producing and trading large masses of used material. Freshly occupied provinces are not yet able to take over the full supply of the army. If one unit is in garrison at the same place for any length of time, it might attempt to become more independent of supplies coming from Gaul or Italy. Some samian ware, for example, was made at the legionary fortress at Haltern;[33] nevertheless this production was not sufficient to supply the whole legion, as is proved by a mass of contemporary Italian samian ware that was found at Haltern.[34]

The beginnings of phase two are to be seen in this production, starting in the final phase of conquest or shortly after that. The province would gradually want to become independent of expensive imports from Italy or Gaul. In this second phase, I think most equipment would be produced in the fabricae of the forts or their vici when the situation became more settled. The mass of the troops are not fighting any longer and they have to hold the occupied parts of the province. During this first period of consolidation, private craftsmen and traders come to the new province slowly. We have to count further with native traders and craftsmen. At this point, imports from Italy and Gaul are still necessary to supply the needs of the troops fully. With continual romanization, phase three slowly begins with more or less full economic independence of the new province. The main burden of supply was now taken over by smaller or larger local workshops or manufactories, or by the fabricae of the forts and fortresses. Phase four is reached in late Roman times, when private industry was no longer able to supply the army. From that point onwards, the government takes over the supply of the army with weapons and equipment.

As far as the dating of the different phases is concerned, I can give only a general review. Phase one covers the period from Augustus to the Claudian/Neronian period, phase two from the Neronian to the Flavian period; phase three was from the late Flavian period up to the middle of the third century, whilst from the middle of the third century, phase four slowly began, and is obvious in the Diocletianic era.

NOTES

1. I would like to thank Mr. David P. Davison M. Phil., Oxford for reading my English manuscript and for some interesting discussions.

2. OLDENSTEIN, 1976, 68ff., cited below as Ausrüstung.

3. Notitia dignitatum Or. XI, Oc. IX.; HOFFMANN, 1969, 61ff.; Ausrüstung 83f. esp. notes 161-7.

4. For these finds see Ausrüstung passim.

5. Ausrüstung 68ff., FISCHER, 1984, 299ff.; MACKENSEN, 1983, 565ff. esp. 571ff.

6. For the German limes see OLDENSTEIN, 1982, 117 Nr. 8 s.v. Geräte zur Herstellung von Bronzeguß.

7. Ausrüstung 80ff.

8. See the new paper of M. Feugère (1983, 45ff).

9. See notes 5 and 6.

10. MACMULLEN, 1960, 12ff; NUBER, 1972, 483ff. with older literature.

11. ULBERT, 1971, 44ff.

12. Ausrüstung 82 note 155, CIL VI 1952, IX 3962, X 3986 (gladiarii), CIL VI 9043, X 3971 (scutarii).

13. Ausrüstung 82, CIL XIII 6677.

14. See the article by E. Ritterling in the RE s.v. "legio" passim.

15. OLDENSTEIN-PFERDEHIRT, 1983, 303ff.

16. Ausrüstung 80f., Tacitus, Historien 2,12 and Vegetius, Ep. rei milit. 4,8.

17. Ausrüstung 80f. with notes 135 and 136, CIL XIII 6763.

18. Ausrüstung 81 and Vegetius Ep. rei milit. 4,8.

19. Ausrüstung 152ff. Nrs 353-87 Taf.39, 40 for example.

20. Ausrüstung 160f. Nrs 425, 426 Taf.43 for example.

21. To that problem see MACKENSEN, 1983, 571ff.

22. _Ausrüstung_ 78 Nr. 951 and 1058 Taf. 72 (Pfünz) Taf.79 (Weißenburg), or Nrs. 1110 (Heftrich) 1111 (Zugmantel) Taf.84; Nrs. 1139 (Niederbieber) 1138 (Zugmantel) Taf.88.

23. For the career of Roman officers see MAXFIELD, 1981, 24ff. esp. Map 4.; DOBSON, 1978, passim, esp. 165ff.; SPEIDEL, 1978, esp. 67ff.

24. A. Böhme in SCHÖNBERGER, 1978, 211ff. Abb.73.

25. See distribution map _op. cit._ 225 Abb.79 and list of the finds 223, 6.

26. See distribution map _op. cit._ 224 Abb. 78 and list of the finds 222, 5.

27. See note 25.

28. See note 25.

29. See note 25.

30. _Ausrüstung_ Taf. 29.

31. _Ausrüstung_ 83ff.

32. See the article by H. Comfort in the RE s.v. Terra Sigillata.

33. VON SCHNURBEIN, 1974, 77ff.; VON SCHNURBEIN, 1981, 69ff.

34. VON SCHNURBEIN, 1982.

BIBLIOGRAPHY

DOBSON, 1978: B. Dobson Die Primipilares. Entwicklung und Bedeutung, Laufbahnen und Persönlichkeiten eines römischen Offiziersranges, Beihefte Bonner Jahrbücher (Köln 1978)

FEUGÈRE, 1983: M. Feugère, L'équipement militaire romain dans le Département de la Loire. Cahiers archéologiques de la Loire 3, 1983, 45-66

FISCHER, 1984: Th. Fischer, Eine Bronzegießerei im Lagerdorf des römischen Kastells Pfünz, Gde. Walting Ldkr. Eichstätt, Oberbayern. Bayerische Vorgeschichtsblätter 49, 1984, 299ff

HOFFMANN, 1969: D. Hoffmann, Das spätrömische Bewegungsheer und die Notitia Dignitatum. Epigraphische Studien 7.1 (Düsselforf 1969)

MACKENSEN, 1983: M. Mackensen, Eine limeszeitliche Scharnierfibel aus Resafa, Nordsyrien. Germania 61, 1983, 565-78

MACMULLEN, 1960: R. MacMullen, Inscriptions on armor and the supply of arms in the Roman empire. American Journal of Archaeology 64, 1960, 23-40

MAXFIELD, 1981: V. Maxfield, The Military Decorations of the Roman Army (London 1981)

NUBER, 1972: H.U. Nuber, Zwei bronzene Besitzermarken aus Heddernheim. Chiron 2, 1972, 483-507

OLDENSTEIN, 1976: J. Oldenstein, Zur Ausrüstung römischer Auxiliareinheiten. Ber. Römisch-Germanische Kommission 57, 1976, 51-284

OLDENSTEIN, 1982: J. Oldenstein, Der Obergermanisch-Raetische Limes des Römerreiches, Fundindex (Mainz 1982)

OLDENSTEIN-PFERDEHIRT, 1983: B. Oldenstein-Pferdehirt, Die römischen Hilfstruppen nördlich des Mainz - Forschungen zum obergermanischen Heer I. Jahrbuch des Römisch-Germanischen Zentralmuseums 30, 1983, 303-48

VON SCHNURBEIN, 1974: S.v. Schnurbein, Bemerkenswerte Funde aus einer Töpferei des Hauptlagers von Haltern. Germania 52, 1974, 77-88

VON SCHNURBEIN, 1981: S.v. Schnurbein, Untersuchungen zur Geschichte der römischen Militärlager an der Lippe. Bericht der Römisch-Germanischen Kommission 62, 1981, 7-101

VON SCHNURBEIN, 1982: S.v. Schnurbein, Die unverzierte Terra Sigillata aus Haltern. Bodenaltertümer Westfalens 19, 1.2 (Münster 1982)

SCHÖNBERGER, 1978: H. Schönberger, Kastell Oberstimm, die Grabungen 1968-1971. Limesforschungen 18 (Berlin 1978)

SPEIDEL, 1978: M. Speidel, Guards of the Roman Armies: An Essay on the Singulares of the Provinces, Antiquitas Reihe 1, 27 (Bonn 1978)

ULBERT, 1971: G.Ulbert, Gajus Antonius, der Meister des silbertauschierten Dolches aus Oberammergau. Bayerische Vorgeschichtsblätter 36, 1971, 44-9

'BELL-SHAPED STUDS'?

Lindsay Allason-Jones

One of the most common finds from Roman sites, particularly in the north of Britain, is the 'bell-shaped' stud. It is also one of the more puzzling finds with almost as many suggestions for its use as there are examples known. Because of this lack of clear identification the present author coined the phrase 'bell-shaped stud' on the grounds that they could all be described as roughly bell-shaped and had shanks.[1]

The object in question consists of a cast circular bronze head with a countersunk face, in the centre of which there is either a dimpled boss or cone (Fig.1). Some of the central bosses do not project beyond the lip whilst others project several millimetres. The studs themselves vary remarkably in size from 13mm to 57mm in diameter. The heads are decorated with lathe-incised concentric circles on the faces and occasionally around the waist and skirt.

The studs divide into two clear types: those with an iron shank secured in a trough in the head by lead, and those with a bronze shank cast in one with the head. The method of caulking an iron shank to a bronze head with lead is also known on the dome-headed studs from northern military sites, particularly the 4th century fort at Piercebridge.[2] The iron shanks of Type 1 tend to be roughly square in section and, as far as one can tell after corrosion, seem to be quite short. The bronze shanks of Type 2 are much longer (up to 60mm) and twice as wide as they are thick. Complete examples of this type have a circular hole near the end of the shank and some have pieces cut from the edge of the shank.[3]

Types 1 and 2 have been regarded until recently merely as variations of the same object having the same function, and there have been numerous suggestions as to the character of that function. Curle in 1911 called the Newstead examples 'bolts for fastening lockplates'.[4] Kenyon referred to the Leicester studs as 'handles from boxes or drawers'.[5] Webster, in his catalogue of military bronzes at the end of the article 'The Advance under Ostorius Scapula', described them as 'probably decorations for a vehicle or piece of furniture although the suggestion has been made that they are the ends of keys or latch-lifters'.[6] This latter suggestion had been made by Jacobi in 1897.[7] In 1979 the present author interpreted them as doorstuds of the type commonly

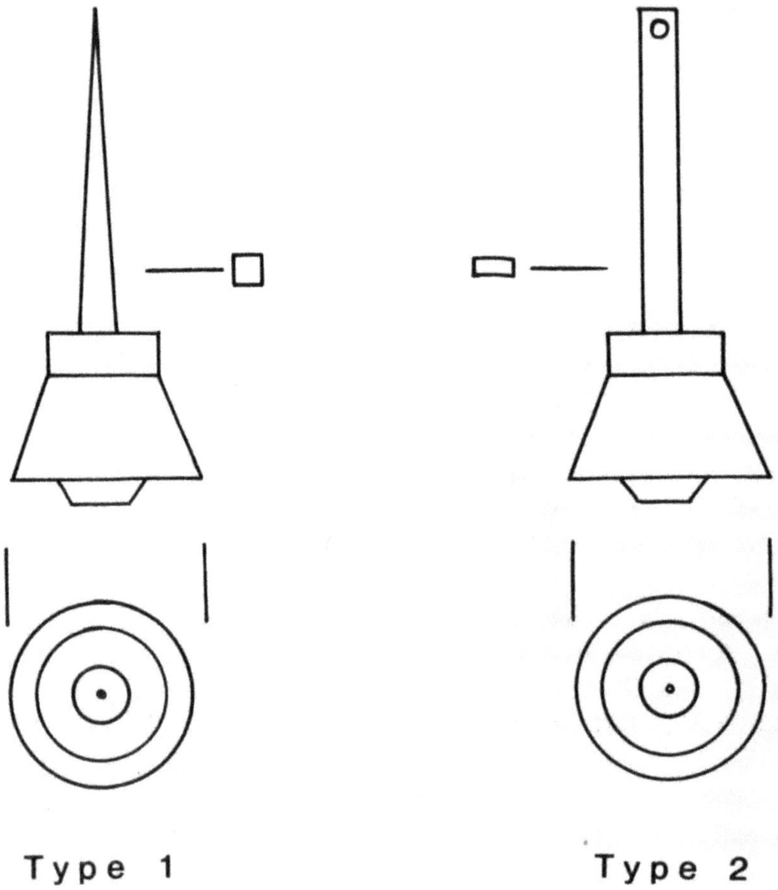

Fig. 1: Bronze bell-shaped studs. Drawn by W. Hubbard.

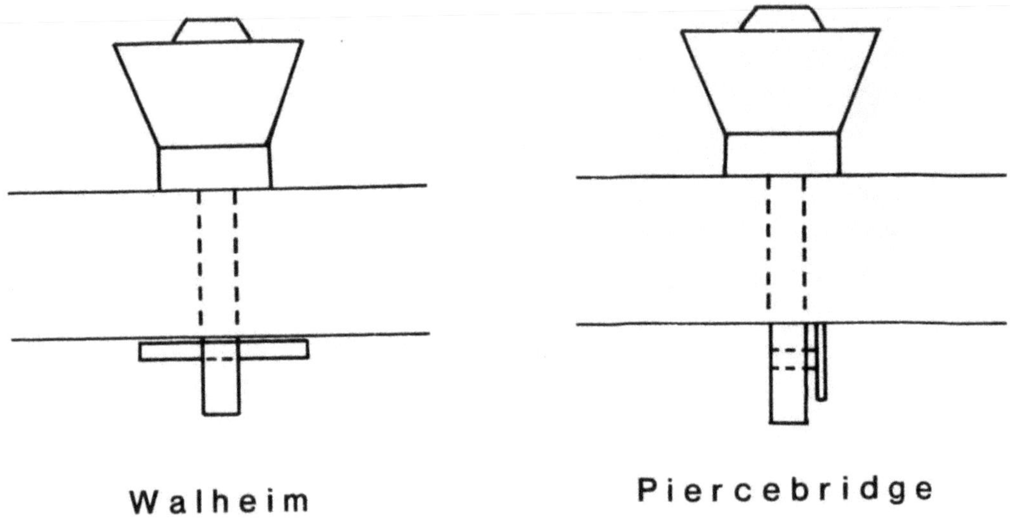

Fig. 2: Methods of fastening the studs. Drawn by W. Hubbard.

found on medieval church doors and known in modern building parlance as 'Spanish studs'.[8] That such studs were being used during the Roman period can be seen on coins, e.g. a sestertius of Nero showing the closed doors of the temple of Janus on the reverse.[9]

In 1982 Planck published a group of eight studs found in position on a plate from a wooden chest or box found at Walheim in West Germany (Pl.I).[10] This may not appeal to modern taste yet it appeared to solve the problem of how bell-shaped studs were used. Two factors complicate this identification. Firstly the length of the shanks. It would have been a massive box to warrant a stud shank of 60mm, the length of several examples from Piercebridge and thicker than a door to a modern room. It is therefore possible that they were used, not only for fixing a plate over the entire surface of a casket but also to fix a lockplate to a substantial chest. A bronze bar piercing the chest wall and fixed by a metal rod would ensure that the lock could not be wrenched or levered off. The second factor is the small disc-headed stud found in position in the hole at the end of the shank on a Type 2 stud from Piercebridge (Fig.3).[11] If this was used as a box decoration with the smaller stud acting as a chock, 20mm of shank would project inside. Several of these would limit the internal area of a small box and would catch on the contents. The limited length of the small stud's shank which barely projects through the hole coupled with the diameter of the head (10mm) makes it unsuitable for securing the shank through wood - to hold the main stud firmly a rod which could be held tightly against the box wall would be more efficient (Fig.2). The implication is that this particular stud may not have been used in the same manner as the Walheim studs. The shank has obviously been intended to be covered as the surface has been roughly filed and the patination is of a different quality to that of the head. It is possible that the shank was attached to a second material rather than intended to penetrate wood and the limited length of the smaller stud might suggest leather or cloth. However, it is equally possible that this bell-shaped stud was used as a box fitting and the small disc-headed stud used to replace a missing rod - a 'Heath Robinson' repair.

Despite these two factors there seems to be little doubt that the majority of Type 2 studs were intended to be used as box fittings of one sort or another. It is possible that they were designed to allow the boxes to be stacked, holding each box clear of the locks or decorative panels of the others. However, when one looks at the Type 2 studs in the light of this identification several problems arise. Firstly, none of the Type 1 studs have been found attached to boxes and the majority of them appear to come from military contexts rather than from civilian sites. The

Pl. I: Box plate from Walheim. Taken from PLANCK 1982.

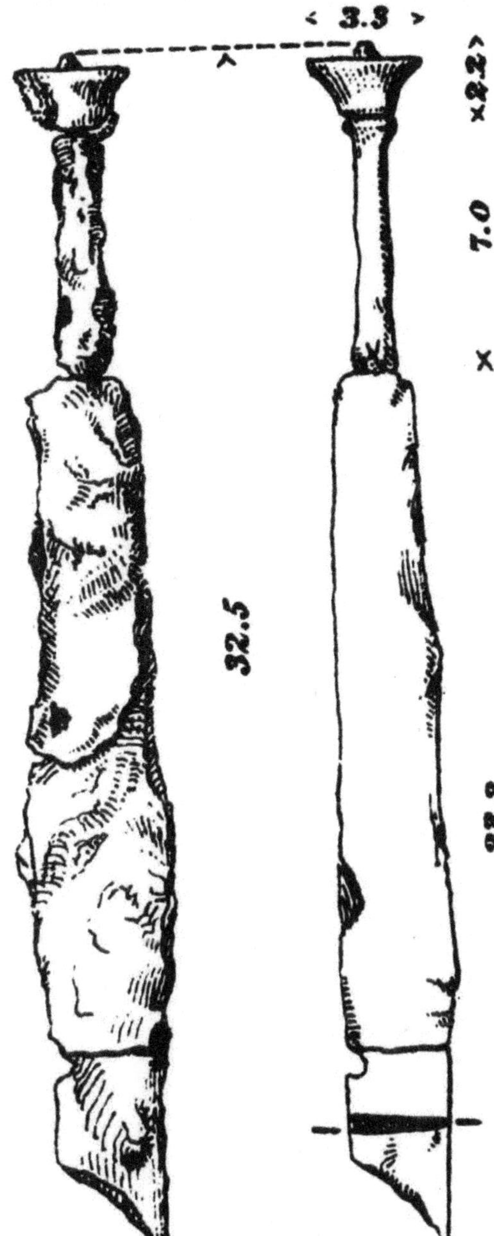

Pl. II: Bell-shaped stud from Carnuntum used as a dagger pommel. Taken from RLÖ III, 1902, Taf.X.

two from Shakenoak and one from Newport (Isle of Wight) seem to be the exception rather than the rule.[12] Secondly, all eight of the Walheim studs are exactly the same: they are a matching set of the same size with the same decoration and apparently come from the same mould. A similar casket from Pfünz published by Winkelmann in 1926 also shows a set of six matching studs on the upper plate.[13] In contrast there are forty-nine examples of Type 1 studs from the 19th and 20th century excavations at South Shields, not one of which can be found to match another. It would be inconceivable that on one site forty-nine boxes would each lose a single stud, particularly as the excavated area was limited in size.[14] The recently excavated material from Housesteads, Piercebridge and Wallsend has also failed to produce any matching studs. Coventina's Well at Carrawburgh produced the most interesting group in this respect: the Well consists of a stone trough 8'6" X 7'9" internally and 7' deep, which was sealed in the early 4th century by large slabs of stone.[15] Seven bell-shaped studs were found in the votive deposit in the Well - four of Type 1 and three of Type 2, all varying in size and appearance. The diameter of the Type 1 studs range from 25mm to 35mm and the Type 2 studs from 42mm to 57mm. This must suggest that these particular examples were used singly rather than in sets. There would be little point in throwing a single stud from a box into a votive deposit.

In 1902 a bell-shaped stud from Carnuntum was published which had been found in situ as the pommel of a short sword or dagger (Pl.II).[16] The text states that many more 'Schwertknäufe' were found but omits to say how many. In 1904 Groller showed a second stud in diagram form used in the same way but did not make it clear if the stud was found in situ or if the author was merely explaining its use on the evidence of the 1902 stud.[17] In 1909 another, found at Lauriacum, was identified as a dagger pommel although attention was drawn to their use as box decorations, unfortunately giving no parallels.[18] A single stud found in a male grave from Fischamend (Aquinoctium) on the R. Leithen in Austria, adds to the evidence.[19] Their use as dagger pommels might explain the individual examples found on military sites although the quantities involved imply an unmilitary carelessness.

Close scrutiny of the article by Groller in 1902 reveals a possible second use for the Type 1 studs as a hinge for a bronze border or guard, apparently a dolabra-sheath (Pl.III).[20] Such sheaths were fitted over the cutting edge of military axes and held in place on either side of the blade by hooks which were tied by cords to the wooden hafts.[21] Some of the studs from Piercebridge which have fragments of thick bronze sheet under the heads may have been used in this way.

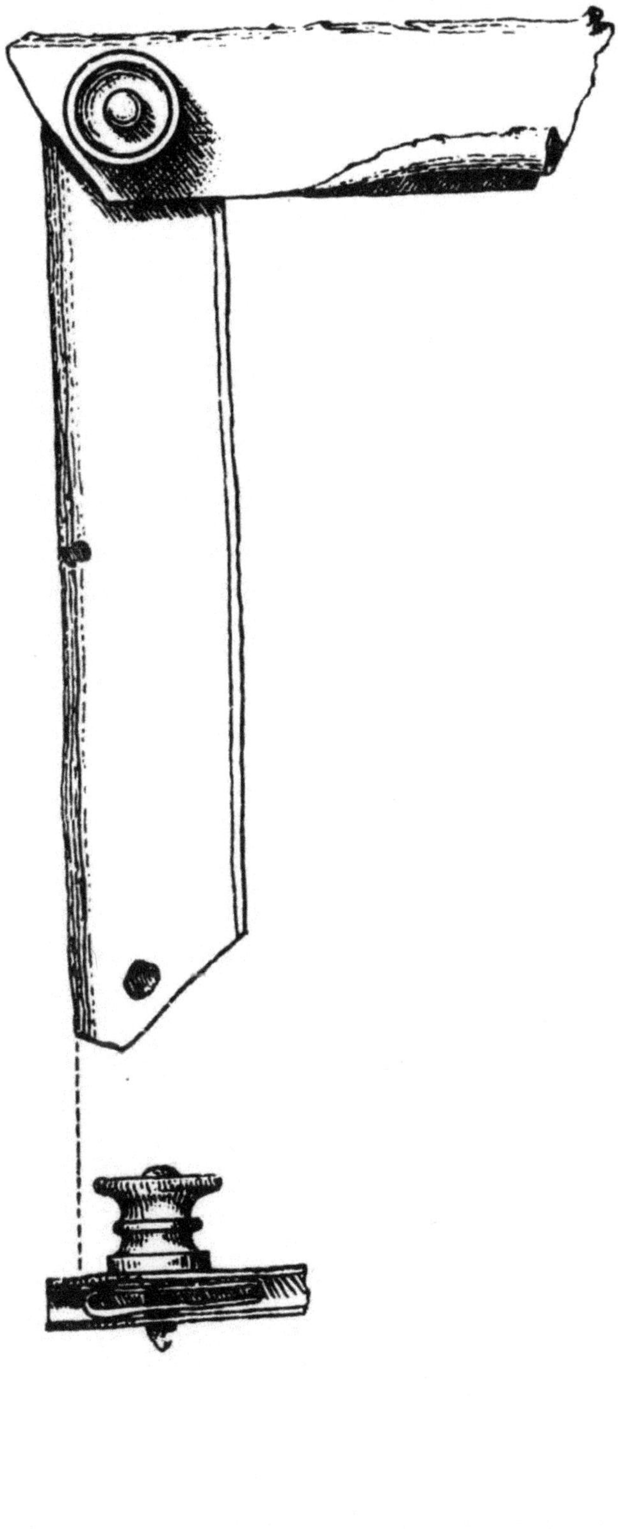

Pl. III: Bell-shaped stud used as a dolabra-sheath. Taken from RLÖ III, 1902, Taf.X.

The studs of both types which come from dated contexts in Britain suggest that they were used over a long time span. They are present at the Agricolan supply base at Corbridge Red House which cannot have continued in occupation long after A.D.90[22] but none have been found at the main site at Corbridge which was built in A.D.90 and continued in some form of occupation into the 4th century. Why there should be a dearth at Corbridge is unclear particularly when most of the other excavated forts along the line of Hadrian's Wall and the Stanegate have produced them.[23] Their use in the 2nd century A.D. is implied by their discovery at Sewingshields, a milecastle on Hadrian's Wall which had limited use in the 2nd century before being abandoned and reused as a metalworking factory in the 4th century,[24] and at the wall forts of Housesteads, South Shields and Wallsend - sites which had no 1st century occupation. Their survival into the 4th century is indicated at Piercebridge which is believed to have been built in the time of Constantius or Constantine.

On the Continent they have been found at Kapersburg which was built around A.D.100 and abandoned in the middle of the 3rd century,[25] and at Stockstadt, Saalburg, Zugmantel and Pfünz, which were all built under Domitian and abandoned in the mid 3rd century.[26] The box from Walheim has been dated to the second half of the 2nd century.[27] So far no reference has been found to their discovery in Gaul or Italy.

A group of ten Type 1 studs from Piercebridge were analysed semi-quantitatively by energy dispersive X-ray fluorescence by Justine Bayley and Brian White (see table). They proved to be of leaded bronze or gunmetal which is somewhat surprising in view of the suggestion that military fittings were normally brass.[28] If they were domestic fittings then their composition, a mixture of leaded alloys, is what would normally be found for massive castings where mechanical strength was not important. This seems to contradict the evidence that most Type 1 studs are found in military contexts but the date range is so extensive that the apparent contradiction may not be significant.[29]

To conclude one can suggest that the Type 2 studs were used for attaching decorative panels or lock-plates to boxes or chests but that the Type 1 studs were used for a variety of purposes. Although both can be found throughout the Roman period Type 1 studs tend to cluster on military sites whilst Type 2 studs can be found in either military or civilian contexts. To test these hypotheses more information is required as to the context and association with other finds of each bell-shaped stud found in future excavations.

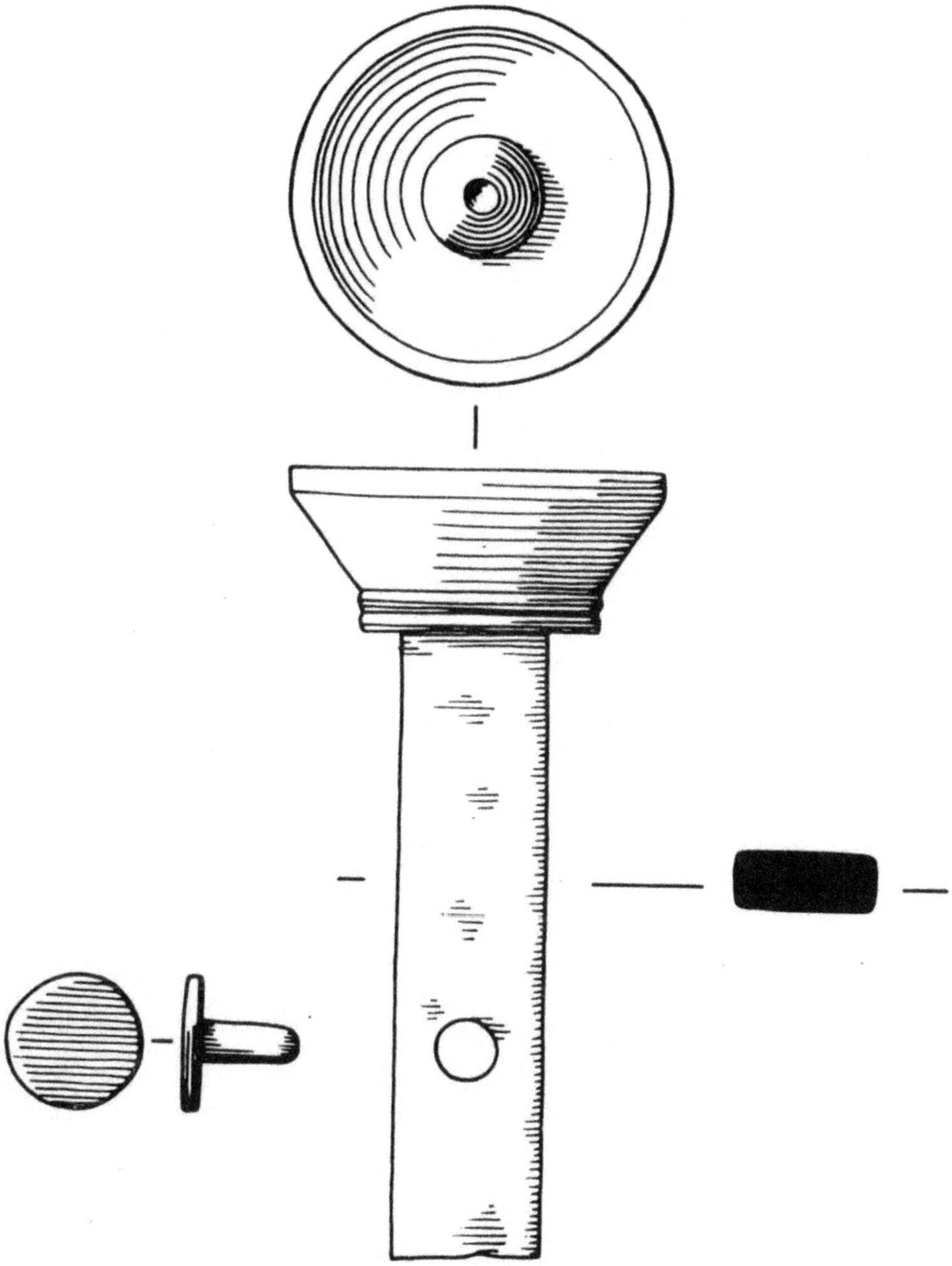

Fig. 3: Bell-shaped stud from Piercebridge. Drawn by S. Morris.

FOOTNOTES

1. HANSON _et_ _al_., 1979, 62.

2. Allason-Jones, in preparation.

3. _Saalburg_ _Jahrbuch_, VII, 1930, Taf.X, nos.15, 18.

4. CURLE, 1911, 306 & Pl.LXXVIII, No.10.

5. KENYON, 1948, fig.88, No.23.

6. WEBSTER, 1960, 94.

7. JACOBI, 1897, fig.76, nos.49, 56, Taf.XXXXV, nos.14, 15, 18, 19.

8. HANSON _et_ _al_., 1979, 62-64.

9. SUTHERLAND, 1974, no.309.

10. PLANCK, 1982, 123.

11. SF No.129 v38.

12. BRODRIBB _et_ _al_., 1968, Part I, fig.3, no.53; Part II, fig.50, no.124. STONE, 1929, 148, fig.2, no.12.

13. WINKELMANN, 1926, p.103, Abb.22.

14. ALLASON-JONES & MIKET, 1984, nos. 3.889 - 3.995; DORE & GILLAM, 1979, fig.2.

15. ALLASON-JONES & MACKAY, forthcoming.

16. RLÖ III, 1902, Taf.VIII, nos.23, 24, 25.

17. RLÖ V, 1904, Fig.33.

18. RLÖ X, 1909, fig.36, no.7.

19. RLÖ V, 1904, fig.11, p.23.

20. RLÖ III, 1902, Taf.X, fig.1.

21. See examples from Vindonissa illustrated in CURLE, 1911, p.279, fig.39.

22. HANSON _et_ _al_, 1979, 62.

23. E.g. Benwell: Museum of Antiquities of the University and Society of Antiquaries of Newcastle upon Tyne Acc.No. 1926.23; Chesters: Chesters Museum 1467, 2538, 2984, 2989, 2542, 2982, 1492, 1469, 887; Wallsend: SF Nos. 125, 839, 1053, 1135, 1340; Vindolanda: BIDWELL, P., forthcoming; Carlisle: Blackfriars SF No.77.

24. Archaeologia Aeliana[5] XII, 1984, p.78, fig.12, nos.13, 14.

25. ORL 12, Taf.VII, no.24.

26. ORL 33, Taf.VII, no.43; JACOBI, 1897, Taf.XXXXV, nos.14, 15, 18, 19; ORL 8, Taf.XIII, nos.60, 67, 68, 71; ORL 73, Taf.XIV, nos.72, 79.

27. Personal comment from D. Planck.

28. Bayley, in preparation; Craddock, P.D., pers. comm.

29. Cf. brooches: Bayley, J., 'Copper alloys and their use in Iron Age and Roman Britain', a paper presented to the 1984 Symposium on Archaeometry, Smithsonian Institute, Washington.

TABLE OF RESULTS

AM Lab. No.	XRF peak heights				Alloy
	Copper	Zinc	Lead	Tin	
841146	65500	-	5550	1070	Leaded bronze
841147	65500	-	19800	2000	Leaded bronze
841148	25500	-	3600	400	Leaded bronze
841149	55200	-	9400	570	Leaded bronze
841150	65500	8000	1500	400	(Leaded) gunmetal
841151	61300	-	3500	590	(Leaded) bronze
841152	56600	8800	4800	230	Leaded gunmetal
841153	32300	3600	5500	600	Leaded gunmetal
841154	65500	6350	380	405	Gunmetal
841155	65500	1250	2730	275	(Leaded) bronze/ gunmetal

NOTES TO TABLE

'(Leaded)' alloys probably contain less lead than 'Leaded' ones; all contain more than a few percent of the metal. Bronzes are copper-tin alloys while gunmetals also contain significant amounts of zinc. Pure copper-zinc alloys are called brasses. Different elements fluoresce more or less strongly so the absolute figures cannot usefully be compared. It is the values of the ratios of each element to copper which are compared with ratios obtained from objects of known composition and so provide an indication of the types of alloy used.

ABBREVIATIONS

RLÖ - *Der römische Limes in Österreich*
ORL - *Der obergermanisch-raetische Limes der Römerreiches*

BIBLIOGRAPHY

ALLASON-JONES & MACKAY, forthcoming: L. Allason-Jones & B. Mackay, *Coventina's Well: a Shrine on Hadrian's Wall* (forthcoming)

ALLASON-JONES & MIKET, 1984: L. Allason-Jones & R.F. Miket, *The Catalogue of Small Finds from South Shields Roman Fort* (Newcastle 1984)

BIDWELL, forthcoming: P.T. Bidwell, *The Roman Fort of Vindolanda* DOE Monograph 12, (forthcoming)

BRODRIBB *et al.*, 1968: A.C.C. Brodribb, A.R. Hands, D.R. Walker, *Excavations at Shakenoak Farm, near Wilcote, Oxfordshire* (Oxford 1968)

CURLE, 1911: J. Curle, *A Roman Frontier Post and its People. The Fort of Newstead in the Parish of Melrose* (Glasgow 1911)

DORE & GILLAM, 1979: J.N. Dore & J.P. Gillam, *The Roman Fort at South Shields: Excavations 1875-1975* (Newcastle 1979)

HANSON *et al.*, 1979: W.S. Hanson, C.M. Daniels, J.N. Dore, J.P. Gillam, 'The Agricolan supply base at Red House, Corbridge', *Archaeologia Aeliana*[5] VII, 1979, 1-88

JACOBI, 1897: L. Jacobi, *Das Römerkastell Saalburg bei Homburg vor der Höhe* (Homburg v.d.H. 1897)

KENYON, 1948: K.M. Kenyon, *Excavations at the Jewry Wall Site, Leicester* Society of Antiquaries Research Committee Reports No.15 (Oxford 1948)

PLANCK, 1982: D. Planck, Grabungen in Kastellvicus von Walheim, Kreis Ludwigsburg, *Archaölogische Ausgrabungen in Baden Württemberg 1982* (Stuttgart 1982)

STONE, 1929: P.G. Stone, 'A Roman villa at Newport, Isle of Wight', *Antiquaries Journal* IX, 1929, 141-51

SUTHERLAND, 1974: C.H.V. Sutherland, *Roman Coins* (London 1974)

WINKELMANN, 1926: F. Winkelmann (ed.), *Eichstätt. Sammlung des* historischen Vereins Katalogue West- und Süddeutscher Altertumssammlungen VI (Frankfurt 1926)

A CHASED CHEEK-PIECE FROM STANWIX

Julian Bennett

The cheek-piece which forms the basis of this note was discovered at Barn Close, Brampton Road, Stanwix, during construction work before 1936.[1] It was donated to the Tullie House Museum, Carlisle, in that year, and was accessioned as a single item.[2] There is no archaeological record of its discovery, and no other items with this provenance were donated to the museum in this or later years, apart from three Roman coins,[3] and it is therefore not possible to say anything concerning the precise circumstances of its recovery. Nonetheless, although the precise find spot is not known, Barn Close itself is within the south-east corner, and thus the praetentura, of the Hadrianic fort of Stanwix, known to have been garrisoned by the ala Petriana milliaria probably from its foundation to the closing years of Roman occupation, and it is a reasonable presumption that the cheek-piece came from a stratified or disturbed Roman deposit within the fort, and was not associated, for example, with a burial or other non-specifically military deposit. Apart from brief notices,[4] the Stanwix cheek-piece has not been published, presumably on account of its rather unspectactular appearance when compared with other items of the same general class. This is regrettable, as not only can the item be ascribed with some confidence to a known unit, and does provide interesting information concerning its production, but there is every reason to believe that cheek-pieces of this type were more common than generally supposed: moreover, the previous publications contain certain errors in describing the manufacture and decoration of this piece. It is hoped that publication of the Stanwix find might prompt identification of similar examples now surviving only as fragments.

The cheek-piece, which comes from the wearers' left-hand side of the helmet, survives about seven-eighths complete, with some damage to the upper margin and to the stylised ear, the former caused by cutting and the latter by breakage and loss through corrosion. It measures 150 x 110mm, and is 1mm thick. It is formed from copper alloy, with some tinned areas, a golden surface surviving on the outer side, where not tinned, contrasted with a chestnut-brown patina on the reverse, suggesting that there may have been some post-discovery cleaning. The leading and rear edges are both folded over, the leading edges slightly cusped and the rear edge rounded. The position of the ear indicates that this cheek-piece was only slightly shaped to fit into a protector, although as currently displayed at Tullie

House, it is restored with a prominent shape on its leather backing (Pl.1). There is no surviving evidence for the hinge, nor for the original backing of this piece, although an iron backing may be confidently assumed, necessitated by the thinness of the metal and suggested by the 10mm wide folded over-edges: such iron backing-plates rarely survive, or at least are rarely reported, but have been noted on other examples of the same specific class, e.g. that from Heddernheim[5] and indeed on other plainer versions and some of the more highly decorated examples.[6] It is reasonable to assume that decorated cheek-pieces of this general type, even certain of the more elaborate examples, with or without ear protectors, come from service helmets rather than parade-gear, as can be seen from the Heddernheim and other helmets, or even those depicted on early cavalry tombstones.

The decoration is simple, if not naive, in conception and execution, and consists of a stylised ear, of which only the lobe survives, with a pearled border surrounding a design of two crossed oval shields on a punched background. Pearled borders are a common feature on cheek-pieces of this type, whether highly-decorated or plain, although cabled borders are perhaps equally common. The stylised ear also requires no further comment, being present on the majority of cheek-pieces of auxiliary cavalry origin. The principal design, of crossed shields on a punched background would, however, appear to be unusual, the overwhelming majority of published cheek-pieces being of the highly decorated type, commonly with mythological scenes, such as the Dioscuri, the Imperial Eagle, Mars, Victory, etc. This type of cheek-piece, however, is surely that represented on several early cavalry tombstones, for example that of C. Romanius, from Mainz, with a rosette, or one from Worms, with feathering.[7] Actual examples are rare, or perhaps more accurately are rarely published. Leaving aside those of probably legionary or at least infantry origin, there are a number of low-relief or plain auxiliary cavalry cheek-pieces, which belong to the same general class as the Stanwix piece. Plain examples have been discovered at Valkenburg and Nijmegen, both with a raised central area demarcated by a plain border, and no decoration other than the stylised ear.[8] The surviving cheek-piece of the Witcham gravel-pit helmet is a development stylistically on these, with an embossed semi-circle beneath the hinge together with a stylised ear, and with traces of five bosses, now missing, but probably plain, in view of the other bosses on this helmet. More akin to the Stanwix example is one of the Heddernheim helmets, which retains a right cheek-piece, with a chased rosette in place of the stylised ear, and a chased pearled border: at the top of the piece is a chased design representing locks of hair, surmounting a half-rosette. Finally, there are fragments from Dormagen and Heddernheim, the former with stylised ear, and a pattern of overlapping scales contained

Pl.1: The Stanwix cheek-piece (scale 1:1).

within a double cabled-border, the latter with a stylised ear, and a scale pattern beneath a half-rossette, with a scroll pattern further down.[9]

Owing to the circumstances of its disovery, there is no evidence to date the Stanwix piece by, other than through stylistic analysis. It has been claimed that the piece is probably third century,[10] presumably on account of its poor decoration. Consideration of the other pieces within this general group, however, would allow a date-range from the first century to the later second, and further speculation on this point would be unfounded, beyond noting the Hadrianic date for the foundation of Stanwix. It should be noted, however, that there is an early type cavalry tombstone from Stanwix (RIB 2030) which might imply an pre-Hadrianic foundation on the site (if it does not derive from a cemetery associated with the earlier fort in Carlisle), although the Dis Manibus formula in full, as on this tombstone, also occurs on the tombstone of Flavinus at Hexham, of late Flavian, more properly, Trajanic date (RIB 1172).

Detailed examination of the low-relief decoration on the Stanwix piece indicates that this was produced entirely by chasing, rather than embossing or engraving, as is often claimed for this and other cheek-pieces. Chasing involves the hammering of the surface of the item with shaped punches, for example a beader, to produce shallow or deep depressions (pointille), or a hollow beader, with its open circular or oval end, to produce a raised area by driving down the surrounding metal, or a straight- or curved-edged beader to produce lines. This particular method of working is well suited to the production of low relief decoration. The other processes are quite distinct: embossing involves the hammering of the piece from behind while it is either supported in a soft bed (as in much silverwork) or, as in repousse work, while it is hammered into a metal die (as used to produce the sheet bronze decoration on toilet boxes and mirrors); engraving, on the other hand, requires the actual removal of strips of metal by cutting the surface with a graver, with a 'V'- or 'U'-shaped blade. The production of the more elaborate cheek-pieces evidently involved a combination of all three techniques, although chasing was by far the most commonly used one: embossing in nearly all cases was restricted to the central motif, chasing to the border and other low-relief areas, with engraving restricted for background detail or for sharpening the embossed areas. That chasing was the sole method used in the manufacture of the Stanwix and other pieces of the same class is clear from the absence of hammer marks on the reverse, and the sharp indentations made by the beaders used to form the pearled border and the other decorative motifs.[11]

Decoration of the Stanwix cheek-piece was completed by

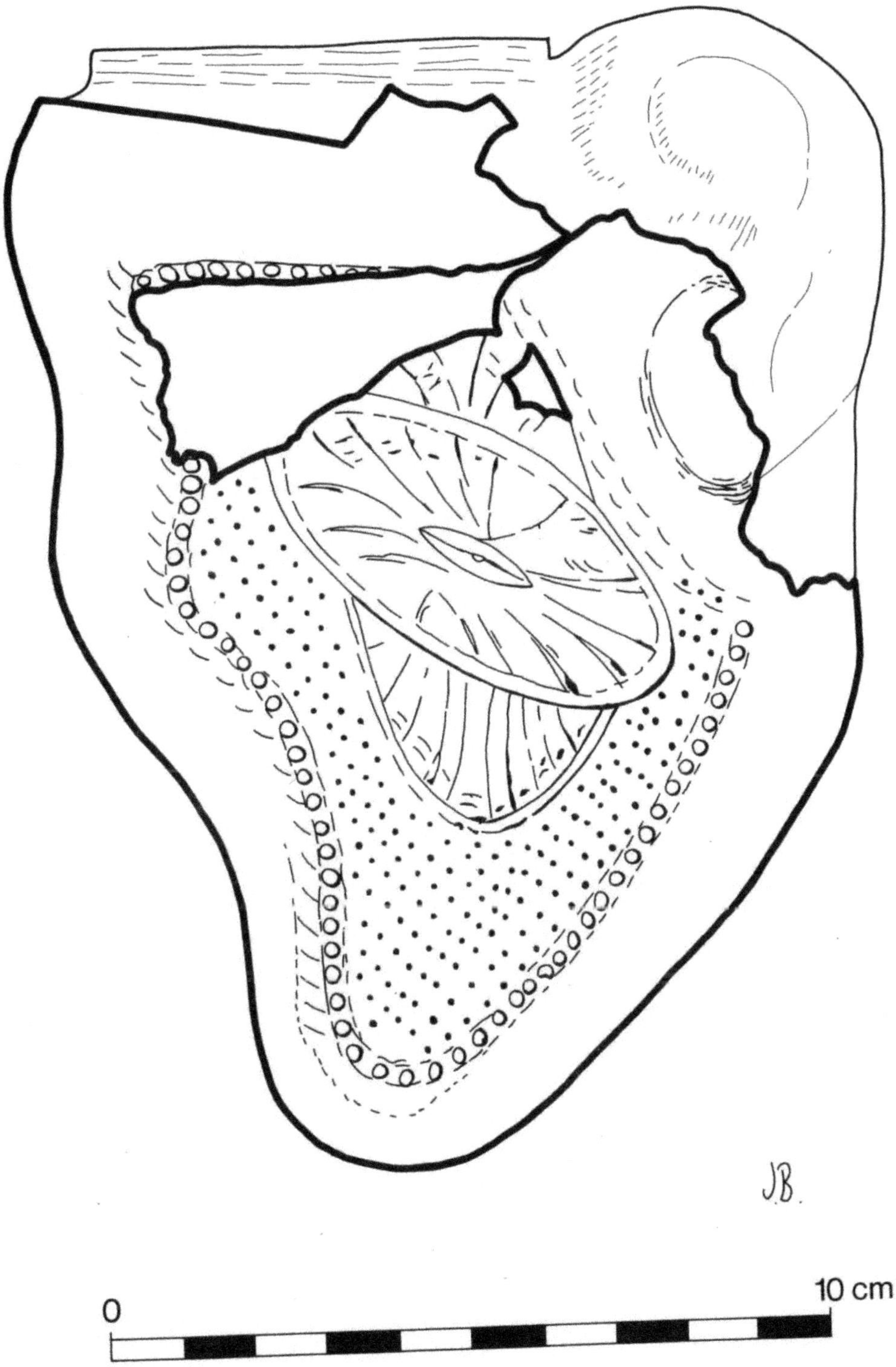

Fig.1: The Stanwix cheek-piece (scale 1:1).

tinning. This is restricted to the shields, ear, pearled border and outer edges of the cheek-piece, and was intended to make them stand out from the punched background: as has been noted, the golden appearance of the background might owe its present appearance to cleaning after discovery, but the absence of any visible tinning in the depths of the punched depressions confirms that this area was not originally tinned, and such surface treatment perhaps having been subsequently removed through injudicious cleaning. Tinning is a common technique on many Roman military bronzes, and several civilian items also, and was the final stage in the finishing of the piece in question. The method is recorded by Pliny, and was recommended by him - with particular reference to drinking receptacles - to prevent the formation of verdigris: the process used by the Romans was presumably to wax those surfaces that were not required to be tinned, whilst flushing the remaining areas with the molten metal.

It would be most unwise to speculate on the organisation behind the production of the Stanwix and other decorated auxiliary cheek-pieces, whether they were official issue from central government, or produced uniquely by each auxiliary regiment, for individual issue or purchase. My feeling is for the latter, given the variety in styles and techniques that has been recorded in the equipment, such variety perhaps reflecting the amount of cash any one soldier was prepared to pay: certainly, the highly decorated or even full-closed helmets were not restricted to officers, as is shown by the ownership marks they sometimes bear. Localised small-scale production of auxiliary equipment is certainly known at Stanwix, and presumably at other sites, the Stanwix fabrica producing scabbard chapes and mail armour amongst other items,[12] and it is not beyond the bounds of possibility, and certainly within the bounds of practicality, that larger items such as helmets and their cheek-pieces might also have been produced on a regimental basis according to an approved pattern.

ACKNOWLEDGEMENTS

I am most grateful to Colin Richardson, of Tullie House Museum, Carlisle, for his encouragement in publishing the Stanwix cheek-piece, and rescuing it from undeserved obscurity. Mr Richardson also kindly provided the photograph for Pl.1, taken by Mr G. Paule, the copyright of which remains with Tullie House Museum. My academic supervisor, Charles Daniels, and my colleagues Jon Coulston and Mike Bishop, all of the Department of Archaeology, Newcastle University, have helped with certain details, and I remain grateful to them for these, although none are responsible for any errors of fact or opinion.

NOTES

1. NGR NY403572

2. 29.1936

3. 27.1944.1 and 2, and 55.1945.10

4. TOYNBEE, 1964, 297; ROBINSON, 1975, Pl.403 and 135; ROBINSON, 1979, 9; GARBSCH, 1978, No.76

5. FISCHER, 1973, 98, Abb.22,1

6. For plain examples, see those from Nijmegen and Valkenburg published respectively in KLUMBACH, 1974, No.45, Taf.44, and GROENMAN-VAN WAATERINGE, 1967, Fig.76; for decorated examples, e.g. that from Newstead, ROBINSON, 1970

7. Both conveniently published in ROBINSON, 1975, Pls.302 and 292

8. GROENMAN-VAN WAATERINGE, 1967, Fig.76; KLUMBACH, 1974, No 45, Taf.44

9. MULLER, 1979, 126-127, Taf.21, 12; FISCHER, 1973, 98, Abb.22,1

10. ROBINSON, 1975, 135

11. A fragment of what is probably one of these cheekpieces in the Grosvenor Museum, Chester shows signs of having been embossed: NP 63-4 7 (FU) (50) (information from M.C. Bishop, who inspected this and other pieces by kind permission of the Grosvenor Museum).

12. COLLINGWOOD, 1931

BIBLIOGRAPHY

COLLINGWOOD 1931: R.G. Collingwood, 'Roman Objects from Stanwix' Trans. Cumb. West. Ant. Arch. Soc., Ser.2, 31, 1931, 69-80

FISCHER 1973: U. FISCHER, Grabungen in römischen Steinkastell von Heddernheim, 1957-1959, (Frankfurt am Main 1973)

GARBSCH 1978: J. Garbsch, Römische Paraderüstungen, (Munich 1978)

GROENMAN-VAN WAATERINGE, W., 1967: W. Groenman-van Waateringe, Romeins Lederwerk uit Valkenburg, z.H., (Groningen 1967)

KLUMBACH 1974: H. Klumbach, Römische Helme aus Niedergermanien, (Bonn 1974)

MÜLLER 1979: G. Müller, Durnomagus: das römische Dormagen, (Cologne 1979)

ROBINSON 1970: H.R. Robinson, 'Fragments of a Roman Cavalry Cheek-piece from the Fort at Newstead', Proc. Soc. Antiq. Scotland, 102, 1970, 290-292

ROBINSON 1975: H.R. Robinson, The Armour of Imperial Rome, (London 1975)

ROBINSON 1979: H.R. Robinson, What the Soldiers Wore on Hadrian's Wall, (Newcastle upon Tyne 1979)

TOYNBEE 1964: J.C. Toynbee, Art in Britain Under the Romans, (Oxford 1964)

NOUVELLES OBSERVATIONS SUR LES CABOCHONS DE BRONZE ESTAMPES DU CINGULUM ROMAIN

Michel Feugère (*)

ABSTRACT

A re-examination of those decorated bronze studs previously studied by G. Ulbert and other scholars, leads to a critical analysis of the scheme generally used in the study of such diffused objects. Both iconography and the distribution of types bring the author to place an official workshop producing these cingulum elements in Besançon during the second half of the 1st century A.D.; the studs then would not have been distributed by commerce, nor by troop movements, but given to the soldiers as part of imperial propaganda on specific occasions.

1. INTRODUCTION

Le thème de réflexion proposé aujourd'hui par M.C. Bishop, "production et diffusion de l'équipement militaire romain", est relativement nouveau dans la démarche des chercheurs qui ont eu à se préoccuper de ce type de mobilier jusqu'à présent. Certes, les archéologues s'intéressent depuis longtemps, par exemple, à l'origine des armes romaines; l'étude des influences subies par tel ou tel type de casque, de poignard, devait inévitablement conduire les auteurs à aborder la question des ateliers. Mais l'interrogation débouchait essentiellement sur des considérations générales, faute de pouvoir s'appuyer sur une documentation précise et, surtout, sur une problématique claire. C'est bien, semble-t-il, la première fois que la question est au centre d'un débat.

Du point de vue de l'archéologue, ce retard peut s'expliquer par le fait que jusqu'à une époque récente, une réponse argumentée à l'alternative objet militaire/objet civil pouvait apporter pleine satisfaction. Peu importait, en fait, de savoir si l'objet était de fabrication locale ou italique, ou comment il était arrivé sur le site en question. L'approfondissement de la démarche archéologique entraîne, c'est bien naturel, un déplacement sensible de la problématique; en fait, ce ne sont pas les seuls archéologues de terrain qui sont amenés à questionner leurs documents sous un autre angle: les chercheurs spécialisés dans l'équipement militaire viennent, eux aussi, à d'autres champs d'investigations. L'identification de la fonction d'un objet ne constitue plus le but ultime de leur recherche; il s'agit maintenant d'intégrer le document militaire dans un système de production et d'échanges; c'est-à-dire d'appréhender, sans négliger aucune source, toute sa signification sur le plan historique.

2. METHODOLOGIE

Cette démarche est, en fait, celle-là même qui a renouvelé, depuis quelques années, les études sur le petit mobilier en général. Il y a bien longtemps que les céramologues, non contents de dater leurs tessons, se sont préoccupés de savoir d'où ils provenaient. En revanche, ce n'est que depuis une date très récente que, principalement sur la base d'études chorologiques, on s'intéresse à l'aspect économique, voire sociologique, des petits objets dans les fouilles d'époque romaine.

Il n'est donc pas inintéressant d'examiner en détail la méthode généralement suivie pour l'étude de ces mobiliers (épingles, fibules, miroirs..., etc.). Nous nous interrogerons ensuite sur les éléments de cette méthode qui sont appliquables à l'étude des mobiliers militaires, avant de tester nos résultats sur l'exemple choisi des cabochons estampés.

Quelle que soit la catégorie considérée, la première tâche consiste toujours à regrouper un corpus des documents utilisables; les objets sont choisis a priori selon les critères pré-existants. Ce préalable, s'il tient compte d'un minimum de bon sens, ne constitue pas un facteur d'erreur significatif dans la mesure où le tri typologique intervient plus tard. Ce n'est, en effet, qu'une fois le corpus initial réuni que l'on peut songer à répartir la population en catégories. A ce stade, en revanche, l'archéologue doit effectuer des choix décisifs, soit qu'il s'agisse d'éliminer certains individus non conformes aux lois de la série, soit qu'il faille définir les facteurs discriminants sur lesquels reposeront toutes les interprétations ultérieures.

Le classement sélectionné en dernier ressort gagne, en principe, à subir le test que constitue pour lui une carte de répartition. Si l'objet de la typologie a été la mise en évidence des productions de différents ateliers, on ne peut guère douter que les origines diverses ne se manifestent par des distributions nettement dissemblables. C'est, d'ailleurs, sur ce point que les opinions des auteurs divergent le plus: sans que l'on précise le but exact recherché, on se défend souvent vivement d'avoir voulu traduire dans la typologie les origines matérielles des objets (comme si la classification constituait une fin en soi).

Quoi qu'il en soit, cette méthode a permis de mettre en évidence, avec un certain succès, des productions aussi différentes que celle de coffrets en Suisse septentrionale, d'épingles en os de IVe s. dans le Nord-Est de la Gaule, de fibules et de miroirs en bronze en Gaule méridionale.[1] Ayant déjà exposé ailleurs les possibilités et les limites de cette démarche, je ne reviendrai pas ici en détail sur son fonctionnement.[2] Il est cependant évident que l'archéologue doit, à chaque étape, effectuer des choix qui sont autant d'occasions de s'engager dans une impasse. L'échelle géographique à laquelle il faut examiner le phénomène est, par exemple, capitale; cet aspect est intimement lié au type de prospection sur lequel repose le corpus, et au cadre fixé a priori dans la problématique. Dans le domaine des mobiliers militaires, toute réflexion d'ensemble se doit d'envisager au minimum l'Europe occidentale, avec deux extensions possibles, vers l'Afrique et/ou vers le Moyen-Orient. Ce champ d'investigations immense constitue, sans aucun doute, une sérieuse excuse pour qui voudrait expliquer le retard des études chorologiques dans le domaine des mobiliers militaires.

3. APPLICATION

Cette méthode peut-elle être appliquée telle quelle à l'étude des cabochons estampés qui nous intéressent ici? Ces objets, d'abord étudiés par G. Ulbert,[3] ont bénéficié depuis quelques années d'une attention particulière de la part des archéologues. Présentant une grande variété iconographique à l'intérieur d'un type bien défini, relativement abondants, répartis sur de nombreux sites, de la Grande-Bretagne jusqu'à la Hongrie, ils constituent un matériau idéal pour le sujet qui nous occupe. Examinons donc un premier lieu les données acquises, ou considérées comme telles.

3.1 Typologie, chronologie et fonction:

Le mérite de la définition du type revient bien sûr à G. Ulbert: ne rentrent dans la catégorie qui nous intéresse que les objets de bronze présentant au dos une petite tige de fixation pointue, et dont le disque en tôle mince porte un motif estampé figuratif (généralement une tête); le rebord replié vers le bas forme tout autour une mince couronne.

De tels objets ont été répertoriés par G. Ulbert sur 25 sites: il était dès lors évident que cette série était liée d'une manière ou d'une autre au problème du <u>limes</u>; G. Ulbert proposait

Fig. 1: Légionnaire et porte-enseigne romains, début Ier s. ap. J.-C.

en 1971 un classement en 17 types, tenant compte aussi bien du thème iconographique (sujet et style) que de la taille de ces cabochons (entre 16 et 43mm). La datation semble assez homogène, que l'on s'appuie sur les figurations elles-mêmes ou sur les contextes archéologiques: deuxième moitié du Ier s. de notre ère, surtout époque flavienne.

Enfin, la fonction des cabochons estampés est suggérée par deux sources complémentaires: d'une part, les nombreux reliefs militaires montrant le 'tablier' de cingulum couvert de semblables boutons; d'autre part, une lanière de cingulum effectivement conservée, à Mayence:[4] on y observe un rangée de 11 boutons lisses fixés sur le cuir. Aucun document ne montre donc avec précision la fonction de nos cabochons estampés. Néanmoins, leur petite taille et leur fragilité les font attribuer à l'équipement individuel plutôt qu'au harnais; et parmi les accessoires utilisés par le fantassin, on ne voit guère que le cingulum qui convienne à ces petits ornements[5] (fig.1).

Il s'agissait donc, dès 1971, d'un mobilier bien étudié et sur lequel il n'y aurait, en principe, pas à revenir. Mais dans la perspective qui nous préoccupe, cette bonne connaissance constitue au contraire un élément indispensable. Avant d'aborder les problèmes liés à la production et à la diffusion de ces boutons, il nous faut cependant revenir sur les questions d'inventaire et de classement typologique.

3.2 Corpus complété:

Depuis le travail de G. Ulbert, divers articles ont fait connaître d'autres cabochons estampés.[6] Ces publications, complétées par des prospections personelles (bibliographie et musées) permettent aujourd'hui de dresser une liste de 39 gisements ayant livré de tels objets (cf. Annexe 2). La carte mise à jour qui découle de cette liste confirme naturellement le caractère militaire de cette série, même si les sites de l'intérieur sont un peu mieux représentés qu'autrefois. A ma connaissance, seul le limes septentrional de l'Empire est concerné par cette distribution, aucun exemplaire n'ayant encore été signalé, ni en Afrique, ni au Proche-Orient.

3.3 Typologie critique:

Le classement proposé par G. Ulbert repose, on l'a vu, sur le sujet figuré, le style et la taille du bouton. Néanmoins, un examen caractéroscopique, du type de celui auquel ont recours les numismates, montrerait que cette typologie doit, par force, regrouper sous une même catégorie des cabochons estampés à l'aide de coins différents. On peut d'ores et déjà dresser, pour chaque type, la liste des poinçons les mieux connus, en les désignant par une lettre suivant le numéro du type (Annexe 1, et fig.2 et 3).

Il n'est, bien sûr, pas toujours facile de distinguer, sur la base des seules publications disponibles, si deux cabochons d'aspect voisin proviennent véritablement du même poinçon;[7] et on peut d'ailleurs s'interroger, comme on le verra plus loin, sur la nécessité archéologique d'une telle distinction: le fait que deux ateliers reproduisent, avec quelques variantes, le même schéma iconographique ne constitue-t-il pas déjà un rapprochement intéressant? En d'autres termes, quel intérêt y a-t-il à morceler à l'extrême la documentation? Le rôle de la typologie n'est-il pas de rapprocher, plutôt que d'isoler?

3.4 Chorologie:

Une réponse à ces questions peut être recherchée dans l'examen des cartes de répartition. La distribution générale des cabochons (fig.4) nous informe surtout sur leur fonction (militaire) et, à un échelon aussi peu précis, sur l'aire géographique concernée (<u>limes</u> septentrional de l'Empire). Force est de constater que la répartition par types fournit une information relativement décevante, chaque série comportant en général trop peu d'individus pour que la diffusion soit significative.

Néanmoins, il est important d'indiquer pour chaque gisement le nombre d'objets attestés. Les données numériques, trop rarement prises en compte dans la cartographie archéologique, sont ici particulièrement frappantes: la trouvaille de Besançon, localisée en retrait du <u>limes</u>, mais à égale distance des points extrêmes de la répartition, a fourni à elle seule 200 cabochons, soit plus de 73% du total des objets de cette catégorie actuellement répertoriés.

En se limitant aux types les mieux représentés, on peut examiner la répartition des types suivants:

- le type 7 (fig.5) est très diffusé sur toute l'aire concernée; cette première carte met en relief la position centrale de Besançon par rapport aux points extrêmes de la distribution. On chercherait vainement, cependant, des regroupements géographiques suggérant la présence d'un atelier distinct pour chaque variante.
- le type 8 (fig.6) connaît une distribution assez semblable. Besançon semble y jouer un rôle central, sans que cette position se traduise néanmoins par une diffusion centripète.

- les types 9 et 14 (fig.7 et 8) reproduisent à peu de choses près le même schéma; si, dans le premier cas, la trouvaille de Káloz déséquilibre la carte vers l'Est, la diffusion du type 14 ne concerne (sauf l'exception de Fremington Hagg), que le Nord-Est de la Gaule, et le limes jusqu'à Regensburg.

- le type 10 enfin (fig.9) fournit peut-être la carte la plus étonnante: ses 4 exemplaires, répartis sur près de 2000km, ne semblent reliés entre eux par rien, si ce n'est le limes.

4. CONCLUSIONS

Quelles données peut-on retirer de ces observations? Tout d'abord, une remise en cause, dans ce cas précis, de la typologie comme outil d'analyse significatif si on veut l'utiliser pour définir une production. Il est évident, devant plusieurs cartes, que la distinction des variantes, voire même parfois des types, brouille le jeu. La carte la plus lisible reste, paradoxalement, la fig.4 sur laquelle on ne distingue pas les différents types.

Il semble bien que l'on puisse en conclure que les divers types de cabochons ne correspondent pas, comme dans la plupart des objets fonctionnels dont on peut étudier la distribution, à des fabrications successives ou émanant d'ateliers différents. En effet le seul centre qui apparaisse sur les cartes, et qui puisse donc prétendre, dans cette optique, à la qualité d'atelier, est Besançon. Mais à partir de là, les cabochons estampés ne suivent pas les règles 'normales' de distribution des mobiliers domestiques, militaires ou civils. Leur diffusion sur le limes semble obéir à d'autres lois.

Si Besançon est bien, comme je le pense, l'atelier principal, sinon le seul, d'où sont sortis nos cabochons, comment expliquer leur distribution particulière? Il faut, je crois, prendre en compte l'iconographie elle aussi spécifique de ces objets. Il est hors de doute que les motifs, qu'il s'agisse des portraits impériaux, de la représentation d'une Victoire navale ou de figurations mythologiques (?), ressortissent à la propagande impériale généralement confiée, dans le domaine du petit mobilier, à la numismatique, mais également à certaines catégories d'objets personnels.[8] Supports privilégiés de l'image officielle, ces cabochons ont donc pu être distribués dans le cadre d'activités relatives à la vie militaire. Leur abondance interdit d'y voir des dona militaria, récompenses trop précieuses, et en outre pratiquement inaccessibles aux auxiliaires.[9] Ne pourrait-on les considérer comme de simples cadeaux, distribués aux soldats à diverses occasions, par les empereurs successifs, de Néron à Trajan?

Ainsi s'expliquerait la source unique, rendue nécessaire par

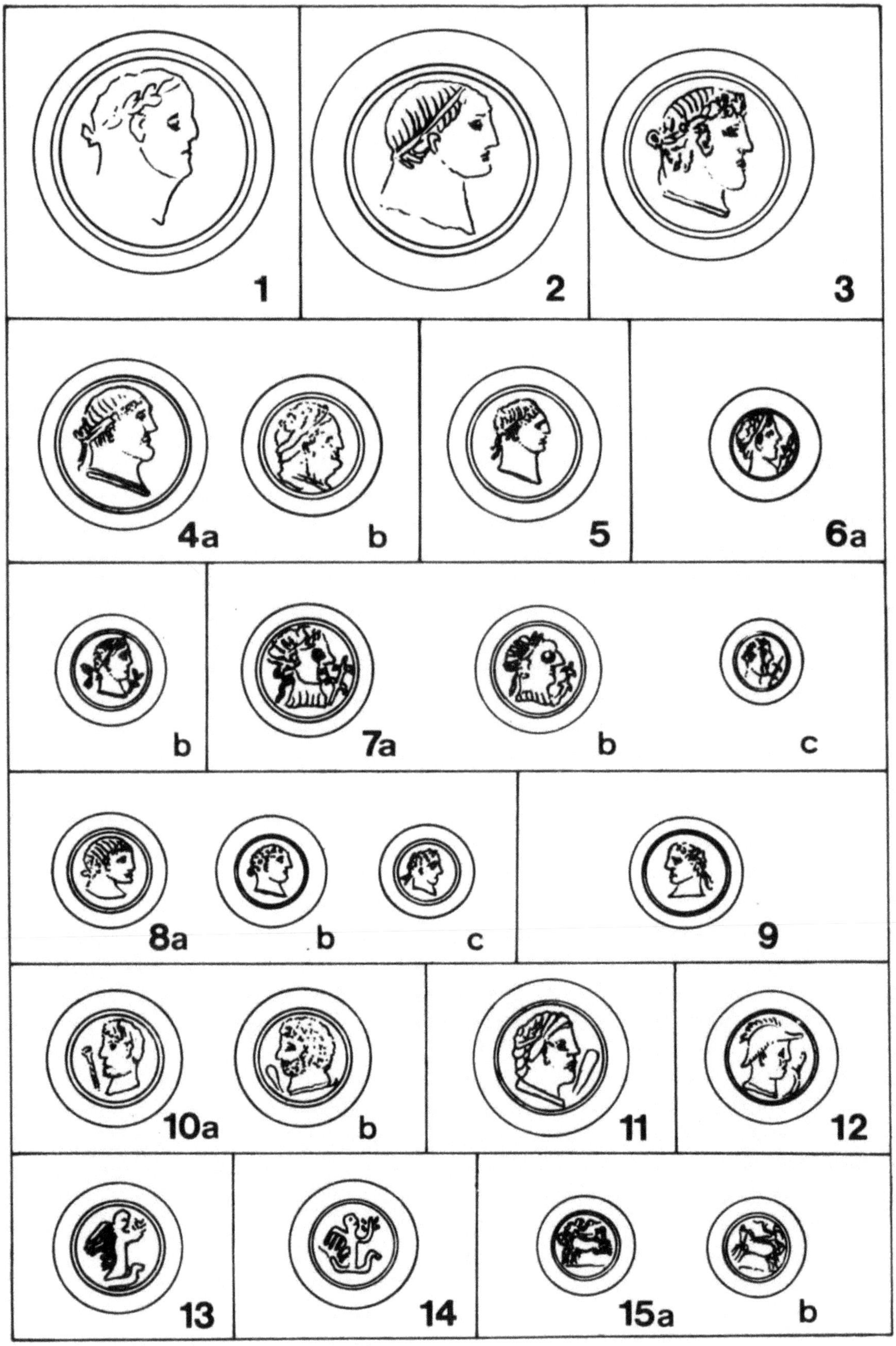

Fig. 2: Typologie affinée des cabochons estampés (suite page suivante).

Fig. 3: (suite en fin).

l'existence d'un atelier officiel, seul garant d'une iconographie conforme au programme impérial. Si tous les types de cabochons ne sont pas attestés à Besançon, il ne faut pas oublier que l'on ne dispose pas d'un décompte exact de tous les exemplaires découverts, la série complète n'étant pas accessible pour l'instant au Musée. La localisation de cet atelier n'est d'ailleurs pas sans rapport avec le problème des ateliers d'équipement militaire proprement dit. On a en effet, depuis longtemps, relevé les influences gauloises qui semblent s'être exercées sur certaines catégories, comme le harnais, ou les casques de l'armée romaine. Celle-ci avait vraisemblablement implanté en Gaule, dont la position centrale dans l'Empire occidental présentait aussi l'avantage d'une riche tradition artisinale, plusieurs *fabricae*. La liste donnée par la *Notitia Dignitatum* pour la fin de l'Empire ne représente d'ailleurs à cet égard, on le sait, que l'évolution d'une situation beaucoup plus ancienne.

Il semble donc raisonnable de considérer les cabochons de bronze estampés, largement diffusés sur le *limes* septentrional de l'Empire, comme des accessoires de *cingula* produits à Besançon dans une *fabrica* impériale, et distribués aux soldats lors de certaines occasions officielles, entre les règnes de Néron et de Trajan. Cette interprétation peut s'appuyer à la fois sur l'iconographie spécifique de ces objets, et sur leur distribution à l'intérieur du *limes*, non conforme aux règles que l'on peut observer à partir d'autres mobiliers.

ANNEXE 1: TYPOLOGIE AFFINEE ET LISTE DES CABOCHONS ESTAMPES

Les renvois indiqués par l'abréviation GU renvoient à l'article de ULBERT (1971 - on se reportera à ce travail pour une description plus complète des types); la bibliographie n'est indiquée in extenso que pour les exemplaires complétant ce premier inventaire.

Type 1: tête d'empereur à dr., couronnée de lauriers (diam. 40mm)
- Camp de Siscia (GU fig.1,3)

Type 2: tête d'empereur à dr., ceinte d'un bandeau (diam. 43mm)
- Pfünz (GU fig.1,1 = 5,6)

Type 3: tête à dr., couronnée de lauriers (diam. 35mm)
- Malton (GU fig.1,16)

Type 4: tête à dr., ceinte d'une couronne de lauriers (ou plutôt d'un bandeau?; diam. 28mm)

.Var. a: tête d'un homme mûr, au long cou:
- Besançon, 4 ex. (GU fig.2,1 à 4 - ici fig.11,1)

.Var. b: tête d'un homme âgé, sans cou:
- Urspring (GU fig.5,3)

.Autre var.?
- Mandeure, diam. 45mm! (Musée de Montbéliard, rens. C. Tchirakadzé - ici fig.11,2)

Type 5: tête à dr., couronne ou bandeau noué sur la nuque (diam. 18 à 22mm)
- Mailhac (Aude), *oppidum* du Cayla, fouilles O. et J. Taffanel (Inv. F.32); 2 ex. inédits, ici fig.10,1 et 2[10]
- Besançon, 3 ex. (GU fig.2,7 à 9)
- Heilbronn-Böckingen? (GU fig.4,1)

Type 6: tête à dr., ceinte d'un bandeau, sceptre devant (diam. 18 à 20mm)

.Var. a: tête remplissant tout le champ:
- Londres, *Tokenhouse Yard* (GU fig.5,10 - ici fig.10,3)
- Malton, 3 ex. (GU fig.1,17 et 18)

.Var. b: tête plus petite, noeud visible sur la nuque:
- Frankfurt-Heddernheim (GU fig.3,1 = 5,10)

Type 7: la même, de style schématique (diam. 18 à 22mm)

.Var. a:
- Londres, 2 ex. (GU fig.5,9 = ici fig.10,4) et un autre ex. (inédit, au Museum of London, Inv.20083 - ici fig.10,5 et 11,3)

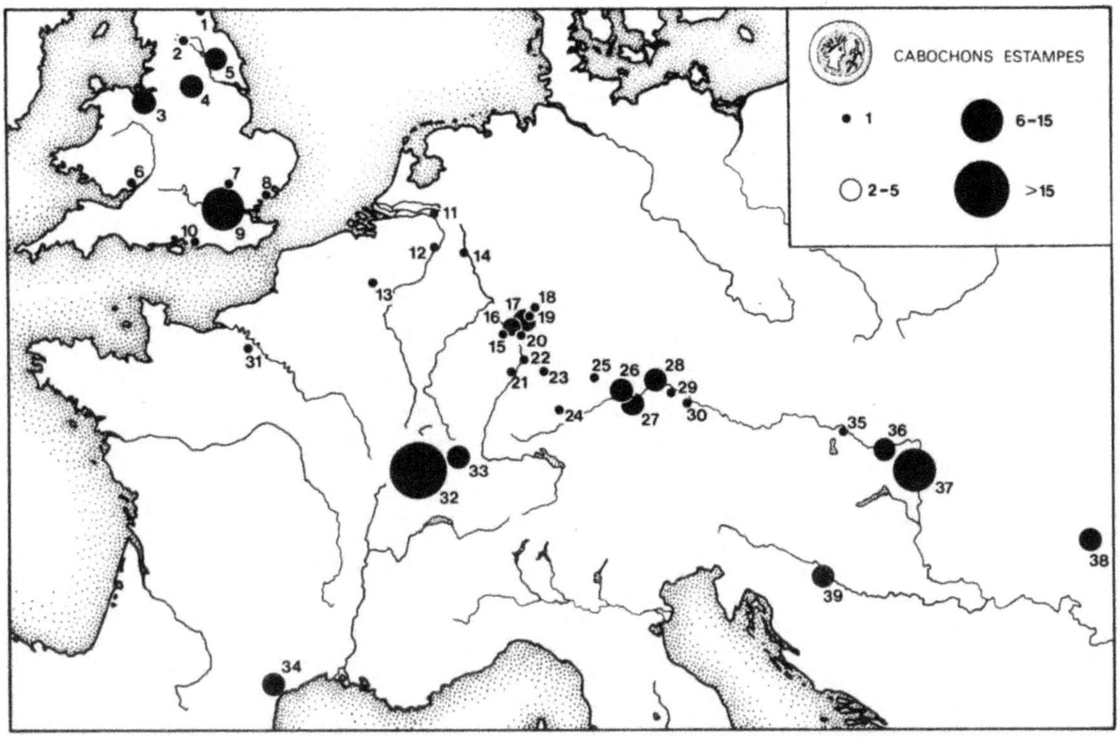

Fig. 4: Carte de répartition des cabochons estampés, tous types confondus (d'ap. G. Ulbert, complétée; pour les localisations, se reporter à l'Annexe 2).

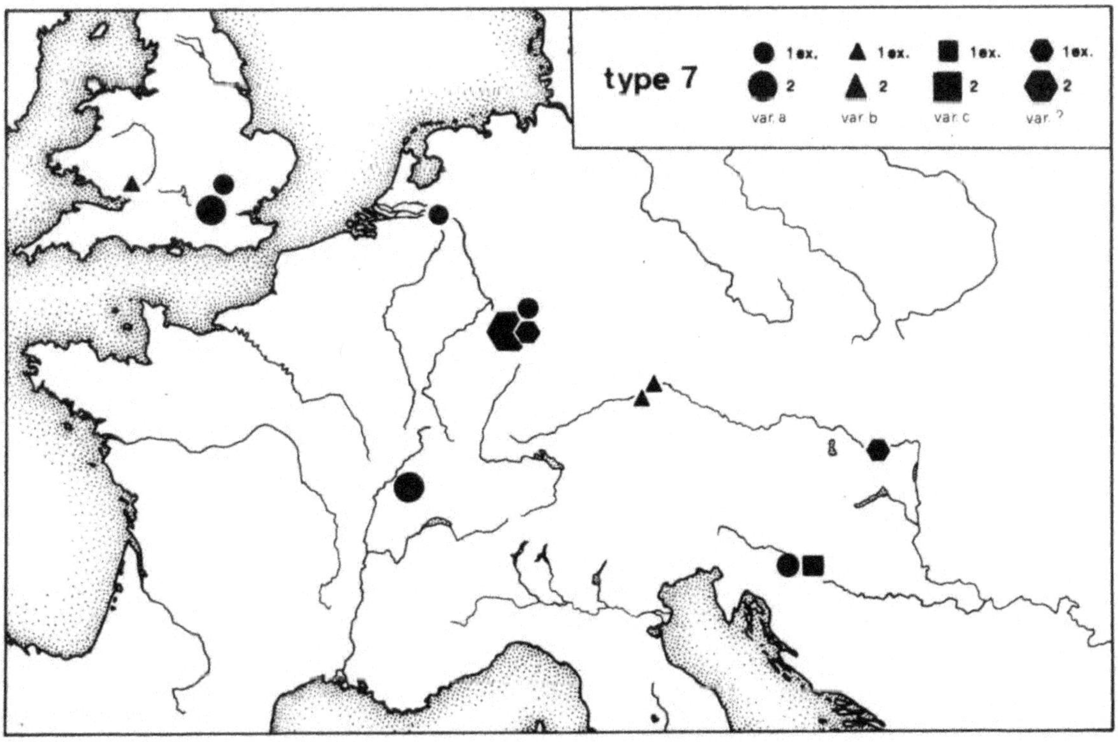

Fig. 5: Distribution du type 7.

- Skeleton Green (PARTRIDGE, 1981, 308,a et fig.113: tombe à coffre n°XXXV; réutilisation probable - ici fig.10,6)
- Nimègue, castra (ZADOKS-JITTA et al., 1973, n°139)
- Besançon, 2 ex. (GU fig.2,5 et 6)
- Okarben (GU fig.3,5)
- Siscia (GU fig.1,2)

.Var. b:
- Regensburg, nécropole (GU fig.4,4 = 5,1)
- Oberstimm, camp (BÖHME, 1978, pl.22, n°174 et pl.43)
- Caerleon, camp légionnaire (GU n°1, non ill.)

.Var. c: tête plus petite et de facture moins barbare:
- Siscia (GU fig.1,4)

.Var. ?
- Gross-Gerau (GU fig.3,7)
- Mayence, camp légionnaire, 4 ex. (GU fig.5,7, 8 et 11; fig.3,6)
- Szöny-Brigetio (GU fig.1,9)

Type 8: petite tête stylisée à dr., cheveux noués (diam. 18 à 20mm)

.Var. a: grosse tête, coiffure peignée:
- Malton (GU fig.1,13)
- Le Vieil-Evreux (GU fig.5,4)

.Var. b: tête plus petite, coiffure bouclée:
- Malton, 2 ex. (GU fig.1,14 et 15)
- Buggenum (Pays Bas) (Musée de Leyde, Inv. GL.147 - ici fig.11,4)
- Besançon, au moins 8 ex. (GU fig.2,12 à 19)
- Straubing, camp (GU fig.4,6)
- Siscia, camp légionnaire, 2 ex. (GU fig.1,5 et 6)

.Var. c: tête encore plus petite, noeud bien visible sur la nuque:
- Gnotzheim? (GU fig.5,12)
- Rheinzabern (Inv. L6 81/39, trouvaille de surface; rens Dr. Reutti - ici fig.10,7)

.Var. ?
- Künzing? (GU fig.4,2)

Type 9: petite tête similaire, mais à g. (diam. 18 à 20mm)
- Londres, 2 ex. (inédits, Museum of London, Inv. 20252 et 20813 - ici fig.10,8 et 9)
- Pfünz (GU fig.5,5) (type 9 et non type 8)
- Besançon, au moins 5 ex. (GU fig.2,20 à 24)
- Szöny-Brigetio (GU fig.1,10)
- Káloz, 12 ex. (SZABO, 1976, 193-7)
- Siscia, camp légionnaire, 2 ex. (GU fig.1,7 et 8)

Type 10: tête à g., de style réaliste, objet allongé devant (diam. 22mm)

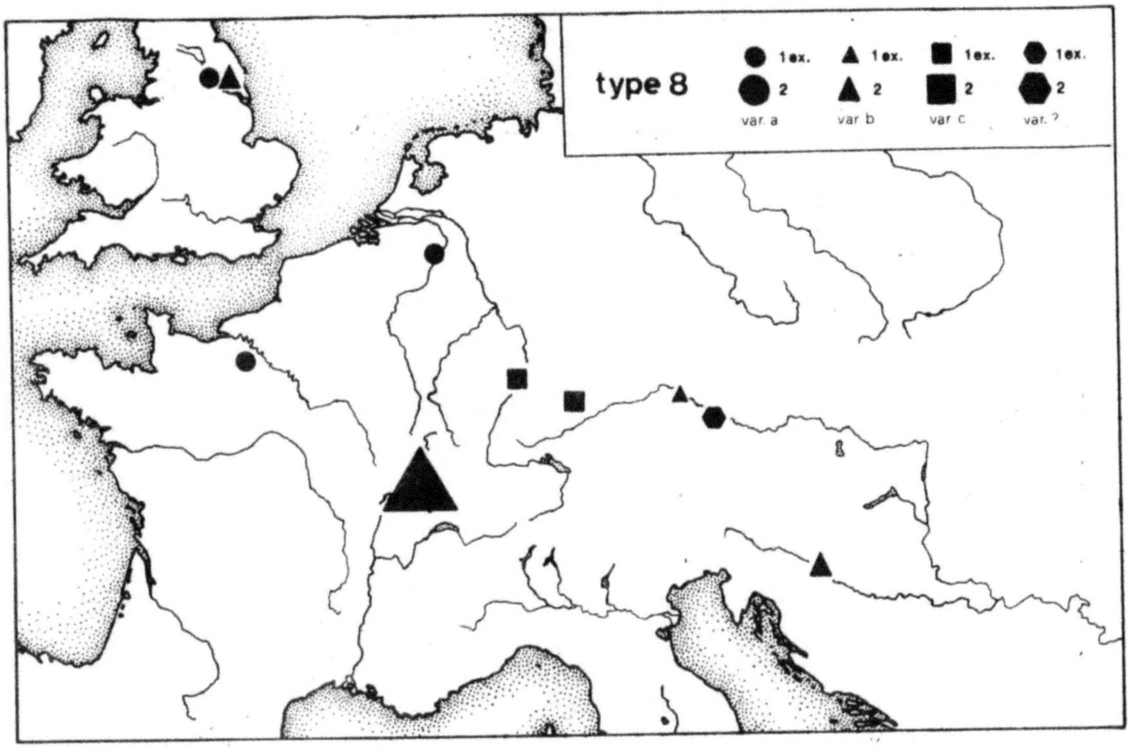

Fig. 6: Distribution du type 8.

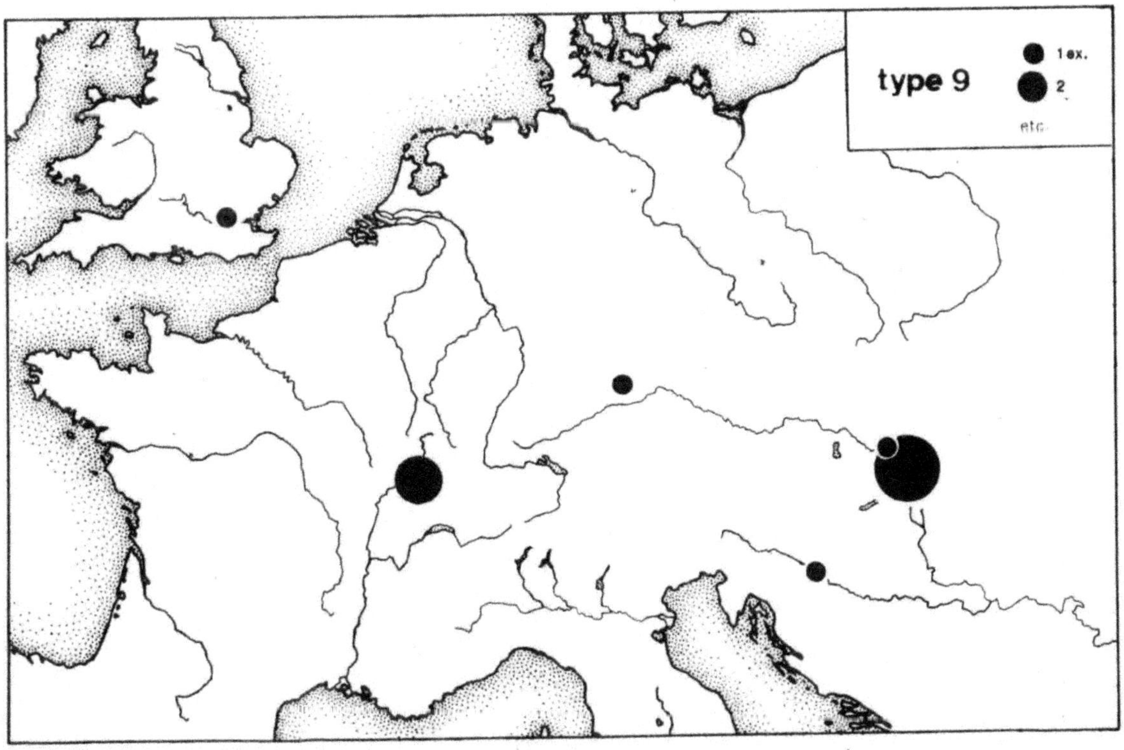

Fig. 7: Distribution du type 9.

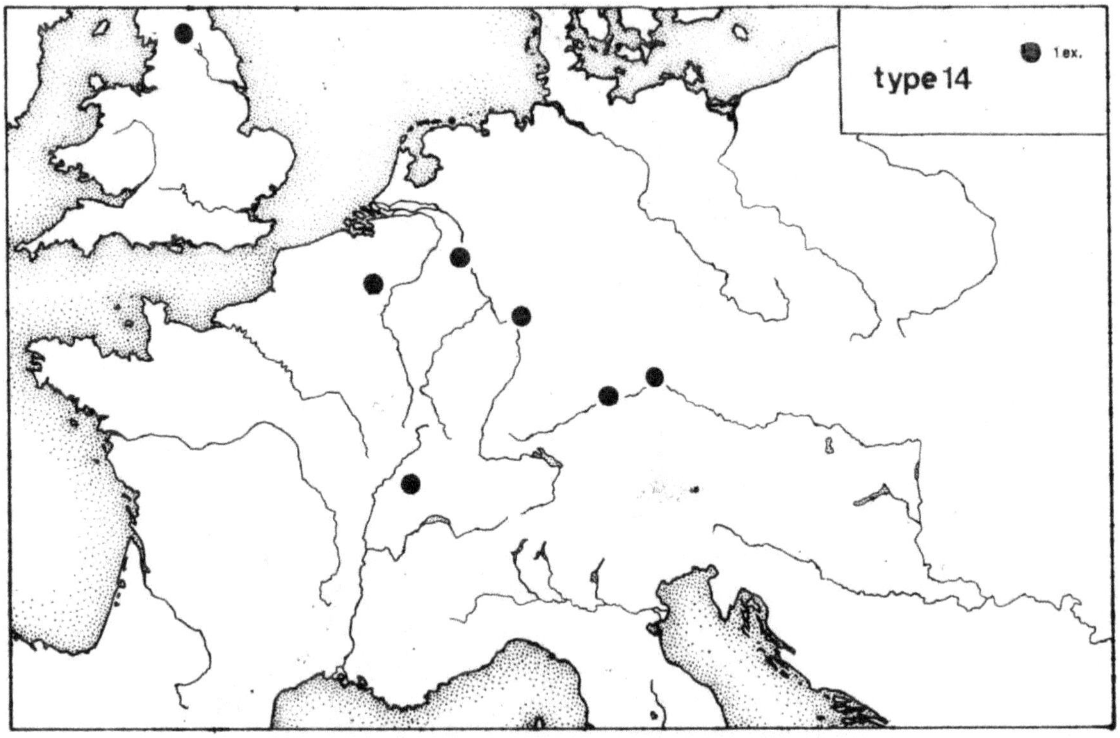

Fig. 8: Distribution du type 14.

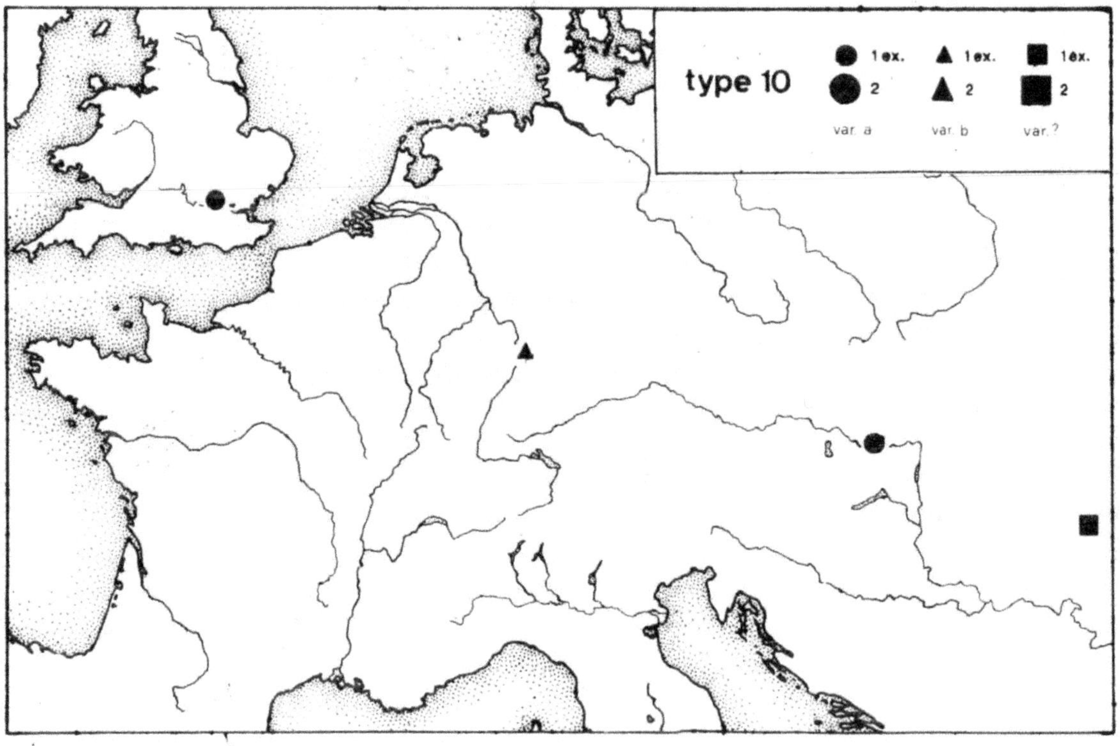

Fig. 9: Distribution du type 10.

.Var. a: tête imberbe, avec sceptre (?)
- Londres (inédit, Museum of London - ici fig.11,5)
- Szöny-Brigetio, camp légionnaire (GU fig.1,12)

.Var. b: tête barbue, avec massue (?) - Hercule?
- Frankfurt-Heddernheim (GU fig.3,4)

.Var. ?
- Sarmizegetusa (rens. D. Alicu - ici fig.10,10)

Type 11: tête à dr. couronnée et barbue; instrument (massue?) devant le visage (diam. 26mm)
- Chester (DROOP et NEWSTEAD, 1931, 133 et pl.XLVI,24; ici fig.11,6)
- Szöny-Brigetio, camp légionnaire (GU fig.1,11)

Type 12: tête casquée à dr., cimier; objet rond devant le visage: bouclier (?) - Mars? (diam. ?)
- Kreuznach, nécr., tombe 10 (GU fig.4,5)
- Rusovce (près de Bratislava), camp romain (GU fig.7,7)

Type 13: Victoire navale à dr., sur proue, très schématisée, tenant devant elle une palme stylisée (diam. 22mm)
- Frankfurt-Heddernheim (GU fig.3,2)

Type 14: même sujet, plus réduit (diam. 20 à 22mm)
- Fremington Hagg (BÖHME, 1978, liste 4,1)
- Liberchies (GU n°25, non ill. - ici fig.10,11)
- Besançon (GU fig.2,25)
- Neuss, camp légionnaire (GU n°11, non ill.)
- Frankfurt-Praunheim (GU fig.3,3)
- Oberstimm, camp (GU fig.7,6 = BÖHME, 1978, pl.22,B.173 et pl.43)
- Regensburg-Kumpfmühl, camp (GU fig.4,3 = 5,2)

Type 15: bige (diam. 16mm)

.Var. a: bige à dr.
- Besançon, au moins 4 ex. (GU fig.2,26 à 29)

.Var. b: bige à g.
- Besançon (GU fig.2,30)

.Var. a ou b?
- South Shields (MIKET, 1983, 112 no.63 fig.71)

Type 16: bige à g., mais diam. plus important (25mm)
- Besançon (GU fig.2,31)

Type 17: bige à dr., stylisé, animal au premier plan (diam. 32mm)
- Chester (NEWSTEAD, 1928, pl.VI,3 - ici fig.11,7)
- Montbéliard (GU fig.4,7)

Type 18: tête à g., couronnée, coiffure de rouleaux

concentriques; objet (?) devant le visage: empereur flavien (diam. 30mm)
- Heidelberg-Naunheim (NUBER, loc. cit. - ici fig.10,12)

Type indéterminé:
- Doncaster, 3 ex. (fouilles P. Buckland, rens. G. Lloyd-Morgan): 1971, DQ/MV, 113: tête à g., bandeau; diam. 18mm; 1971, DQ/NK, 122: tête à dr.? diam. c.18mm; et Baxtergate 1966, DA/EX, S, AB, 57, tête à dr., diam. c.22mm
- Chichester (DOWN, 1981, fig.8.31,31: tête à g.)
- Sarmizegetusa (rens. D. Alicu - ici fig.10,13)

ANNEXE 2: LISTE DES SITES AYANT LIVRE DES CABOCHONS ESTAMPES.

Cette liste reprend la documentation répertoriée ci-dessus, mais sous la forme d'un classement géographique. Il n'y avait donc pas lieu de redonner à chaque fois les références bibliographiques détaillées (voir pour cela l'Annexe 1). Les gisements sont présentés du Nord au Sud et d'Ouest en Est; les numéros correspondent à la carte de répartition générale des cabochons estampés, fig.4.

Grande-Bretagne:

1. South Shields, 1 ex. du type 15 (a ou b?);
2. Fremington Hagg, 1 ex. du type 14;
3. Chester, 1 ex. du type 11, 1 du type 17;
4. Doncaster, 3 ex. (type?)
5. Malton, 1 ex. du type 3, 1 du type 8a, 2 du type 8b;
6. Caerleon, 1 ex. du type 7b;
7. Skeleton Green, 1 ex. du type 7a;
8. Colchester? 1 ex. atypique;
9. Londres, 1 ex. du type 6a (Tokenhouse Yard), 2 du type 7a, 2 du type 9, 1 du type 10a;
10. Chichester, 1 ex. du type indétérminé;

Pays-Bas:

11. Nimègue, 1 ex. du type 7a;
12. Buggenum, 1 ex. du type 8b;

Belgique:

13. Liberchies, 1 ex. du type 14;

R.F.A.:

14. Neuss, 1 ex. du type 14;
15. Kreuznach, nécropole, 1 ex. du type 12;
16. Mainz, 4 ex. du type 7 (var.?);
17. Frankfurt-Heddernheim, 1 ex. du type 6b, 1 du type 10a, 1 du type 13;
18. Frankfurt-Praunheim, 1 ex. du type 14;
19. Okarben, 1 ex. du type 7a;
20. Gross-Gerau, 1 ex. du type 7 (var?);

21. Rheinzabern, 1 ex. du type 8c;
22. Heidelberg-Naunheim, 1 ex. du type 18;
23. Heibronn-Böckingen, 1 ex. du type 5;
24. Urspring, 1 ex. du type 4b;
25. Gnotzheim, 1 ex. du type 8c;
26. Pfünz, 1 ex. du type 2, 1 du type 9;
27. Oberstimm, 1 ex. du type 7b, 1 du type 14;
28. Regensburg, 1 ex. du type 7b, 1 du type 14;
29. Straubing, 1 ex. du type 8b;
30. Künzing, 1 ex. du type 8 (var.?);

France:

31. Le Vieil-Evreux (Eure), 1 ex. du type 8a;
32. Besançon (Doubs), 200 ex. dont au moins 4 du type 4a, 3 du type 5, 2 du type 7a, 8 du type 8b, 5 du type 9, 1 du type 14, 4 du type 15a, 1 du type 15b, 1 du type 16;
33. Mandeure (Doubs), 1 ex. du type 4 (var.), 1 du type 17;
34. Mailhac, oppidum du Cayla (Aude), 2 ex. du type 5;

Tchécoslovaquie:

35. Rusovce, 1 ex. du type 12;

Hongrie:

36. Szöny-Brigetio, 1 ex. du type 7 (var.?), 1 du type 9, 1 du type 10a, 1 du type 11;
37. Káloz, 12 ex. du type 9;

Roumanie:

38. Sarmizegetusa, 1 ex. du type 10 (var.?), 1 de type indéterminé;

Yougoslavie:

39. Siscia, 1 ex. du type 1, 1 du type 7a, 1 du type 7c, 2 du type 8b, 2 du type 9.

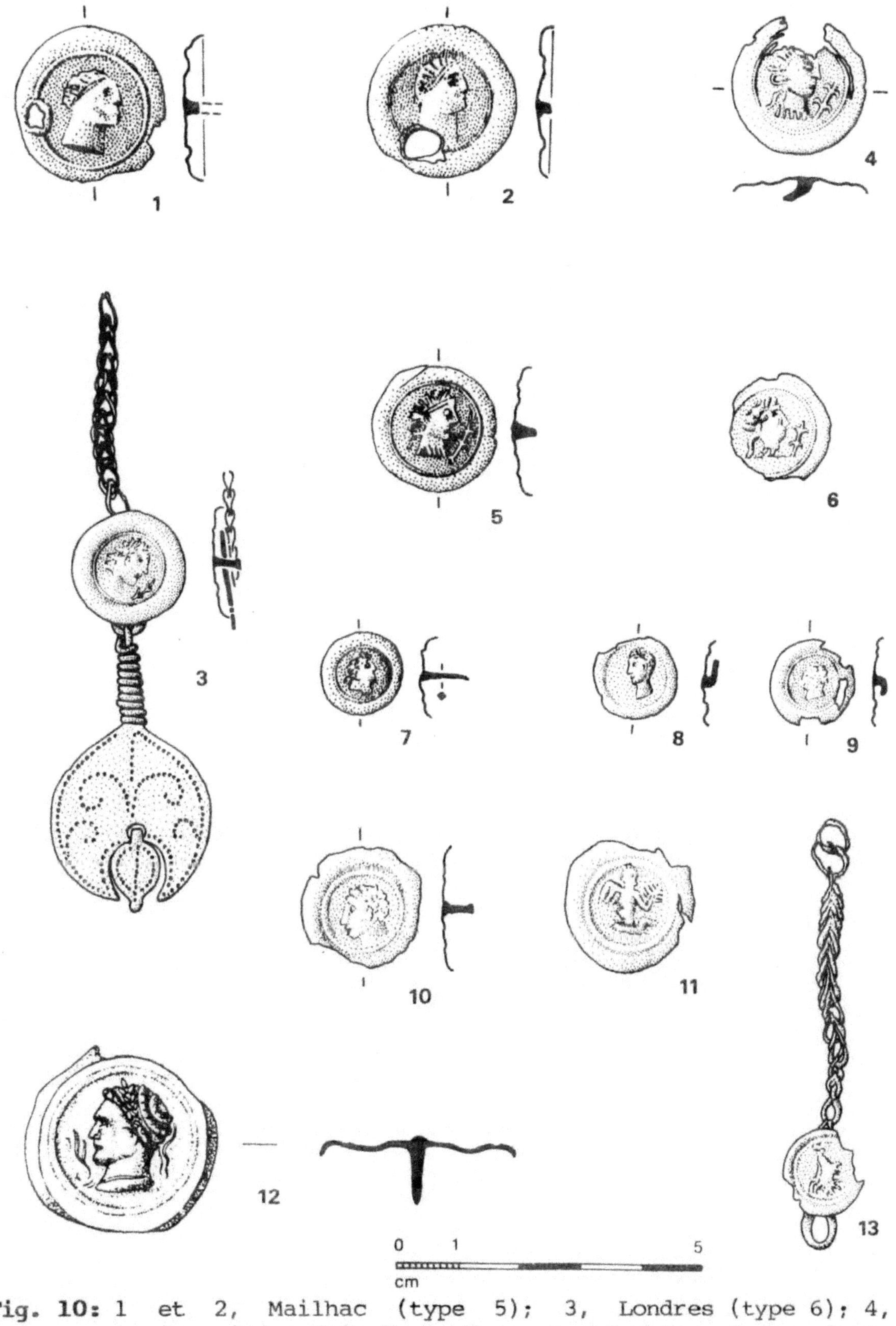

Fig. 10: 1 et 2, Mailhac (type 5); 3, Londres (type 6); 4, Londres (type 7a); 5, Londres, Bucklesbury House (type 7a); 6, Skeleton Green (type 7a); 7, Rheinzabern (type 8c); 8 et 9, Londres (type 9); 10, Sarmizegetusa (type 10); 11, Liberchies (type 14); 12, Heidelberg-Neunheim (type 18); 13, Sarmizegetusa (type ?).

ADDENDUM

Cet article était à l'impression quand trois nouveaux exemplaires ont été portés à ma connaissance:

Type 7:

.Var. a:

Verulamium (GOODBURN, 1984, n°.138)
Canterbury, Marlowe I (ibid., 45, note 13)

Type 9:

Verulamium (ibid., n°.139)

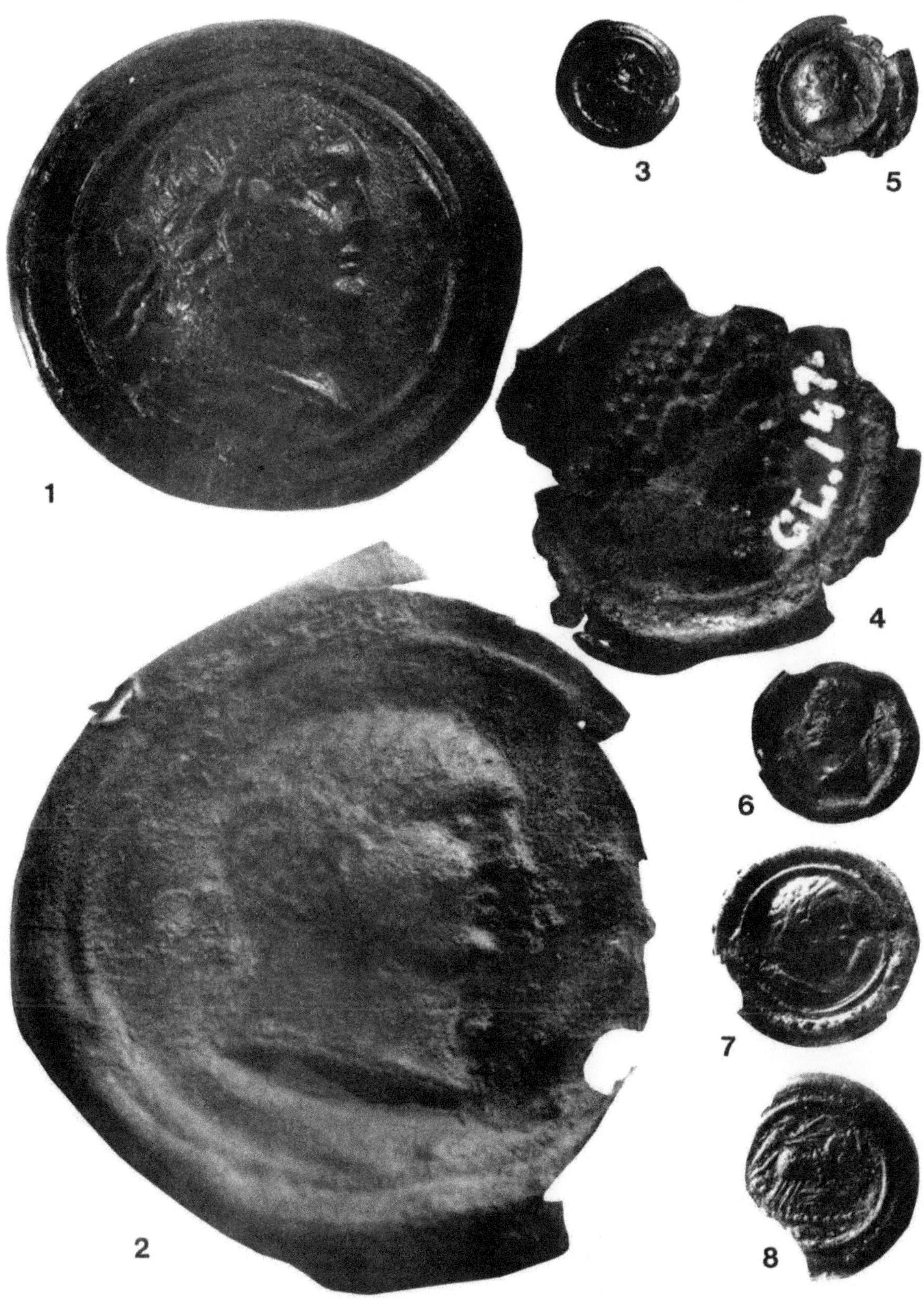

Fig. 11: 1, Besançon (type 4); 2, Mandeure (type 4, var.); 3, Londres (type 7a); 4, Buggenum (type 8b); 5, Londres (type 10); 6, Chester (type 11); 7, Chester (type 17).

NOTES INFRAPAGINALES

* Attaché de Recherche au C.N.R.S., Centre Camille Jullian, Université de Provence, 29 Av. Robert Schuman, 13621 Aix-en-Provence, France.

Je tiens à remercier tous les fouilleurs et responsables de collections qui m'ont autorisé à étudier les cabochons estampés qu'ils avaient découvert ou dont ils assuraient la conservation: en Grande-Bretagne G. Lloyd-Morgan, P. Buckland et H. Chapman; au Pays-Bas P. Stuart; en Allemagne A. Böhme, S. Rieckhoff-Pauli et G. Reutti; en France C. Tchirakadzé, P. Lagrange, O. et J. Taffanel; en Roumanie, D. Alicu. Je dois une attention particulière à G. Lloyd-Morgan, du Musée de Chester, dont l'aide efficace m'a été précieuse pour l'inventaire des cabochons de Grande-Bretagne.

1) Coffrets: MARTIN-KILCHER, 1976; épingles en os, *cf.* BÉAL, 1983, notamment p.40-3; fibules, *cf.* FEUGÈRE, 1984; miroirs rectangulaires dans le midi de la France, voir *Doc. Arch. Mérid.* 4, 1981, 148, fig.12.

2) ADAM et FEUGÈRE, 1982, 149-57 en particulier.

3) ULBERT, 1971, 278-97.

4) *Ibid.*, fig.7,1.

5) C'est le parti adopté dans la plupart des reconstitutions sérieuses de l'équipement du fantassin disponibles à ce jour: ici fig.1, légionnaire et porte-enseigne (Auguste-Tibère), dessins de P. Connolly pour le Römisch-germanisches Zentralmuseum de Mayence.

6) Publications de fouilles comme PARTRIDGE, 1981, 308, fig.113,a; d'objets isolés, comme NUBER, 1981, 517, fig.12; mais aussi de séries, comme BÖHME, 1978, 210, et liste mise à jour des types Ulbert 6/7 et 14, p.222 (carte fig.77, p.223); également SZÁBO, 1976, 193-7.

7) L'existence même de certaines variantes, comme par exemple ici la var. c du type 7, pourrait être suspectée: une mauvaise reproduction photographique n'a-t-elle pas fait prendre un bouton de type 6 pour une variante du type 7? Bien qu'une erreur de ce genre soit toujours possible, dans l'étude d'une série d'objets dispersés, le classement proposé ici s'attache à traduire les variantes les plus apparentes.

8) Fibules: *cf.* ADAM et FEUGÈRE, *loc. cit.*, 171-8; également

VIDAL, 1982, 234 s.

9) MAXFIELD, 1981, notamment 121 sqq.

10) Ces deux exemplaires ont été découverts en même temps, serrés l'un contre l'autre et percés d'un trou qui en faisait une petite boîte, ou un pendentif creux. Cette particularité, outre l'isolement géographique, montre bien que ces cabochons n'ont pas été apportés à Mailhac en tant qu'éléments d'un cingulum, mais en tant qu'objets décoratifs et, peut-être, de souvenir.

ORIGINE DES ILLUSTRATIONS

Fig.1: Reconstitutions de P. Connolly pour le Römisch-Germanischen Zentralmuseum de Mayence, cliché RGZM.

Fig.2 à 9: dessins M. Feugère.

Fig.10: d'ap. les originaux, sauf 6, d'ap. C. Partridge; 10 et 13, d'ap. D. Alicu; 11, d'ap. Y. Graaf; 12, d'ap. H.-U. Nuber.

Fig.11: 1, 4, 5 et 6, photos M. Feugère n°1402, 1798, 905 et 920; 2, photo Musée-Chateau de Montbéliard, C. Tchirakadzé; 3, photo Museum of London; 7, d'ap. R. Newstead.

BIBLIOGRAPHIE

ADAM et FEUGÈRE, 1982: A.-M. Adam et M. Feugère, Un aspect de l'artisanat du bronze dans l'arc alpin oriental et en Dalmatie au Ier s. av. J.-C.: les fibules du type 'de Jezerine', Aquileia Nostra LIII, 1982, 129-88

BEAL, 1983: J.-C. Béal, Epingles en os à tête féminine à Vienne et dans la région Rhône-Alpes, Bull. Soc. Amis de Vienne n°78, 1983, 31-47

BÖHME, 1978: A. Böhme, Metallfunde, in SCHÖNBERGER, 1978.

DOWN, 1981: A. Down, Chichester Excavations 5 (Chichester 1981)

DROOP et NEWSTEAD, 1931: J.P. Droop et R. Newstead, Excavations in the Deanery Field, Chester, 1928, Part II, The Finds, Annals. of Arch. and Anthr. XVIII, 3-4, 1931, 113-47

FEUGERE, 1984: M. Feugère, Les fibules en Gaule méridionale, Suppl. 12 à la Rev. Arch. Narb. (Paris 1984)

GOODBURN 1984: R. Goodburn, dans S. Frere, Verulamium Excavations, vol. III (Oxford 1984)

MARTIN-KILCHER, 1976: S. Martin-Kilcher, Das römischer Gräberfeld von Courroux im Berner Jura (Derendingen 1976)

MAXFIELD, 1981: V.A. Maxfield, The Military Decorations of the Roman Army (Londres 1981)

MIKET, 1983: R. Miket, The Roman Fort at South Shields. Excavations of the Defences 1977-1981 (Newcastle 1983)

NEWSTEAD, 1928: R. Newstead, Report on the Excavations on the Site of the Roman Fortress at the Deanery Field, Chester, Annals of Arch. and Anthr. XV, 1928, 3-32

NUBER, 1981: H.-U. Nuber, Waage mit Kaiserporträts aus Heidelberg-Neunheim, Fundberichte aus Baden-Württemberg 6, 1981, 501-28

PARTRIDGE, 1981: C. Partridge, Skeleton Green, A Late Iron Age and Romano-British Site (Londres 1981)

SCHÖNBERGER, 1978: H. Schönberger, Kastell Oberstimm, die Grabungen 1968-1971 (Berlin 1978)

SZABO, 1976: K. Szábo, Cabochons décorés de Káloz, Alba Regia XV,

1976, 193-7

ULBERT, 1971: G. Ulbert, Römische Bronzeknöpfe mit Reliefverzierung, Fundberichte aus Schwaben Neue Folge 19, 1971, 278-97

VIDAL, 1982: M. Vidal, Les motifs moulés et figurés de deux fibules à charnière de Vieille-Toulouse (Hte-Garonne), Rev. Arch. Narb. XV, 1982, 223-36

ZADOKS-JITTA ET AL., 1973: A.N. Zadoks-Jitta, W.J.T. Peters, A.M. Witteveen, Description of the Collections in the Rijksmuseum G.M. Kam at Nijmegen (Nimègue 1973)

SOME CHANGES IN THE MANUFACTURE AND SUPPLY OF ROMAN BRONZE HELMETS UNDER THE LATE REPUBLIC AND EARLY EMPIRE

John Paddock

I had originally intended to speak about the supply of all aspects of military equipment under the Republic, but I soon realised that this task was quite enormous and would take many hours, and indeed, that for the supply of certain items, such as body armour, the evidence is lacking. I therefore propose to concentrate exclusively on the helmet, principally because there is a large number of these items both provenanced and of known date, and that there is a reasonably large body of evidence to work with. I would like to make some remarks about the question of supply and to look at the changes in quality, form and processes of manufacture from the end of the 4th century B.C. to the mid 1st century A.D., with particular regard to the Montefortino helmet and its immediate bronze successors.

There are, in my opinion, two distinctive breaks in the continuity of the bronze Roman helmet, especially in its manufacture and the quality of the individual items, the first of which would seem to date to about the end of the 2nd century B.C. and the second to the early years of the 1st century A.D. or just before. I believe that both changes are probably related to changes in the nature of the individual within the Roman army and to the nature of the army itself.

I would like, first of all, to examine the question of the supply of equipment to the legions in the Republic. There can be no doubt that at the time of Polybius the individual legionary was responsible for supplying his own equipment, his position within the legion depending upon his age and property qualification amd the armour he could supply. Polybius, in Book VI,26,1-2,[1] leaves us in no doubt that there was a basic standard of equipment laid down for each type of soldier beyond which the individual might purchase more expensive or better equipment. This is confirmed in Book VI,39,15-40 where he tells us that the allies had their rations as a 'free gift', but the Romans had to pay for their own clothes and any additional arms they required. As we know from Tacitus (Annals I,17),[2] this situation continued into the Empire and is listed as one of the grievances of the legionaries in A.D. 14. However, by this time it would seem that equipment was held on a sort of lease-back arrangement. Again, Tacitus states this in his Histories (2,67)[3] when describing the events in Rome in A.D. 69 that the Praetorians "when they were offered in addition the inducement of an honorable discharge, had begun to hand over their arms to the Tribunes when the news of Vespasian's war began to spread."

This is a marked change from the early to middle Republican practice, when the legionary presumably returned home with his weapons upon discharge, and is a probable result of the Lex Julia de Vi Publica and the change from a citizen to a permanent, professional army. In this context it is interesting to note that by far the majority of Republican helmets turn up in civilian rather than military contexts.

Under the Republic the soldier's equipment was his own personal property and this, allied with the stylistic variation and the differing degrees of elaboration and the decoration, would seem to indicate a small scale supply by local metalworkers on an individual basis.

Under normal circumstances this system would seem to have worked adequately, but it must have been totally unequipped to have dealt with military disasters such as Cannae, and indeed the whole Second Punic War, when the demand for men and new equipment to replace the losses was immense. At such times of national emergency it would seem logical to postulate some form of over-riding state control of metalworkers to enable the state to fulfil its commitments.

Likewise, before a major campaign the Consuls would of necessity have to stockpile replacements for equipment likely to be damaged or lost, as well as large quantities of projectiles, including pila.

Like the equipment itself, evidence for areas of supply and manufacture is scarce under the Republic, but if one looks at the Montefortino helmet from the 4th century B.C. onwards, one can discern some possible areas of production and even gain some insight into the identity of the manufacturers. There are two groups of stylistically very similar and almost identical helmets numbering about 20 in all, from graves dating to the 4th and 3rd centuries B.C., which centre around a small area of Etruria and most particularly the vicinity of Perugia. We know from Polybius that when the Consul raised an army, all citizens of military age were summoned to Rome, and after having been selected they were dismissed to arm themselves and to return to their homes. They were required to assemble at a given time and place. Although the recruit might buy his equipment at any of these locations, it seems likely that a good number bought from local craftsmen near their homes. Furthermore, a man is generally also buried near his home and in view of the great similarity in these items, this may indicate a manufacturing centre; however, the evidence is far from conclusive.

There are three helmets (Figs.I-III) which come from Italy and date to the period between the end of the 2nd and the mid 1st centuries B.C. which bear stamped marks that can only be interpreted as maker's or armourer's marks. These marks are quite unlike the usual Republican or early Imperial inscriptions on equipment, which are usually just scratched on or executed in punctile work, and give the soldier's name and usually his unit, but are actually countersunk into the bronze with a die. This was done at the time of manufacture, as can be seen by the stamp on the helmet from Rieti, where the stamp overlies one part of the decoration, but is itself overlain by another part (Fig.Ib).

The fact that an armourer would go to the extent of making or having made a stamp for his work might seem to argue a relatively large scale production. In this connection it is also interesting that these helmets show a marked decline in the standard of workmanship over previous examples, which might be taken as further evidence of large scale production. I will return to this point later, but will say now that two of these helmets date to about the beginning of the Civil Wars and the Social War, when the large scale increase in army size would demand huge increases in the numbers made and the urgency of this demand would perhaps account for the decline in the standard.

The first of these three helmets, now in the Staatliche Museum, Munich,[5] bears a stamp 'Q Cossi Q' of Quintus Cossius the son of Quintus, probably a private artisan or entrepreneur, and seems to date to between the end of the 3rd and begining of the 2nd century B.C. It is possible that the two other helmets which bear maker's marks, that from Loreto Apruntino[6] and that from the Merrick Collection, now in the British Museum[7] carry government marks, respectively a monogram (FR) and RON (Figs.IIb, IIIb), which could stand for Rom Fecit and RON may have in fact been intended to read ROM, but I think it is more probable that once more these are of a private nature, as there is some indication that the stamp on the British Museum helmet had a letter, possibly an F, before the RON, and could therefore stand for Frontinus, as could the monogram FR.

I would now like to turn to the methods of manufacture and changes in quality.

For those who are not conversant with the raising of bronze, I would like, briefly, to outline the process. The armourer would start with an annealed bronze sheet; the metal would be formed into the desired shape over a variety of stakes, some ball headed, others not, by concentric, slightly overlapping hammer blows spiralling downwards. The metal, as a result, would become work-hardened and require frequent annealing. The hammer used in this process would be of cross-pein form, a type comparable to a modern raising hammer and occurring on both civilian and military sites, e.g. Silchester and Newstead. Finally[8] the surface would be planished to remove the marks of the raising hammer and then decorated, ground, cleaned and finally polished.

As I said at the beginning, there is a sharp drop in quality of the helmets at about the time of the Marian Reforms. Up until then, the quality of helmets had been fairly consistent and the bowls well decorated and finished. However, after the Marian Reforms, with their resultant influx of the poorest citizens into the army, there must inevitably have been a massive demand for cheaper equipment, a situation which can only have been exacerbated by the Civil Wars and the intervening Social War, which led to the vast multiplication of armies.

Practically all surviving helmets from this period are of Montefortino type although a small number of other forms may date to this period, e.g. the Italo Corinthian helmet, and have, in general, a very battered appearance, being left rough from the raising process, the hammer blows being especially obvious on a helmet from Montenerodomo (Fig.IVa). They are often asymmetrical and very crude in form and finish and every short cut, both in time and skill has been exploited. For example, the hinge plates for the cheekpieces are held by one rivet instead of two, and are often made of re-used scrap; decoration has disappeared completely; more often than not the crest knob is off-centre and on four or five helmets are applied and not made in one with the bowl. These crest knobs are cast and soldered on, some retaining flash marks, some have moved off-centre during soldering (Fig.IVb).[9] The edges of the bowl are not thickened, and the bowl itself is noticeably thinner.

The second change that may be detected in the manufacture of bronze Roman helmets seems to occur in the early years of the Principate, possibly as a result of the establishment of a standing army following the reforms of Augustus in 27 B.C. it is unfortunate that there are no known helmets that can be definitely dated to the years immediately following these reforms, but it seems probable, given the decreases in the number of legions, that for the next 20 years or so stocks of old equipment would continue in use.

The first securely datable early Imperial helmets come from the Rhine Frontier, most noticeably around Nijmegen, and show a very marked increase in the quality of the workmanship and general finish, including edge thickening and slightly larger neckguards. And it is about this time that there occurs the widespread adoption of spinning and other methods and techniques of mass production. A good number of the early Imperial Coolus/Hagenau[10] and at least one Imperial Montefortino from Nijmegen (Figs.V-VII)[11] have been spun and not beaten, a large number having spin marks and occasionally a punched centring point still visible. However, this is often obscured by the affixing of a cast or spun crest knob.

As with all ancient manufacturing processes, it is of course impossible to state categorically how a Roman bronze worker would set about spinning a helmet bowl, but thanks to the intuitive work of Dr. Alfred Mutz[12] we can say that it was probably not dissimilar to the technique now used by modern silversmiths.

The process known as spinning is applicable to shapes which are circular in section, although the finished form may be altered by the use of snarling irons, or reshaping on stakes. The principle tools needed are a single geared lathe, capable of 800 revolutions per minute, a forme and a spinning tool. The forme is usually a block of wood of the exact shape required, but smaller by exactly the thickness of the metal. The spinning tool would have to be a heavy bar of hardened and tempered iron/steel, highly polished and with no sharp corners to cut the work.

The annealed sheet of bronze would be placed between the forme and a suitable follower and centred. The tool would be used in gentle stroking motions from the centre to the edge as the metal disc rotated, forcing it against the forme.

Alfred Mutz has shown that this technique was used to produce vessels in the Classical World many years before its adoption by armourers. However, it is possible, and indeed probable, that given their shape and dimensions, Pilos helmets of Classical and Hellenistic Greece were made in this way. The rims of these helmets are completely circular and in all cases 10 dactyls[13] in diameter and height. Unfortunately, the final finishing process and the trimming up which would be necessary often obscure any remaining surface spinning marks, and therefore without structural analysis it is difficult to tell how far this process was used on Imperial helmets. This method of manufacture is obviously quicker, and in the long run cheaper, and is more suited to production on a large scale than is beating, and it is interesting in this context that the majority of Hagenau helmets, although falling typologically into different groups, within these are of very similar dimensions, and the internal diameters of the bowls are almost the same giving near hemispheres, consistent with spinning over a similar shape and size of former.[14]

Although steel is spun today, and this process would have

worked perfectly adequately for bronze helmets, it could not have been used upon the iron Imperial Gallic varieties of helmet since the carbon content of the iron/steel being used could not have been kept constant throughout, and the smith might easily find that the material he was spinning was harder than the tool he was using, and also that any impurities in the iron would be ripped out, ruining the work. Obviously the larger dimensions of the Hagenau type of helmet would allow for, indeed demand, a greater amount of padding at the sides and back of the head and the addition of a brow reinforcement would allow for the reduction in the height of the bowl since it gives more protection to the top and front of the skull, reducing the need for padding there. The higher bowls of the earlier Montefortino helmets, with their small neckguards, would give more chance of a whiplash injury.

The reasons for the change mentioned above seem to be two-fold: first and foremost, obviously a spun helmet is quicker and cheaper to produce and spinning lends itself readily to mass production. The second reason is perhaps less tangible and relates, I believe, to a change in the nature of the army and its personnel.

The reforms of 27 B.C. left Rome with an army of 28 or so legions, designed primarily for the defence of the Empire and composed of citizens serving for a minimum period of sixteen years with the colours and four years as a veteran, and often more than this. It was a regular army with an established system of officer N.C.O. ranks, and offered a recognised career. Dhe nature of the campaigns of this period were, to all intents and purposes, different to those of the Republic. They were no longer the great wars of conquest which yielded booty and easy money to the troops to augment their meagre stipendium, but were either conquests which yielded little by way of profit, or policing actions. As a result, it proved harder to attract men into the army and obviously greater care was taken of the now trained and valuable individuals. The reign of Tiberius was to find the army taking up permanent garrisons upon the borders and for the first time the army had a period of relative peace and the military mind being what it is, obviously would start to demand greater uniformity, which is well reflected in the plan of the camps of this period, and indeed would have time to think up improvements in equipment. I cannot agree with H. Russell Robinson that this development in the helmets was an ad hoc response on a local level to the use of slashing weapons by the opponents of the Roman army, who had in fact, for the last 300 years, been using these very weapons. No previous attempt has been made by the Romans to adopt such obvious modifications, and it may be that the new helmet became a necessity with the reduction in size of the scutum at this period.

Finally, although it cannot be doubted that the new helmets show Gallic features, and are of probable Gallic origin, there is no evidence for the spinning of helmets by the Gauls, especially those of the Coolus type. It may well be that this second change in the quality and design of Roman bronze helmets is part of a larger design reflecting Imperial policy and the changing role of the Roman army, as these helmets seem to date from early in the Principate and very rapidly supercede all the old types; or perhaps, quite simply, the widespread adoption of the improved technique of spinning by entrepreneurs supplying the army.

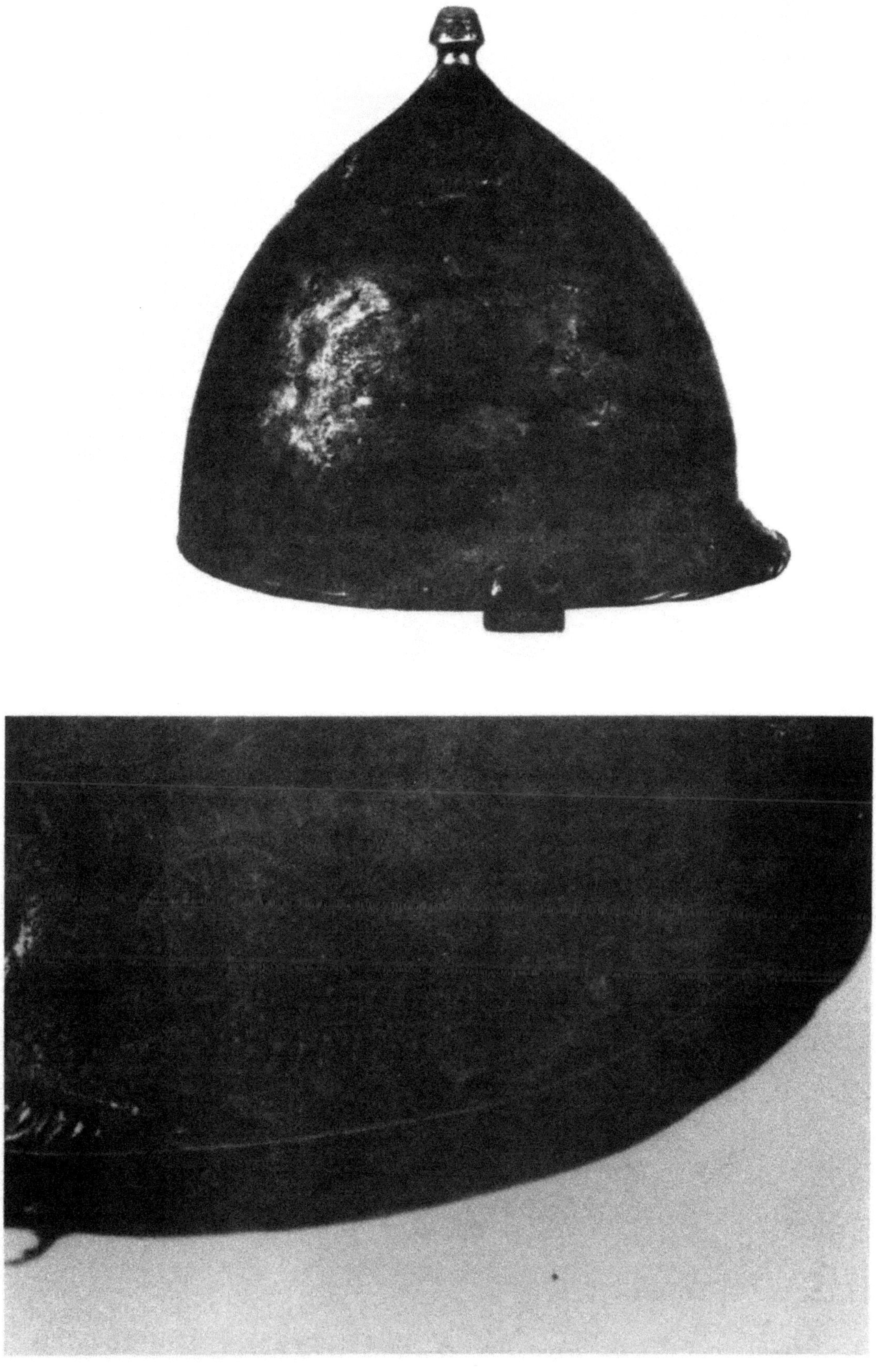

Fig. I: a) Montefortino helmet from Rieti
b) Maker's mark stamped into the neckguard

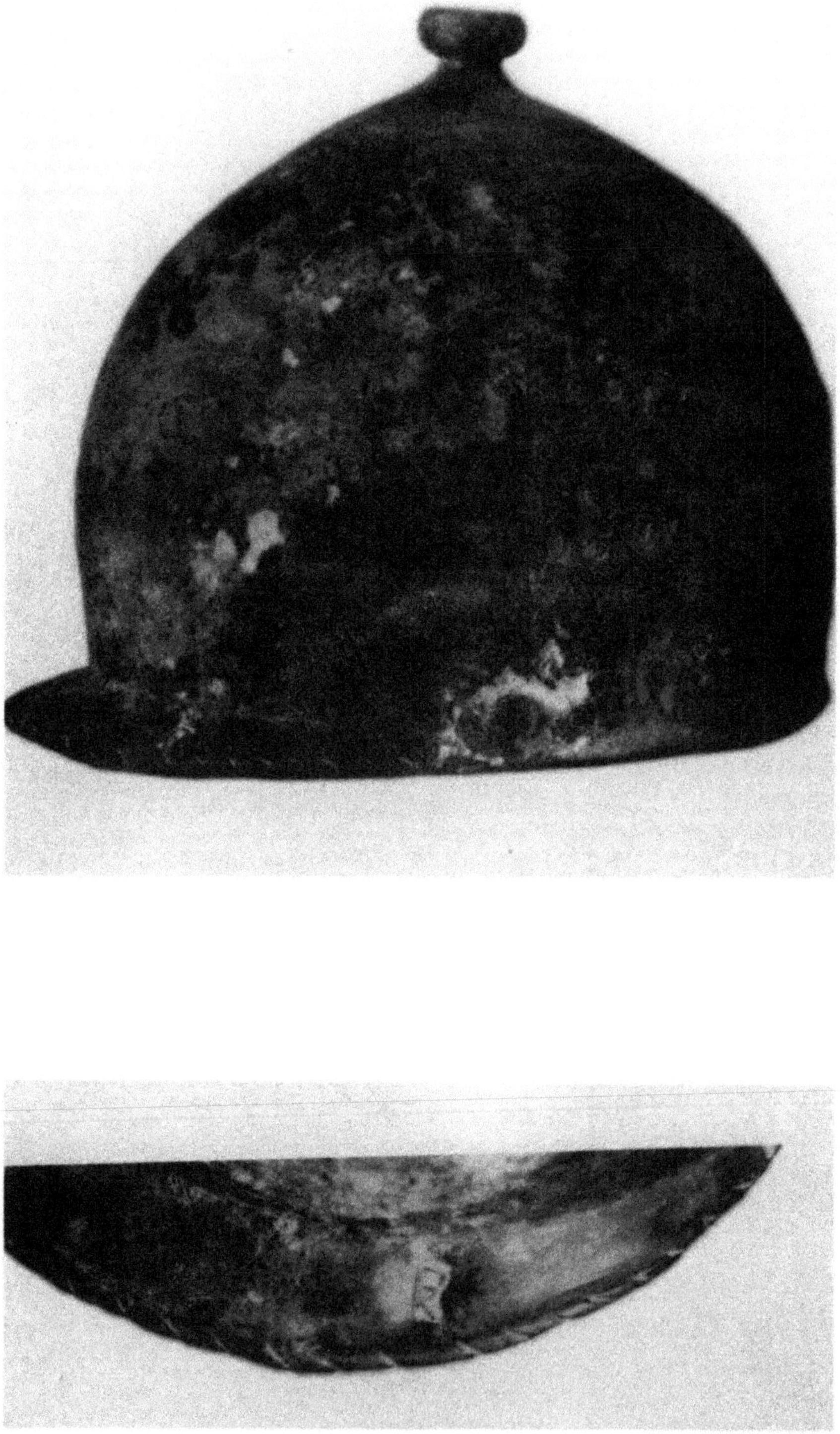

Fig. II: a) Montefortino helmet from Loreto Apruntino
b) Maker's mark stamped into the neckguard

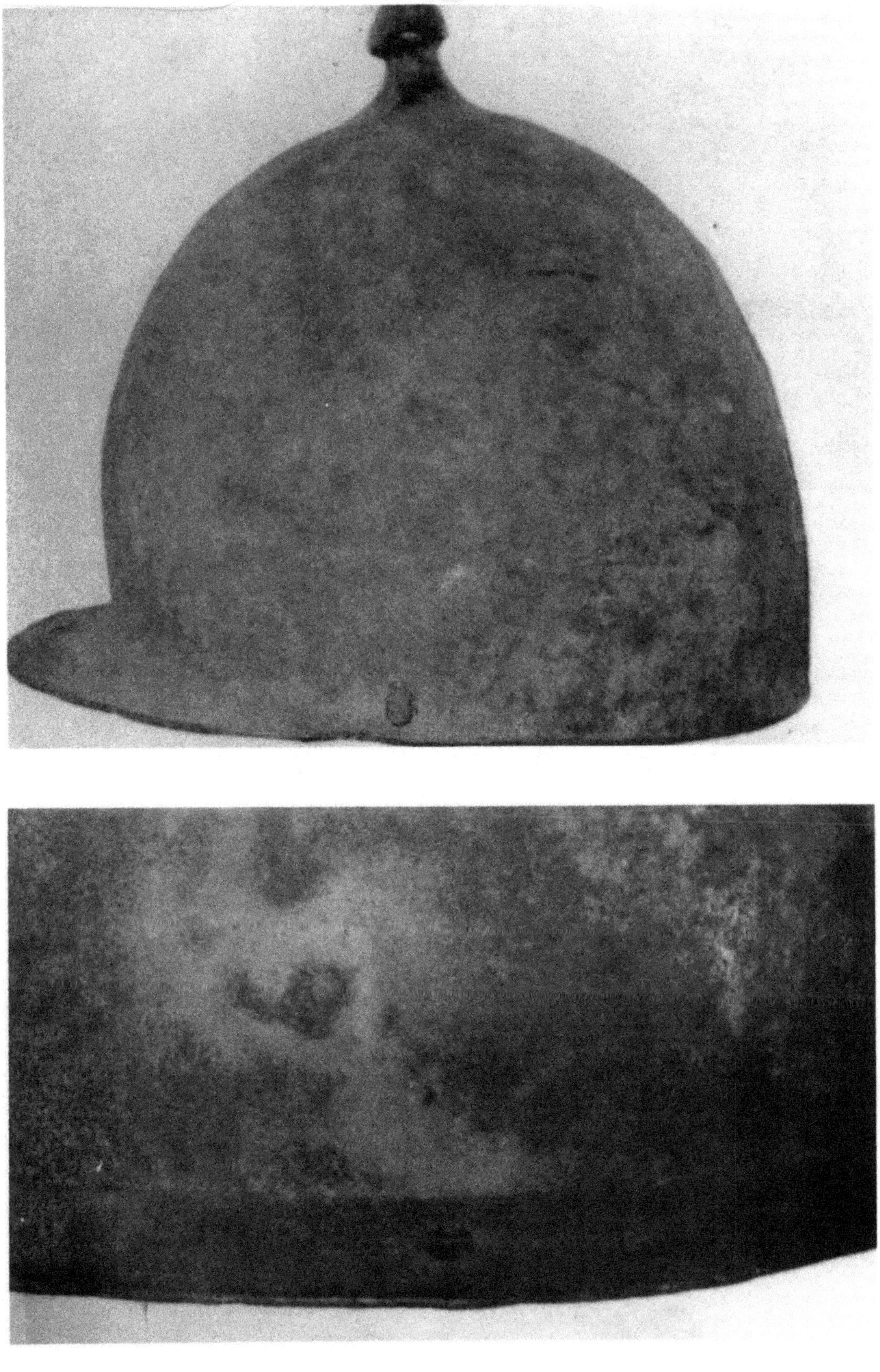

Fig. III: a) Montefortino helmet in the British Museum (81,7-25.2)
b) Maker's mark stamped into the neckguard

Fig. IV: a) 1st century B.C. Montefortino helmet from Montenerodomo
b) 1st century B.C. Montefortino helmet in the Vatican Museum

Fig. V: a) Spun early Imperial Montefortino helmet from Nijmegen

b) Spun Coolus/Hagenau helmet from Bosham Harbour

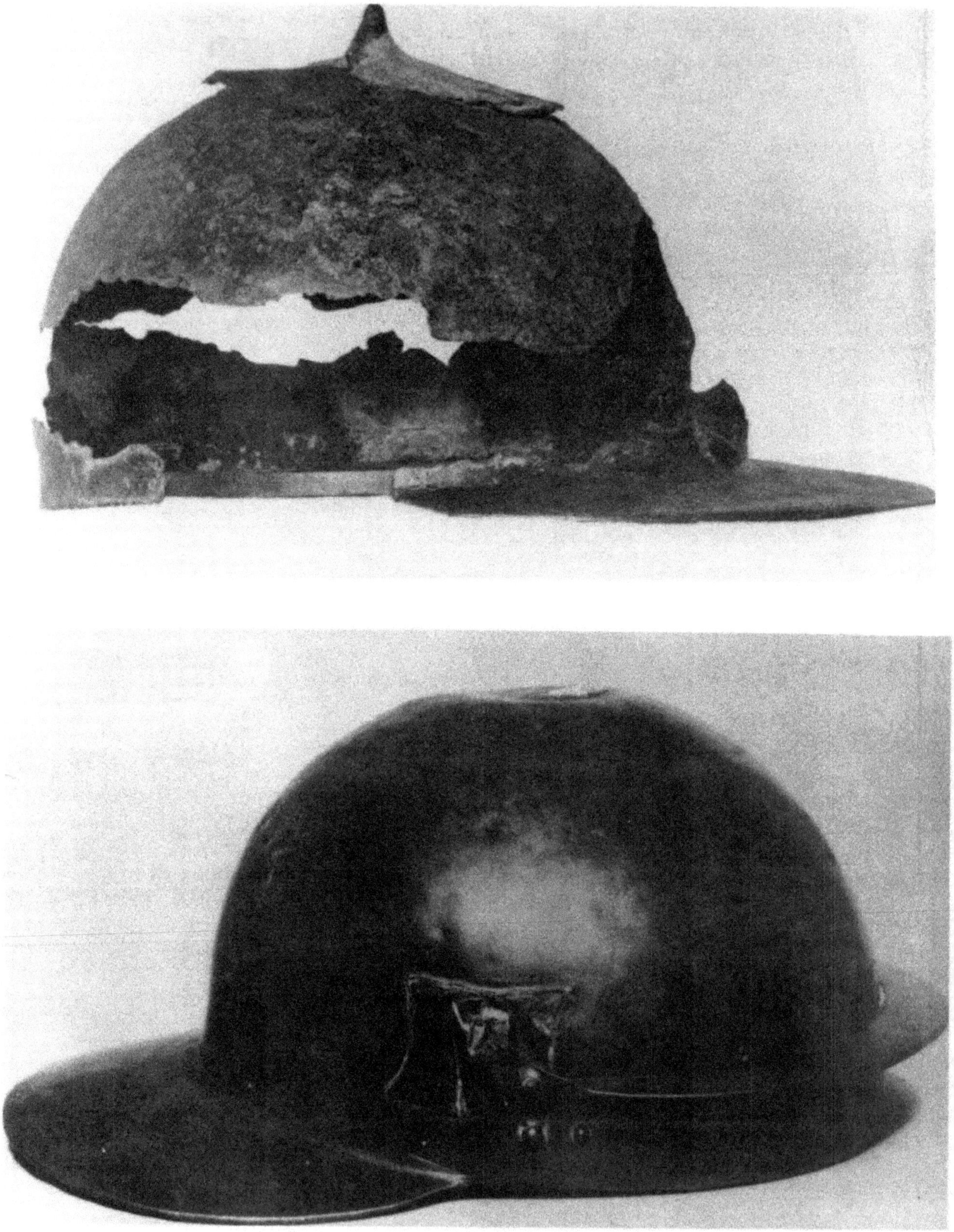

Fig. VI: a) Spun Coolus/Hagenau helmet from Bosham Harbour
b) Spun helmet from the Walbrook, London

Fig. VII: a) Spun helmet from Berkhamstead
b) Spun helmet from Nijmegen

Fig. VIII: a) Spun helmet from the Sava, Yugoslavia
b) Spun helmet from Drusenheim near Hagenau

NOTES

1. Polybius Book VI,39,15-40
"Of the allies the infantry receive the same..., these rations being a free gift to the allies; but in the case of the Romans the Quaestor deducts from their pay the price fixed for their corn and clothes and any additional arms they require."

2. Tacitus, Annals I,17
"ten Asses a day was the assessment of body and soul: with that they had to buy clothes, weapons and tents."

3. Tacitus, Histories II,67
"At first they were kept apart; later the offer of an honourable discharge was employed to soothe their feelings, and they started to turn over their arms to their tribunes, until the report that Vespasian had begun war became common."

4. Polybius Book VI,26,1-2
"The Tribunes thus having organised the troops and ordered them to arm themselves in this manner dismiss them to their homes."

5. Now in the Staatliche Museum. Munich. Inventory No.69. Published REINECKE, 1951, 42; MACMULLEN, 1960, 35.

6. Now in the private collection of Signor G.B. Leopardi in Penne. No inventory no. Published CIANFARANI, 1970.

7. Now in the British Museum. Inventory no.81,7-25.2. Published ROBINSON, 1975, 21.

8. PITT-RIVERS, 1897-1905; CURLE, 1911.

9. Helmet of late second - early 1st century B.C. date from Central Italy, now in the Museo Gregoriano in the Vatican. Inventory no.10652.

10. The following bronze Hagenau/Coolus helmets all show the tell-tale signs of spinning:

    a) a helmet from Bosham Harbour, now in the Sussex Archaeological Society Museum, Lewes. Inventory no.E.1.169. Published ROBINSON, 1975, 37.

    b) helmet from the Walbrook, now in the British Museum. Inventory no.1950.7-6. Published <u>ibid.</u>, 33.

    c) helmet from Berkhamstead, now in the British Museum.

Published *ibid.*, 33.

d) helmet from the Legionary barracks at Neuss, now in the Clemens-Sels Museum, Neuss. Inventory no.R 5016. Published KLUMBACH, 1974, 25.

e) helmet from the Waal at Nijmegen, now in the Rijksmuseum van Oudheden, Leiden. Inventory no.NS 429. Published *ibid.*, 39.

f) helmet from Hagenau near Drusenheim, now in the Nonin Historisches Museum. Inventory no.20714. Published ROBINSON, 1975, 39.

g) helmet from the river Sava at Rugviga, now in Zagreb Archaeological Museum. Inventory no.9228. Published POPOVIC, 1969, 120.

These helmets are the only ones where spin marks are still visible, but there are at least seven others of this type which are possibly spun out of a total of 26.

11. Now in the Rijksmuseum, Nijmegen. Inventory no.BE XIV 8B, 27. Published KLUMBACH, 1974, 21. It has recently come to the author's attention that there is at least one spun helmet of Republican date, c.100 B.C., i.e. the time of the Marian Reforms. This helmet comes from a shipwreck in Grand Basin B at Gruissan, France. But in all other respects it conforms to the usual shoddy workmanship of helmets of this period.

12. MUTZ, 1972, 14-52.

13. Ten dactyls are equivalent to 193mm, and the overwhelming majority of Greek Pilos helmets conform to between 190-195mm in diameter and height.

14. Dimensions of helmets listed in note 10

Helmet	**Inventory no.**	**Internal Diameter Front to Back (mm)**	**Transverse Diameter (mm)**
a	E.1.169	212	208
b	1950.7-6	212	184
c		216	208
d	R 5016	208	208
e	NS 429	204	204
f	20714	216	216
g	9228	205	200

15. ROBINSON, 1975, 26.

BIBLIOGRAPHY

CIANFARANI, 1970: V. Cianfarani Culture Adriatische d'Italia (Rome 1970)

CURLE, 1911: J. Curle A Roman Frontier Post and its People. The Fort of Newstead in the Parish of Melrose (Glasgow 1911)

KLUMBACH, 1974: H. Klumbach Römische Helme aus Niedergermanien (Kunst und Altertum am Rhein Nr.51) (Köln 1974)

MACMULLEN, 1960: R. Macmullen 'Inscriptions on armor and the supply of arms in the Roman Empire' in American Journal of Archaeology 64, 1960, 23-39

MUTZ, 1972: A. Mutz Die Kunst des Metalldrehens bei der Römern (Basel 1972)

PITT-RIVERS, 1897-1905: A.H.L. Pitt-Rivers Excavation in Cranbourne Chase I-V (London 1897-1905)

POPOVIC, 1969: Lj. Popovic Greek and Roman and Early Christian Bronzes in Yugoslavia (Belgrade 1969)

REINECKE, 1951: P. Reinecke 'Ein fruhkaiserzeitliche Bronzehelme aus der Donau bei Straubing' in Germania 29, 1951, 34-44

ROBINSON, 1975: H. Russell Robinson The Armour of Imperial Rome (London 1975)

All plates are by the author, except Fig.I (Staatliche Museum, Munich) and Fig.V (P. Connolly).

FIRST CENTURY MILITARY DAGGERS AND THE MANUFACTURE AND SUPPLY OF WEAPONS FOR THE ROMAN ARMY

Ian R. Scott

In this paper I will discuss the chronology and typology of first century Roman daggers and their sheaths. I shall then attempt to draw, from that evidence and from other historical and archaeological material, answers to two questions. Firstly I want to identify where these daggers and sheaths were made, and whether the location of manufacture changed during the first century. Secondly I wish to see what evidence there is that could identify who made the daggers. Again I want to see whether this also changes during the century. There are three main problems that we must keep in mind when we consider the evidence that we shall use to answer these questions. The first problem is the fact that our sample is very small. We only have just over seventy inlaid sheaths and I have listed about sixty daggers (see Appendices 1 & 2). This is all that remains of several tens of thousands of weapons that must have been in use at any one time.[1] The second problem, which relates to the first, is the small proportion of our sample that can be dated at all closely. The final problem, which may not affect the arguments in this paper but should be borne in mind, concerns the limited geographical distribution of the weapons. This, in fact, may simply reflect the large number of troops in Illyricum, in the Rhineland, and in Britain at various times in the First century A.D.[2]

I

The establishment of a chronology and some form of working typology is clearly central to the discussion of any artifact, particularly when it was in use over a period of time and over a wide geographical range. Because of the importance of typology and chronology, and because of the problems associated with them, I propose to look at them in detail.[3] Secondly having established a chronological framework we want to see whether geographical distribution had a bearing upon the form and decoration of daggers and sheaths independent of any chronological changes we might observe, for this may be a pointer to place of manufacture.

There are some practical problems involved in establishing a chronology, besides the obvious problem of the paucity of dated examples. Let us for a moment think about how and when any particular dagger or sheath came to be in its archaeological context. Firstly they were costly pieces of equipment, and, therefore, it was in the soldiers' interest to look after them. This applies regardless of whether daggers and sheaths were private property or government issue; either way the soldier would lose if weapons were mislaid.[4] We must not, of course, completely discount the possibility that weapons were lost whilst still in use. It is reasonable to assume, nonetheless, that most of the weapons extant today were old or broken weapons that had been discarded at the end of their useful life. Secondly, because they were durable pieces intended to last, their useful life might have been quite long, lasting well after the date of their manufacture. The effect of these factors, for our purposes, will be to blur or smudge the dating in one direction giving it a later bias. We will be able to establish the date of the introduction of a new type, or of a new feature, with some certainty, after taking due account of our meagre evidence, but we will have greater difficulty establishing the date at which an old feature, or type, ceased to be manufactured. For example, in the earlier part of the first century daggers were made with flat tangs; we can say with confidence, on presently available evidence, that daggers with rod tangs were introduced no earlier than the very end of the reign of Claudius, or, more probably in the the principate of Nero. We cannot say whether the older type with flat tangs ceased to be made at the same time as the newer type came into use, or whether it continued to be manufactured for some time after; it certainly continued in use to judge from the archaeological record, but this may merely reflect its useful life, not that it was still being made.

II

In this paper, my particular interest is not with the origins of the Roman military dagger, nor with their introduction into service with the Roman army. I am concerned with their use in the early principate, when they were associated with elaborately inlaid sheaths, and with their final disappearance in the later first century A.D. The fact that military daggers were found at Castillejo, near Numantia, dating to the second century B.C., serves to confirm their long history prior to the principate of Augustus.[5] Because of my concern with the daggers of the principate, I have concentrated my attention almost exclusively on those daggers that can assigned to the period from Augustus onwards. For convenience I have listed all the daggers of this period known to me, with a brief description and references, in Appendix 1. The first part of the list consists of those daggers, which can be dated, placed in chronological order. The second part lists the remaining daggers, those that cannot be dated at all closely. A study of the dated examples brings to light certain features that it is possible to date.

These are i) the form of the midrib, ii) the form of the tang, and iii) the presence, or absence, of rivet holes through the shoulders of the blade. Three further features appear to be characteristic of the latest daggers: iv) very slim blades, v) blades with very little waist, and vi) blades with very slight midribs.[6]

i) Midribs. There are basically two forms: simple upstanding midribs, and midribs flanked and defined by grooves. The latter is found in its most developed form on daggers such as those from Hod Hill (App.1, No.11) and from Leeuwen (App.1, No.54). The earliest dated occurence of the midrib defined by grooves is on a dagger from the Auerberg (App.1, No.8), which is of Tiberian or early Claudian date. The dagger from Augsburg-Oberhausen with a grooved blade and a rod tang could also be early in date (App.1, No.7).[7] The dagger from Dunaföldvár (App.1, No.39) found in the River Danube - and so unprovenanced - seems to have a midrib defined by grooves. It was associated with a sheath of Type A (App.2, No.7), which, possibly, could be dated as early as the principate of Augustus. The dating is not certain, and so we must not place too much reliance upon its testimony. Simple upstanding midribs are found on daggers with early provenances; for example those from Dangstetten (App.1, No.2), dated c.15 to c.10 B.C., and from the Titelberg (App.1, No.1), dated c.30 to c.10 B.C. This form of midrib is also found on the Castillejo daggers already referred to. There is a dagger, with a simple midrib, from a Flavian context at Nijmegen (App.1, No.18). As the authors of the published report on this dagger state, it is possible "daß er schon vor 70 n. Chr. in klaudisch-neronischen Zeit hergestellt worden ist".[8] There is an undated dagger, possibly from Nijmegen (App.1, No.51) which has a very similar blade form, but with a mid rib defined by grooves. This warns us to be wary of placing too much reliance on the blade form for dating, until we have more evidence.

ii) Tangs. We have already noted above that the earliest daggers that are of interest to us, have flat tangs pierced to take the rivets that fastened their handles. The Castillejo daggers appear to have flat tangs. There is little doubt that this feature is found on the earliest daggers: see for example Appendix 1, Nos.2-4, 6 and 10-14. The earliest example of a dagger with a rod tang that can be closely dated is from a Neronian pit at Usk (App.1, No.15). There is a dagger from Kingsholm (App.1, No.16) which could as early in date. There is also the dagger from Augsburg-Oberhausen (App.1, No.7) noted above, which has a rod tang, and which could be even earlier in date. The problems in dating the material from this site, make me loathe to place any great weight on this evidence.[9] The evidence suggests that the rod tang was introduced at the very end of the principate of Claudius, or at the beginning of Nero's reign. The dagger of Flavian date from Nijmegen (App.1, No.18), mentioned above, had a flat tang, but proves only the continued use of daggers with this feature. It may have been made earlier. The same applies to the unprovenanced dagger from Straubing (App.1, No.20); the site was first occupied in the Flavio-Trajanic period. Whatever the exact date at which the manufacture of daggers with flat tangs ceased, we know that rod tangs were introduced in the principate of Nero, or slightly earlier.

iii) Rivet holes through the shoulders of the blade. This is an early feature. The daggers from the Titelberg and from Dangstetten (App.1, Nos.1 & 2) each have four rivets through their shoulders, two either side of the tang. The more usual number, in the first century A.D., is two; one each side. Although an early feature, it does appear to continue in use for some time. The daggers from Nijmegen and Straubing, mentioned above, have rivet holes. They are from Flavian contexts or sites but could have been made earlier. However the dagger from Rißtissen (App.1, No.19), which had a rod tang as well as rivet holes through its shoulders cannot have been made earlier than the end of the principate of Claudius. It was found with a sheath of Type A (App.2, No.32), which I shall argue is itself to be dated, on the basis of its decoration, to a later period (see below). The earliest securely dated dagger without rivet holes through its shoulders is that from a Neronian pit at Usk (App.1, No.15). It has a rod tang as we have already noted. Possibly of the same date are daggers from Kingsholm (App.1, No.16), and from Mainz (App.1, No.14). The latter may have had a flat tang, and is very possibly later in date; the former has a rod tang, and is more certainly of pre-Flavian in date. The dagger from Augsburg-Oberhausen, referred to above, lacks rivets through its shoulders, and could be earlier in date. We have noted the problems of this site. The evidence strongly suggests that the omission of rivet holes from the shoulders of daggers coincided with the introduction of rod tangs, and should be dated to Nero's principate and after. The Rißtissen and Mainz daggers show that some of the older traits lingered on after new features had been introduced.

To summarise briefly the evidence we have looked at so far: the introduction of the grooved blade seems to have occurred in the reign of Tiberius, but blades with simple midribs continued in use up until the Flavian period. Flat tangs, like simple midribs, are an early feature, but it seems that they were not replaced by the rod tang until, at the earliest, the end of the principate of Claudius, or, more probably, in the principate of Nero. The third new feature, the omission of rivet holes through the shoulders of the blade, was contempary with the introduction of the rod tang. The earliest certain examples without rivet holes also have rod tangs. The dagger from Rißtissen with its rod tang and pierced shoulders shows that some old traits lingered on. None of the new features discussed above would have altered the overall appearance of the dagger very much. Even the discontinuation of the flat rivetted tang did not herald a new form of dagger handle, to replace the old composite handle, as the extant, but badly corroded, handles of the Usk and Caerleon daggers (App.1, Nos.15 & 21) show. The daggers from Vindonissa and from London (App.1, Nos.32 & 59) have wooden handles that are clearly secondary. The main effect of the change, from flat, rivetted tangs to rod tangs, would have been to make the fixing of handles less secure; it may also have made them easier to put on. Since it seems unlikely that the Roman army would have deliberately purchased daggers with "built-in obsolesence", it seems likely that it was easier - and therefore cheaper? - production that was the reason behind the change.

The remaining three features that can be dated are all found on daggers with the later features defined above; ie. on daggers with rod tangs, no rivet holes in the blade, and midribs defined by grooves. Daggers do not necessarily display all of these features.

iv) Slim blades. These blades are very clearly distinguished from wider blades; they are all under 4.5cm wide at their maximum, whereas the wider blades are all 5cm or more wide. There is a distinct gap into which scarcely any daggers fall; one exception being a dagger from London (App.1, No.59). Many of the slim daggers are under 3.5cm wide - as in examples from Vindonissa that are between 3.4cm and 3.1cm wide (App.1, Nos.31 & 32), and an example 3.5cm wide from Gelligaer (App.1, No.36). The earliest example, of a blade as narrow as this, may be that from the Auerberg (App.1, No.8), which appears to be c. 3.5cm wide at the shoulder, and could date from as early as the principate of Tiberius.[10]

v) Little or no waist. Many of the slimmer daggers have very little waist - see examples from Vindonissa (App.1, Nos.25, 26 & 31) - but other slim blades have a marked waist - again from Vindonissa (App.1, Nos.27 - 30). Equally, there are broad blades with little or no waist, as for example the blade from Chester (App.1, No.23).

vi) Reduced or vestigial midribs. Most of the Vindonissa

daggers already referred to have much reduced midribs (App.1, Nos.24, 25-26 & 28), one example (No.27) has the midrib replaced by a groove.

The dating of these features is fairly certain, for, with the possible exception of the Auerberg dagger, all the daggers with these features are of Neronian or later date. What is uncertain is the extent to which these daggers are found with, or without, piled cores.[11] On the basis of a purely visual inspection I would say that none of the Vindonissa ones has a piled core. Nor, would it seem does the Chester dagger (App.1, No.23). The Augsburg-Oberhausen dagger already referred to (No.7) has a grooved blade which appears to have no midrib; it clearly has no piled core.[12] Its blade width is uncertain, because it is badly corroded. It could date to any time between c.10 B.C. and c.A.D. 50. Of the features just described the one that would have had the most obvious effect on appearance is the considerable reduction in width found in the very slim blades. However, the most probable reason for the reduction in width of the blade is, not to change the appearance of the weapons, but to reduce the quantity of iron required in their manufacture. Now I want to turn attention to a consideration of the sheaths.

III

Our approach to this body of material has to be different from our treatment of the daggers, for the simple reason that so few dated examples are available, and because the variables are more complex; not only are there differences in the construction of the sheaths, but also variations in the decoration to consider. I will look first at the structure of the sheaths, and then at the decoration. An initial inspection of the sheaths shows that there are two clearly distinguishable forms, which I have labelled Types A and B. Type A sheaths are made of iron, with a wood or leather liner, and are engraved on the front to take inlay. Type B sheaths, on the other hand, were made of wood and leather, and had decorated iron plates fixed to their fronts. The plates were engraved for inlay, and they are the only part that usually survives in an archaeological context. The two forms of sheath can be dated in relation to each other. The evidence for dating Type B is better than that for dating Type A, and so, contrary to normal practise, I shall start with these sheaths, and work backwards in time, and look at Type A last. I shall conclude my comments with a consideration of the date at which these inlaid sheaths were introduced. The evidence from British sites is especially instructive. Of the 21 sheaths, presently known from Britain, sixteen are of Type B and only five of Type A.[13] This fact alone points to the probability that Type B sheaths continued in use for longer than Type A, and is confirmed by the evidence for individual weapons. Of the five sheaths of Type A, four came from sites associated with the earliest stages of the Claudian conquest, namely: from Colchester, from Waddon Hill, and two examples from Hod Hill (App.2, Nos.23, 19, 18 & 28). The sole exception is the sheath from Lincoln, which comes from a Flavian context, and which, as we shall see, is of a form that can be assigned a later date (App.2, No.31). There is a third sheath from Hod Hill, again dating to the earliest years of the Conquest, this time of Type B (App.2, No.43), which shows that this form of sheath was current by A.D. 43. The Richborough sheath is also probably of early date (App.2, No.42). However the majority of Type B sheaths from Britain have been found on sites, or in contexts, of later date.

The evidence is unequivocal: Type B sheaths are found in later contexts, and on sites with later occupation, than Type A sheaths. With the exception of the Lincoln piece there is no evidence for Type A sheaths from contexts dating later than the reign of Claudius. The evidence from the Continent supports this view, and also enables us to put a closer date to the introduction of Type B sheaths. The earliest securely dated example comes from Velsen, where it was found in a well with a body, and a military belt.[14] The occupation of the site at Velsen I is dated to between c.A.D. 15 and c.A.D. 30. We can therefore date this dagger and its sheath to the principate of Tiberius (App.2, No.41). Confirmation is provided by the Tiberian or early Claudian date of the sheaths from Kempten and the Auerberg (App.2, Nos.33 & 34). Since there are no sheaths of either kind that can be securely dated earlier than this we must not be too dogmatic when we say that Type B sheaths were introduced in Tiberius' reign. Turning to the Type A sheaths, we find that the dating evidence is even poorer than that for Type B. The evidence from the continent does not contradict that from Britain, so far as the date of their disappearance from circulation is concerned. As in Britain, there is only one Type A sheath that is securely dated after the Claudian period. As with the Lincoln sheath, the example from Rißtissen (App.2, No.32) has features that mark it out as a late piece. It is of Vespasianic or later date. Other than this the sheaths with the latest dating are three poorly preserved examples from Mainz (App.2, Nos.24-26) from Claudio-Neronian contexts. Our main difficulty is finding evidence to help date the introduction of inlaid dagger sheaths. We have noted above that there is evidence for the use of the military dagger long before the early Principate. There is a shortage of comparable evidence for the sheaths, but there is a little, which may help us to fix the date at which metal or inlaid sheaths were introduced. There is a dagger, from Oberaden (App.1, No.3) with an inlaid handle, which originally probably had an inlaid sheath. There must remain some doubt about the date of this dagger, because inlaid handles are not common with Type A sheaths. They are not unknown, as witnessed by the sheaths from Allériot and Colchester (App.2, Nos.22 & 23), but these are of Claudian date. Since the camp at Oberaden was occupied briefly between 11/10 B.C. and 8/7 B.C., we must question whether this dagger comes from the Augustan camp; or whether inlaid handles made an earlier appearance than other evidence would lead us to believe. Fortunately we do not have to rely upon the evidence of Oberaden alone. From Dangstetten, another Augustan military camp, occupied between about 15 B.C. and 10 B.C., there is a cast bronze suspension loop of the form used on Type A sheaths, which suggests that this type of sheath was in use in the principate of Augustus.[15] There is some evidence that this type of metal sheath was Augustan innovation; it is slight, and there is no solid evidence before A.D. 9 at the earliest as we shall see when we look at the inlaid sheaths. The Iron Age site on the Titelberg, in Luxembourg, produced, as a stray find, a Roman military dagger and sheath (App.1 & 2, Nos.1). The sheath had originally been of leather and wood with an open metal binding. It is this binding that has survived. The exact nature of the Roman military presence on the Titelberg is in doubt, but the relevant fact for our present purposes is its existence as testified by the dagger and other Roman military fittings on the site. There was clearly some form of military presence, which the excavators would date to the period between the campaigns of M. Nonius Gallus against the Treveri, and the campaigns of Drusus in Germany. In other words between 29 B.C. and about 12 B.C.[16] A similar sheath binding was found at Haltern a site occupied between c.10 B.C. and A.D. 9. This evidence, such as it is, suggests, that until the middle of the principate of Augustus, daggers with metal bound leather and wood sheaths were in use; it does not prove that sheaths with metal shells were not already in use by that date. But, in view of the lack of sheaths from very early contexts, I am not prepared to date the introduction of the inlaid sheath much before the end of Augustus' principate.

IV

Having established a basic chronology for the sheath types I want now to look at the decoration found on the sheaths and to attempt to refine the chronology. The way I approached the problem was, firstly, to list all of the Type A sheaths and the motifs and features found on them; and then to do the same for the Type B sheaths. A comparison between the two lists revealed those features and motifs that were typical of each type of sheath, and also isolated certain sheaths which had mixed sets of motifs. The evidence is summarised in Appendix 2. In the following section I intend to consider the validity of the groups defined, and where, possible, to look at the dating evidence available for them. In a later section I will go on to discuss the implications of their geographical distribution, in order to see whether or not we can localise any of the groups, and, in this way, identify, however tentatively, possible manufacturing centres.

The first distinction, that can be made between the two sheath forms, is that, while most Type A sheaths are inlaid with brass, or yellow metal, some are also found with silver inlay (Appendix 2, Sections A, B & C). Type B sheaths on the other hand are always inlaid with siver, or white metal, and never with brass. There is only one exception to this rule, from northern France (App.2, No.68). The second point that can be made is that each type of sheath has its own range of motifs, and that there is remarkably little overlap between those employed on Type A and those found on Type B. As we shall see, there are small groups of each type of sheath that use motifs from the repetoire of the other form (Appendix 2, Sections D, E & F). As is to be expected in such a small sample, there are a few sheaths that defy categorisation. These are remarkably few in number and I will deal with them as and when the need arises; for our present purposes the groups are more important.

Type A sheaths can be divided into three main groups, on the basis of the metal, or metals, used in their decoration: sheaths with brass and enamel inlays, those with silver and enamel inlays, and finally those with silver and brass, but no enamel inlays.

Type A sheaths with brass and enamel inlay (Appendix 2, Sections A & B). This is the biggest single group of sheaths, and remarkably homogeneous. One division can be made however. There are three sheaths (Section B), which can be distinguished readily from the others, because their decoration is not divided into the usual "four zone" scheme, but rather consists of a narrow single zone. They also employ a much smaller range of motifs than the other sheaths. This small group is securely dated to the principate of Claudius: the examples from Hod Hill, and Waddon Hill can be of no other date; that from Moers-Asberg is probably Claudian. For convenience I have labelled these the Moers-Asberg type. The sheaths listed in Section A of the Appendix, which display the more common "four zone" scheme, and a wider range of motifs, cannot be so easily dated. Only two examples, both from the Auerberg, can be given any date at all. They are probably to be assigned to the principate of Tiberius. The typology of the decoration suggests that the sheaths listed in Section A are of a different date from those listed in Section B. The Auerberg evidence points to an earlier date. Their distribution centres on the area south of the Upper Danube, in southern Germany and Illyricum, and on the Rhineland north of

Map 1

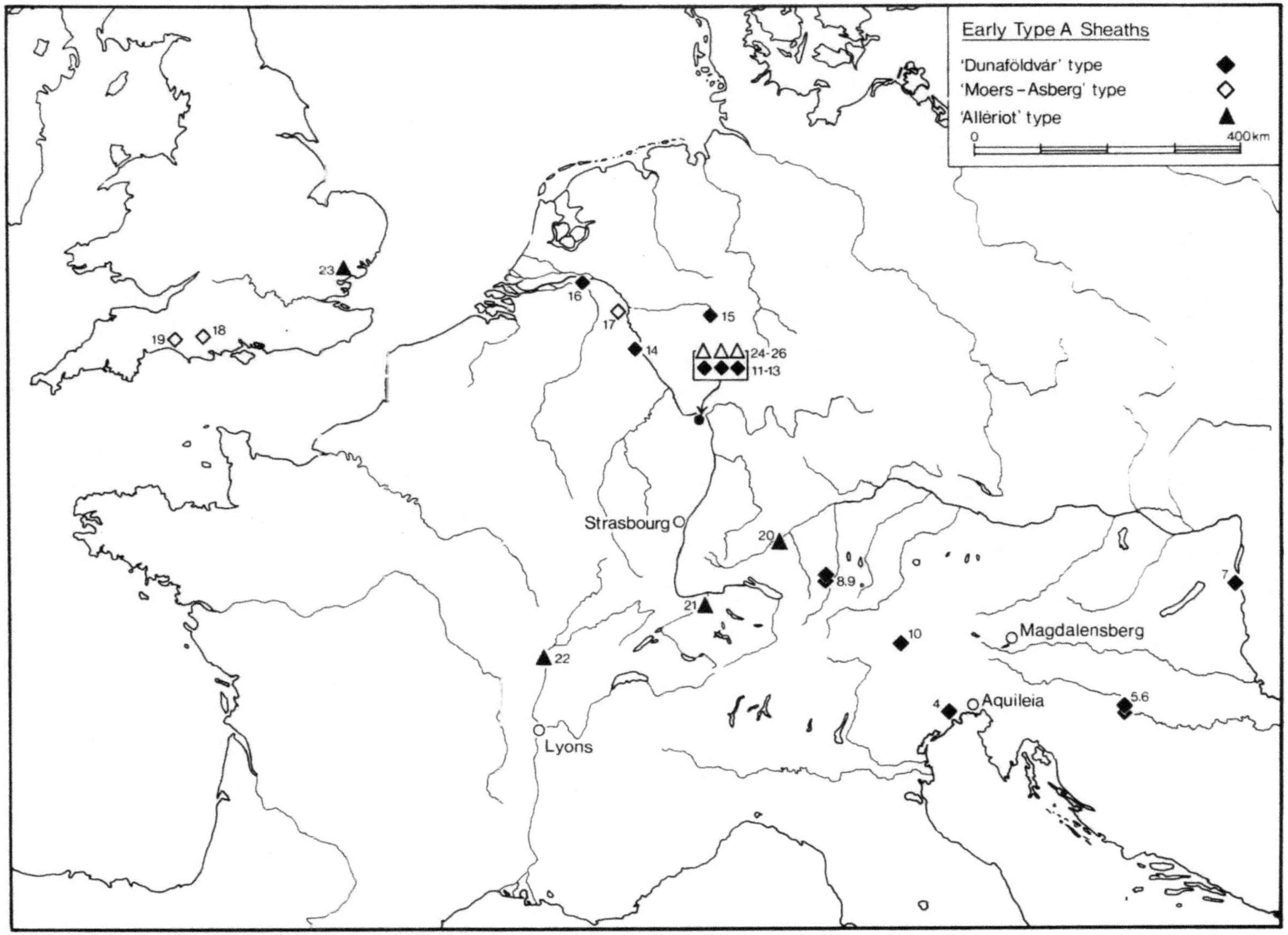
Early Type A Sheaths
'Dunaföldvár' type
'Moers - Asberg' type
'Allériot' type
0
400km
23
16
15
17
14
19
18
24-26
11-13
Strasbourg
20
8.9
7
21
10
Magdalensberg
22
4
Aquileia
5.6
Lyons

Mainz. If we are correct in assigning an early date to these complex sheaths, it would be tempting to date them as early as A.D. 9, for it was at that date that large numbers of troops were transferred from Illyricum to the Rhine frontier in the aftermath of the destruction of legions XVII, XVIII and XIX, under Varus, in the Teutoburg Forest. The legions transferred were XIII and XIV Gemina, and XX Valeria. It may be no more than coincidence that the first base of XIV Gemina, in the Rhineland, was at Cologne. I would not date them much earlier than A.D. 6 to A.D. 9. I have followed Edit Thomas in referring to these as the Dunaföldvár group (Map 1).[17]

Type A sheaths with silver and enamel inlays (Appendix 2, Section C). This group of daggers is much smaller than the preceding group, and less uniform in its character. Having said that, there is no denying, that the motifs used within the group are all similar. The differences may simply reflect different workshops rather than any major difference in date or typology. For example the main difference between the Rißtissen sheath and the Allériot sheath is the fact that the former does not appear to have any engraved hatching inlaid with fine lines of silver filling the fields within its decorative scheme; the latter clearly does (App.2, Nos.20 & 22). The sheaths from Vindonissa and Colchester both have this inlaid hatching (Nos.21 & 23). The sheaths from Mainz are too poorly preserved for us to be certain (Nos.22-4). The dating of these sheaths is unambiguous, for all those that can be dated are Claudian. The Vindonissa sheath may be later in date, but we cannot be sure. This dating distinguishes them from the brass inlaid sheaths quite as much as the different metal. Their distribution is also different being concentrated on the upper Rhine and adjacent areas. These sheaths I have called the Allériot type (Map 1).

Type A sheaths with decorative motifs of late Type B sheaths (Appendix 2, Section E). This small group of sheaths are inlaid in silver and brass with no enamel, and form our third main group. What distinguishes them, apart from the use of metal alone in their decoration, is the fact that they make use of exclusively late Type B decorative motifs. We shall see that Type B sheaths that use these motifs are all Neronian or later in date. The date of the sheaths under discussion is comparable. We have already noticed the sheaths from Lincoln and from Rißtissen, which fall into this group and are of Flavian date (App.2, Nos.31 & 32). The Rißtissen sheath is assigned to this group, because its inlay is all metal, and because the use of a single zone is more often found on late sheaths than on early ones; its decoration is not comparable to that on the other sheaths. The Magdalensberg sheath must date no later than the beginning of the principate of Claudius, for that is when the settlement was abandonned (App.2, No.29).[18] The sheath from

Map 2

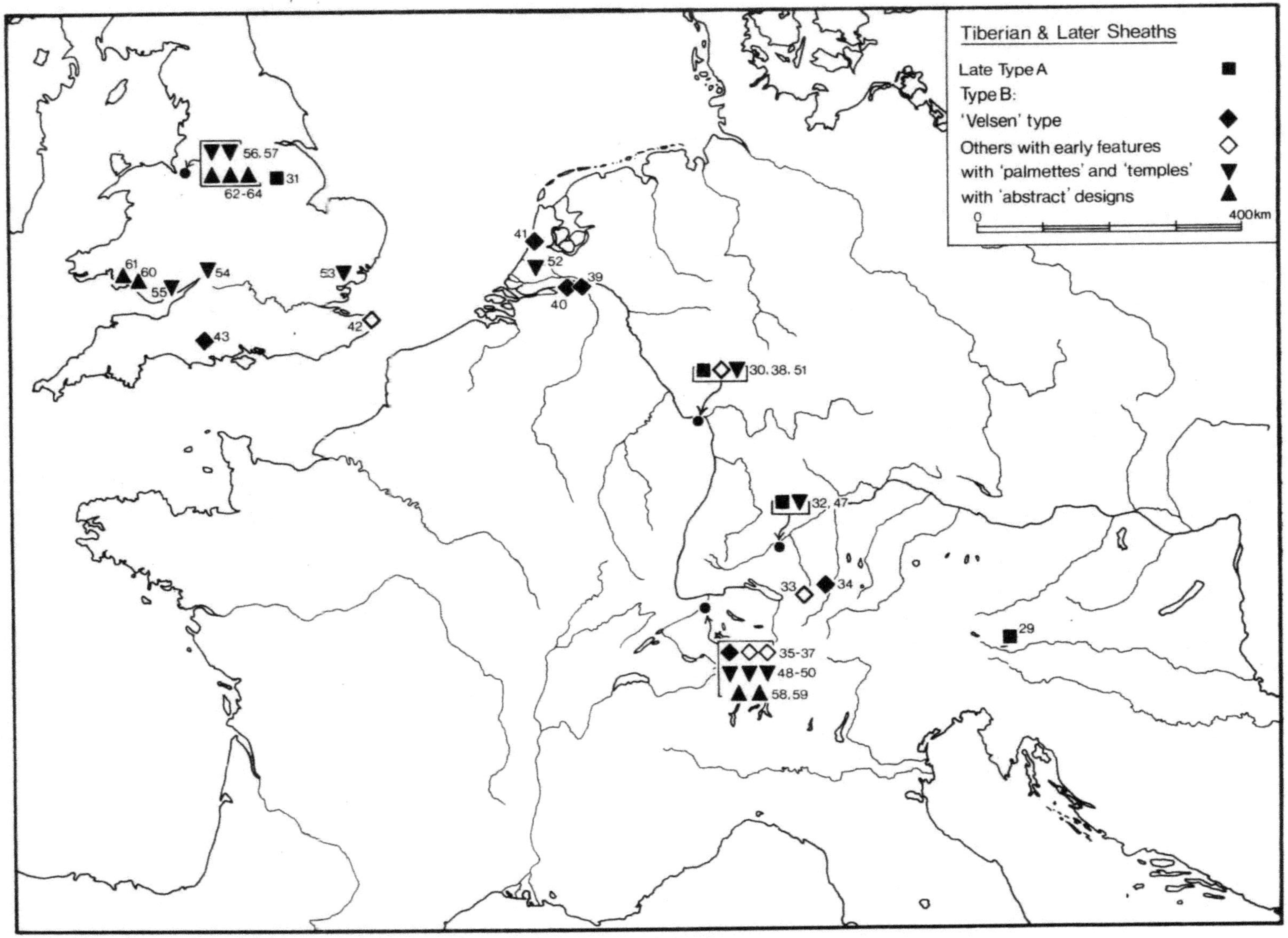
Tiberian & Later Sheaths
Late Type A
Type B:
'Velsen' type
Others with early features
with 'palmettes' and 'temples'
with 'abstract' designs
0
400km
56.57
62-64
31
61
60
55
54
53
42
43
41
52
39
40
30.38.51
32.47
33
34
35-37
48-50
58.59
29

Mainz (No.30) is not dated. As can be seen this tiny group is widely scattered, making it well nigh impossible to make any useful comments on their distribution (Map 2).

This is the best point at which to note two other sheaths, that cannot be categorised. One sheath from Vechten with silver and brass decoration, has a quite complex scheme of decoration. It is unfortunately undated, but typologically is probably late (App.2, No.27). The other sheath is from Hod Hill and has decoration that includes enamel and brass. It is included here, rather than being included under Sections A or B, because it has amongst its motifs a rosette with diamond shaped petals, inlaid with enamel; a motif usually found on Type B sheaths, of the Tiberio-Claudian period.

Type B sheaths can be divided into three main groups on the basis of the designs found on them. The first group are distinguished by the use of various features also found on Type A sheaths, including enamel; the second group use no enamel inlay, and employ a new range of motifs that are distinct from those found on the first group; the final group use the minimum of decorative motifs, and their designs I have labelled as "abstract". There a small number of Type B sheaths that do not form a coherent group, and will be briefly dealt with at the end.

Type B sheaths with Type A features (Appendix 2, Section F). It is possible to distinguish one group of sheaths within this group on the grounds of style, and the use of enamel. The sheaths from the Auerberg (No.34), Nijmegen (No.39), Leeuwen (No.40), Velsen (No.41) and Hod Hill (No.43), I would place in this group, which I would call the Velsen type. I would also include one of the Vindonissa sheaths (No.35). The other sheaths do not seem to form a coherent group stylistically, but they have features in common with each other and with the Velsen type sheaths. The dating of these weapons is securely in the Tiberian and Claudian periods. The examples from Vindonissa (Nos.35-37) and from Nijmegen may be later in date, but the likelihood is that they were made in the Claudian period or earlier. The examples from Britain are from Claudian sites (Nos.42 & 43). Typologically we should not date them later than the Claudian period, because the Type A sheaths, from which they copy some motifs, went out of use then. The distribution of the Velsen type concentrates mainly in the Netherlands; the exceptions being from the Auerberg, Vindonissa, and Hod Hill. The other sheaths have a southern concentration with the exception of the Richborough example (Map 2).

The three sheaths, that I have listed as possibly belonging to this group (Nos.44-46), are later in date, and, although sharing some features such as silver headed nails and rosettes, should be considered as a seperate group. Two of them are from Britain, the third from Vindonissa.

Type B sheaths with "Palmettes" and "Temples" (Appendix 2, Section G). As the Appendix shows these sheaths display a new range of motifs, quite unlike anything found on earlier sheaths. The dating evidence for these artifacts points to their currency from the Neronian period until the end of the first century. They are thus later in date than the preceding group, and are found as far north as the fortress at Chester. They were clearly in use in the late first century. Their distribution centres on Britain in the north and on Vindonissa in the south, with a few on the Rhine. There is no one predominant concentration. One further point should be made, and that is that there is a considerable variation in the quality of the workmanship to be seen in these sheaths. It is not just a matter of seeing different hands at work, but also, of seeing that some of those hands were manifestly less skilled than others (Map 2).

Type B sheaths with abstract designs (Appendix 2, Section H). None of our extant examples need be earlier than the Flavian period. They all, with exception of the Vindonissa examples (App.2, Nos.58 & 59) come from sites first occupied in the late first century. And, with same exceptions, they come from Britain. Like the previous group they show differing quality workmanship (Map 2).

Type B sheaths with miscelllaneous designs (Appendix 2, Section I). These sheaths do not form a coherent group, but they are all probably late in date; all but, that from northern France (No.68) which is a stray find, are of Neronian or later date. The example from France is unusual in having brass inlay, and in a form that is unique. The brass forms two complete motifs, one a bust, the other an eagle, which are applied to the centre of roundels. The remaining inlay appears to be of silver.

V

Having considered at length the dating and typology of both daggers and sheaths, now is the appropriate point to draw the evidence together before proceeding to the next stage in the discussion. There are very few close associations between sheaths and daggers where the blade of the dagger is visible. Most of the associated daggers are corroded into their sheaths and so their blade forms are not known. Most daggers and sheaths are unassociated. The first major change that we have identified is the transition from Type A to Type B sheaths. We find that there are very few daggers associated with Type A sheaths. The daggers, which survive with their blades and are associated with Dunaföldvár type sheaths, number four. They are from Dunaföldvár itself (App.1, No.39), with a midrib defined by grooves, and possibly with a piled core; from Mainz (No.45) and from Rösebeck (No.50), both with simple midribs; and finally from Nijmegen (No.51) again with a simple midrib. The Nijmegen dagger has a distinctive blade form similar to others from The Netherlands (App.1, Nos. 18 & 55). They may be a type that was made there. The only other early Type A sheath to be found with a dagger with its blade visible from was the one from Allériot (App.1, No.41). This had a simple midrib.

If we look at the evidence for the early Type B sheaths we find a similarly confused situation. This time there are five extant daggers, of which four are associated with Velsen type sheaths. The Auerberg dagger (App.1, No.8) is slim, has a midrib defined by grooves and a marked waist; the Nijmegen dagger (No.18) a simple midrib and a piled core; the Velsen dagger (No.9) is probably similar. The Leeuwen piece (No.53) has a blade with a marked waist and strongly defined grooving. Its blade has a piled core. The dagger from Mainz (App.1, No.43), associated with the fragments of a sheath which is not of Velsen type, has a simple midrib. No clear picture of associations is emerging. We cannot at present say that any particular dagger form is associated with a particular sheath type. With more evidence the picture may become more clear. What I think it does bring out is the fact that the sheaths we have been looking at so far were all in use over a comparatively short period. Although we have distinguished between those of possibly Augustan date and those of Claudian date, we are actually talking of a period of no more than fifty years from about A.D. 6 until the death of Claudius in A.D. 54. That is assumming that my early dating of the Dunaföldvar sheaths is accepted. It also suggests that daggers were not necessarily produced in the same workshops as sheaths. The different skills involved in the manufacture of daggers and of sheaths point to the same conclusion.[19] The second change we have identified is the introduction of a new range of decorative motifs - "palmettes", "temples", and so forth - on sheaths in the principate of Nero. Only one dagger can be directly tied in with a sheath of this later type. It is the one from Mainz with the legionary name inscribed in its decoration (App.2, No.51). The dagger blade is hidden, but the handle was fixed by a rod tang. Since we have argued that rod tangs date from Nero's principate and after, this association is no more than we would expect. The daggers and sheaths from Vindonissa, although not directly associated, must surely be considered as one group. Late dagger and sheath forms are predominant amongst this material. It is my contention that this second change, under Nero, is as important, perhaps more important than the earlier change, because it effected both daggers and sheaths.

I want lastly in this section to consider when Roman soldiers ceased to use the military dagger. I need not do more than draw attention to the difficulties of dating the disappearance of an artifact from the archaeological record; the pitfalls of negative evidence are well known. The first evidence that we can consider is provided by Trajan's Column and the Tropaeum Traiani at Adamklissi.[20] On neither of these monuments are soldiers shown wearing daggers. There are problems in using the evidence of sculpture: we cannot be certain always of the accuracy of the details of the accoutrements of the soldiers portrayed. In this instance the evidence of sculpture is supported by the archaeological record as few of the daggers can be dated later than the end of the first century A.D.; only the dagger from Gelligaer (App.1, No.36), and the sheath from Colchester (App.2, No.53), could date as late as the reign of Trajan. Another piece of evidence is the absence of daggers or sheaths of this type from northern England, or from southern Scotland, where the Romans first campaigned and garrisoned camps towards the end of the first century A.D. The most northerly sites that have produced daggers and sheaths, to the present date, are Chester and Lincoln. Of course, new finds could alter the picture, but I suspect that they will not. All of the sites in Britain that have produced daggers and sheaths were occupied first in the pre- or early Flavian periods. A similar picture emerges on the German and Raetian limes, where the forts that have produced daggers and sheaths are those that were garrisoned in the First century. The forts on the Taunus ridge and in the Wetterau built after Domitian's campaigns, have produced, to my knowledge, no daggers or sheaths. To the south, where the frontier was pushed east from the Rhine by Vespasian and his successors, no discoveries have been made. The forts built north of the Danube produce the same picture. The cumulative effect of this evidence seems conclusive; the Roman army stopped using daggers in the Flavian period.[21]

VI

Having established the chronology and the typology of our material, I want now to begin to answer the questions that I posed at the beginning of this article. The first question that I want to answer concerns where the sheaths were made. Because of the problem, which we noted above, in trying to identify groups of daggers, I am going to concentrate my attention on the manufacture of sheaths. The sheaths made in the period up to, and including the principate of Claudius, I believe were made in specialist workshops run by civilians and and not in army fabricae. I am including all Type A sheaths, and the early Type B sheaths (Appendix 2, Section F); I would also include the late Type A sheaths (Appendix 2, Section E) in this category. Some of the evidence is subjective; the fact that these earlier sheaths are consistent in their appearance, within their groups; all display competent workmanship, and almost all of them have at least two materials, a metal and an enamel, as inlays. Although the evidence is subjective, and therefore difficult to evaluate, it becomes more convincing when we compare the workmanship of the later Type B sheaths. These have only a small repetoire of motifs - "palmettes", "temples", and cross-hatched diamonds are the main ones - which are rarely mixed with other motifs, and yet the levels of skill shown in the execution of the designs range from the well produced to the barely competent. It is for this reason that I think that they were made in army workshops. We would not expect soldiers to have the levels of skill to be found in specialist shops. There is a danger in using what is potentially a circular argument, but the difference in the workmanship is clear and is best explained in this way. We should also note that there is very little overlap between the motifs found on the earlier sheaths and those found on the later ones; even the early Type B sheaths, which might have been expected to share some motifs, provide only four examples; in each case it is a "temple" motif (Appendix 2, Nos.34, 37, 38 & 41). The distribution of the later sheaths is a problem. Those decorated with "palmettes" and "temples" give no clue to any one production centre. If the army did produce its own daggers and sheaths, we might have expected each legion, or workshop, would have produced sheaths with a distinctive pattern. At first sight we do not have this, but no two of the extant sheaths in this group are similar. Furthermore the Mainz sheath with the name of <u>legio XXII Primigenia</u> in the decoration, might be taken as evidence for legionary production. The problem is that it could be quite the opposite! Whatever the answer, there was most certainly only a limited range of motifs employed. The British concentration of the sheaths with abstract designs might be just the evidence for distinctive designs that we are looking for. Unfortunately, I think that these particular sheaths are later in date than those with "palmette" and "temple" motifs. Also I suspect that the present pattern of distribution may be misleading, reflecting the recent spate of discoveries made in Britain. We should note, in particular, the similarity between two recently found sheaths from south Wales and a plate from Vindonissa (App.2, Nos.58, 60 & 61).[22]

I have argued for an early date for the Dunaföldvár type of sheaths (Appendix 2, Section A), and have suggested that the examples found on the Rhine frontier were carried there by soldiers transferred from Illyricum in A.D. 9. The implication of this being that the daggers were acquired, or issued, during service in Illyricum. If this were the case, there are two possible sources from which weapons could have been obtained: from Cisalpine Gaul, or from Noricum.

The best evidence, for the manufacture of sheaths and daggers in northern Italy, is the Type A sheath from Oberammergau (App.2, No.3). This piece does not belong to the group under discussion: it is richly decorated in silver, with silver fittings, and thus quite unlike our brass and enamel inlaid sheaths. The motifs used are also different. It is relevant nevertheless to the question under consideration, because Professor Ulbert has argued, from the maker's name on the hilt plate, and from the classically inspired decorative motifs, that it was a north Italian product.[23] The fact that it may have been a specially commissioned piece - a suggestion based on its rich decoration, and on the fact that it is unparalleled - does not rule it out of court as evidence for the manufacture of daggers and sheaths for the army in Cisalpine Gaul. Quite the reverse. Where would one go to have a special commission made, but to expert craftsmen with experience of making weapons? The Dunaföldvár sheaths may not be so lavishly decorated, but they are as skilfully made. The discovery of a sheath of Dunaföldvár type at Concordia (App.2, No.4), only a few miles west of Aquileia, may not the case, but does add to the circumstantial evidence. Aquileia was one of the centres of the metal industry in Cisalpine Gaul.[24] It also had well attested trade links with our other possible manufacturing centre at the Magdalensberg, and with the Norican iron mines.[25] The Roman trading settlement on the Magdalensberg had a long history going well back into the first century B.C. A number of Italian traders were represented: under Augustus and Tiberius there were merchants from Aquileia, Bononia, Vetulonia, and Rome at the Magdalensberg. The iron mines, iron smelting and metal working were the key to the trade with Rome. Norican metal goods were well known to Romans in Augustus' day.[26] In view of this long tradition of metal working, and the long history of trading with Rome, it would not be very surprising to find weapons being manufactured at the Magdalensberg to supply the Roman armies campaigning in Illyricum. There is a sheath from the Magdalensberg (App.2, No.29). It is a late Type A sheath, and as such not relevant to the present discussion. The transfer of a large part of the army to the Rhine would have increased the potential market for traders in Gaul, and amongst the items that the army would have required would have been arms. This may have given a boost to the metal industries of Gaul. I suspect that we can detect the products of these new centres in the silver inlaid sheaths that we have dated to the principate of Tiberius. The Allériot type sheaths (Appendix 2, Section C) have a distribution that centres on the upper Rhine above Mainz. The dagger and sheath from Colchester (App.2, No.22) are of Claudian date and must have been carried to Britain during the Claudian invasion, and so they can be discounted in considering the distribution. The sheath found at Allériot, near Chalons-sur-Sâone, may be the clue to the location of the manufacturing centre. Is it asking too much of the evidence to suggest that these sheaths might have been produced at Lyons, the trading centre at the confluence of the Rhône and the Sâone? Another site, that may have been a centre for the production of daggers, sheaths, and other weapons is Strasbourg. In addition to a sword scabbard with a maker's name, the site has produced tantalising evidence for metal working.[27] Although the evidence for metalworking is certain, the details of the site, as published are unclear, as is its date. Both are critical for the interpretation of the site; is it a legionary workshop, or a civilian establishment? Either of these suggested centres - and they are only suggestions - would have supplied weapons to the upper Rhine frontier, if the distribution of Allériot sheaths is a guide. By way of contrast the Velsen type sheaths (App.2, Nos.34, 35, 39-41 & 43) are concentrated in the Netherlands, which inclines me to think that they were made in northern Gaul for the lower Rhine frontier. They could be seen as complementing the Allériot type. Remember also that there are some other early Type B sheaths, not included in the Velsen group, for example from Kempten (App.1, No.33) and from Vindonissa (Nos.36 & 37), which are found in the southern area.

VII

The picture that is emerging very faintly outlined, shows production in the first half of the century, in civilian workshops, either adjacent to the army's theatre of operations, as in the case of the army in Illyricum or actually within its province, as in the case of the Rhine army. There is little evidence that the Rhineland army was issued with inlaid sheaths before A.D. 9. And it is this above all else that leads me to date the daggers used by the army of Illyricum no earlier than the time of the Pannonian revolt. These daggers and sheaths I have argued were procured either in Cisalpine Gaul, or in Noricum via the Magdalensberg. With the removal of a number of the legions from Illyricum to the Rhine, the production of daggers and sheaths in northern Italy, or Noricum, seems to have stopped.[28] In Gaul the production of inlaid sheaths, and presumably daggers too, seems to have been divided between the north and the south, with the southern centres supplying the upper Rhineland possibly from Strasbourg or from Lyons; the northern centre, perhaps in the Low Countries, supplying the lower Rhine. The start of production cannot be dated earlier than the principate of Tiberius archaeologically, but, perhaps, began soon after the arrival of the legions from Illyricum? Subsequently, in the principate of Nero, the manufacture of daggers and sheaths appears to have been taken out of the hands of civilian specialists, and placed in the hands of the army's own craftsmen. It seems probable that the production of daggers and sheaths was the responsibility of legionary workshops. The change is marked by the introduction of the rod tanged dagger, and of the sheaths with new motifs and mixed quality craftsmanship. I believe that the new situation arose in part because army units were moving their camps less frequently than in the past. This was due largely to the fact that the army was no longer engaged in large scale campaigns leading to territorial expansion and the constant movement of troops. With more permanent bases the army was in a position to set up workshops to produce weapons and equipment. Being no longer involved in major fighting meant that time and manpower were available too.[29] I also believe that the new policy came about because of the need for economies.[30] Production by the army would have reduced costs and the new attenuated daggers would have used much less iron than the old broad daggers, bringing a further saving. The Roman administration was run on what amounted almost to a fixed income: during the principate until the end of the third century, there was one recorded tax increase, by Vespasian. The only other addition was the inheritance tax introduced by Augustus. There was insufficient economic growth during the principate to increase the empire's financial resources significantly.[31] During the period of territorial expansion the army had paid for itself with booty, and with new land and resources, but once expansion stopped, the position was transformed and the army became one of the major costs that had to be borne by the state. Hence the need to reduce costs. The final stage in the process that we have been following occured under the Flavian emperors, when the army ceased to use daggers.

ACKNOWLEDGEMENTS

I would like to thank all those museums who have allowed me to photograph and draw daggers and sheaths in their collections. I would particularly like to thank the following for allowing me refer to, and illustrate unpublished material from their excavations: The Glamorgan-Gwent Archaeological Trust, and their finds officer Steve Sell (Caerleon, Loughor, and Neath finds); The Trust for Lincolnshire Archaeology (formerly Lincoln Archaeological Trust), and their finds officer Jenny Mann (Lincoln sheath); and Professor W.H. Manning of University College, Cardiff (Usk sheaths). I would further like to thank the Grosvenor Museum, and in particular Dr. Peter Carrington for information about the unpublished finds from Chester. All of these finds are marked (*) in the appendices. For published information that I would otherwise not have seen I would like to thank E.L. Schimmer (Velsen) and Henk Schoorl of the Archaeologische Werkgemeenschap voor Nederland (AWN) (Zwammerdam). I would like to thank Professor Manning for reading this paper in typescript, and saving me from error on a number of points. Finally I must thank Mike Bishop for patience beyond the call of editorial duty.

NOTES

1) If we assume that every legionary was issued with a dagger, that would mean that there were close on 150,000 daggers in use at any one time. This number would be increased if, as I believe, some auxiliaries also used daggers. I base this conclusion on two pieces of evidence: i) the fact that many of the weapons come from the sites of auxiliary forts; and ii) the sculptural reliefs on the tombstones of auxiliary soldiers showing them wearing daggers (see ÉSPÉRANDIEU, VII, No.5850; VIII, Nos.6125, 6136-7 & 6207 for example). I hope to argue both this point and the case for the issue of daggers with inlaid sheaths, as opposed to purchase by individual soldiers in another paper. In the present article I am assuming these points. I should note, however the existence of a papyrus (P. Vindob. L135) in which it is clear that one of the soldiers concerned owned an inlaid dagger sheath. I believe that the key to explaining this document is the fact that the man in question is an eques (in ala Paullini) and may therefore not have been issued with a dagger, but had to purchase one privately. (This is a point that I shall be taking up in my new paper). For P. Vindob. L135 see HARRAUER & SEIDER, 1977; and GILLIAM, 1981. I would like thank Mike Bishop for drawing these references to my attention.

2) The distribution of tombstones with reliefs showing daggers is very similar to that of the daggers and sheaths themselves. The stele of Belaterus a soldier in cohors VI Delmatarum, found in Mauretania Caesariensis cannot necesarily be taken as evidence for the widespread use of daggers in Africa, since it is probable that the cohort had recently been sent to Mauretania following the revolt of Furius Camillus Scribonianus in A.D. 42. (Illustrated in HOLDER, 1980, Pl.10 B).

3) I have previously discussed chronology and typology in a section in the forthcoming British Museum catalogue of Romano-British ironwork. Unfortunately that section was written over 3 years ago and in part is superceded by the present discussion. MANNING, 1985.

4) If a soldier did not actually buy his dagger, but was issued with it he would still have had pay a deposit, which would have been forfeited in the event of lose. See GILLIAM, 1967, and the note in BREEZE et al., 1976, 93-5.

5) Reproduced in KEPPIE, 1984, fig.18.1-3.

6) A simple way to obtain more information about the structure

of dagger blades without damage to them is to use X-radiography. This might help not only with the chronology of daggers, but might also help to identify the products of different workshops.

7) For the problems of dating the Augsburg-Oberhausen find see WELLS, 1970.

8) BOGAERS & YPEY, 1962-1963, 92.

9) See note 7.

10) The width of this dagger was calculated from the published photograph and length measurement, and may not be completely reliable.

11) See note 6. By a blade with a 'piled core' I mean one that has a core that has been built up and the edges apparently welded to it. They are not elaborately pattern-welded.

12) HÜBENER, 1973, Pl.39.

13) The published information on the Richborough sheath does not make it clear what type of sheath it is. I have not been able to see the object, and have identified it on the basis of its decoration alone.

14) The Velsen dagger and sheath were found with the body of their presumed owner, and as a consequence we can say with some confidence that this is set of equipment was disposed off while still in use. As a result we can be equally confident about the dating information that it gives us.

15) FINGERLIN, 1972, Abb.13,9.

16) For the dating of the Titleberg see METZLER & WEILLER, 1977, 39; for the military equipment ibid., Abb.31-3. Also of interest is the tombstone of P. Flavoleius Cordus of legio XIV Gemina, which shows a dagger sheath with a metal edge binding (ÉSPÉRANDIEU, VII, No.5835).

17) THOMAS, 1969.

18) ALFÖLDY, 1974, 78.

19) We can note two possible groups of daggers: i) Those found in The Netherlands - Appendix 1, Nos.18, 51 & 55; ii) daggers with very pronounced midribs and grooves, and well finished - Appendix 1, Nos. 11, 46, 47, 49, 53, 54, 59 & 61. The distribution of the latter is concentrated on the lower

Rhine, and around Mainz, and in Britain.

20) For the Tropaeum Traiani see FLORESCU, 1965; for Trajan's Column see CICHORIUS, 1896 & 1900.

21) The tombstone of Castricius Victor of _legio II Adiutrix_ from Aquincum, which shows him wearing a dagger, may be important. _II Adiutrix_ was based at Aquincum only from the last years of Domitian's reign. See SZILÁGYI, 1956, Pl.XXXVI. If daggers ceased to be issued as late as this it would be tempting to link their discontinuation with the rise in army pay granted by Domitian.

22) The recent British discoveries have nearly all been of Type B sheaths and made by means of X-radiography. They have included the identification of previously unnoticed finds from old excavations. As more finds are published the predominance of the British sheaths with abstract designs may disappear. If the plates with abstract designs do prove to be a variant mainly limited to Britain, even with more evidence, we will have to explain the abstract plates from Vindonissa. Professor Ulbert, when publishing the Vindonissa sheaths, suggested that their disposal might date to A.D.69, and linked it with the troop movements of that and subsequent years. This may be the context in which we should explain the apparent link between Britain and Vindonissa (ULBERT, 1962, 16). If they are to be viewed as a type used by the British garrison, the fact that they date later than the plates with "palmettes" and "temples" may simply reflect the fact that Britain was occupied and developed as a province later than the provinces on the Rhine and Danube frontiers.

23) ULBERT, 1971, 45-8; he also draws attention to the fact that the name of the maker of the Rheingonheim sword - L. VALERIUS - is a common one in Gallia Cisalpina, see CHILVER, 1941, 75.

24) Strabo, v.2,12, where he comments that the mines are not as active as they once were. Pliny, _Nat. Hist._, xxxiii,78; xxxiv,2. In general see CHILVER, 1941, 167-173.

25) On links between Aquileia and the Magdalensberg see ALFÖLDY, 1974, 45-6. For the strong links with the Norican iron mines see CHILVER, 1941, 171, and CIL iii, 4788 (= ILS, 1466), and v, 810; see also CIL iii, 4809 & 5036

26) ALFÖLDY, 1974, 72-3.

27) For the gladius see FORRER, 1927, Vol.II, 529 & Taf. LXXV A; see also ULBERT, 1971, 48 for the possiblity that this could have been manufactured elsewhere. For the metalworking site

see FORRER, 1927, vol.II, 500ff.

28) We have mentioned already the possibility that the Moers-Asberg type may be a late Italian product, on the basis of their decoration and not their distribution. The late Type A sheath from the Magdalensberg may have been made on the site, but I think it unlikely. The distribution of these late sheaths is not helpful in locating where they might have been made.

29) Although the idea of fixed frontiers may not have been accepted until late in the first century, in reality the frontiers had been static since the death of Augustus (see MANN, 1974, 508-514 for the development of the concept of the static frontier). We first find forts with a regular layout being built as a matter of course in the principate of Claudius (JOHNSON, 1983, 234-245). This may indicate a change in the policy on troop dispositions.

30) The Type B sheath may have been introduced as an economy measure for it required less iron to make it than the Type A sheath. This is paralleled by the scabbard used with Ulbert's Pompeii type <u>gladius</u>, which was introduced in the mid first century. This scabbard used less metal than its predecessor (ULBERT, 1969). The use of silver as the inlay material, if it is proven for most late dagger sheaths, goes against the argument of economy, because was valuable, and needed for coining.

31) On tax changes see JONES, 1974, 189-190. Economic development was limited by the available technology, which effected both production and land transport. The main source of wealth was land, and the land tax remained the government's major source of revenue. Not only was revenue restricted in this way, but there was often an actual shortage of coin to make payments, for, as the late Professor Jones has stressed, the Roman administration was run on a strictly cash basis; everything had to be paid for in coin. Which brings us back to a previous point about the use of silver for inlay.

APPENDIX 1

DAGGERS

Only daggers that survive with their blades, or blades that are largely complete have been included in this listing. Daggers which are corroded inside sheaths have been excluded, with the sole exception of the example from Usk (No. 15). Handles and handle plates have not been included.

A. Dated Daggers

1. TITELBERG (c.30 to c.10 B.C.?)
?Slim blade; simple midrib; ?flat tang, handle, not inlaid, intact; 4 rivets through shoulders
Provenance & Associations: Stray find; open bronze sheath binding (App.2,No.1).
Length overall, with sheath: 34.5cm
Reference: METZLER & WEILLER, 1977, Abb.31

2. DANGSTETTEN (c.15 to c.10 B.C.)
Small broad blade; simple midrib; flat tang; 4 rivets through shoulders
Provenance: ?
Length of blade: c.16cm
Width of blade: c.5cm
Reference: FINGERLIN, 1972, Abb.14.5

3. OBERADEN (11/10 to 8/7 B.C.)
Blade form uncertain, piled core; flat tang, inlaid handle; 2 rivets through shoulders
Provenance: ?
Length overall, extant: 23.6cm
Length of blade, extant: 14.3cm
Width of blade: 5.5cm
Museum: Dortmund: E101
Reference: dating - WELLS,1972, 216-8

4. HALTERN (c.10 B.C. to A.D. 9)
Slim blade with extended shoulders, incomplete; simple midrib; ?flat tang; ?no rivets through shoulders
Provenance & Associations: ?; open bronze sheath binding (App.2,No.2)
Length overall, extant: c.13cm
Length of blade, original: c.22cm
Length of blade, extant: c.12cm
Width at shoulders: c.5cm
Museum: Röm.-Germ.Mus., Haltern

5. LORENZBERG (c.10 B.C. to c.A.D. 50) (Fig.1)
Broad blade; simple midrib; ?flat tang, inlaid handle; ?rivets through shoulders
Provenance: ?
Length overall: c.29cm
Length of blade: c.19cm
Width of blade: c.5.6cm
Reference: ULBERT, 1965, 46, Taf.3,1 & 24

6. AUGSBURG-OBERHAUSEN (c.10 B.C. to c.A.D. 50)
?Broad blade, piled core; simple midrib; flat tang, handle not ? inlaid; 2 rivets through shoulders
Provenance & Associations: Found with large deposit of Roman material, no known structures
Length overall: 30.1cm
Length of blade: c.21 cm
Width of blade, extant: c.4cm
Reference: HÜBENER, 1973, No.445, Taf.8,9 & 39,1; WELLS, 1970

7. AUGSBURG-OBERHAUSEN (c.10 B.C. to c.A.D. 50)
Blade form uncertain; flat with two grooves; rectangular cross-section rod tang; no rivets through shoulders
Provenance & Associations: see previous entry & references
Length overall, extant: 23.2cm
Length of blade: c.17.5cm
Width of blade, extant: 2.9cm
Reference: HÜBENER, 1973, No.446, Taf.8,13 & 39,2

8. AUERBERG (c.A.D. 10/15 to c.A.D. 40/45)
Slim blade, marked waist; ?midrib defined by grooves; ?flat tang, handle not inlaid; ?rivets through shoulders
Provenance & Associations: ?; Type B sheath (App.2,No.34)
Length overall: 28.5cm
Length of blade: c.20cm
Width of blade: c.3.5cm
References: ULBERT, 1975, Abb.17:2a;

9. VELSEN (A.D. 15 to A.D. 30)
Incomplete blade, shoulders missing; ?simple midrib, ?piled core
Provenance & Associations: In well, with Type B sheath (App.2,No.41), belt fittings, and skeleton
Length extant: ?
Width of blade: ?
Reference: SCHIMMER, 1979, 111-4 & fig.6

10. COLCHESTER (Claudian/Neronian)
Heavily corroded, broad blade; simple midrib, ?no piled core;

flat tang, handle not inlaid; ?rivets through shoulders
Provenance & Associations: Region 5, area C: few finds or features later than Neronian
Length overall, extant: 30.1cm
Length of blade, extant: 20.7cm
Width of blade: 6cm
Reference: HAWKES & HULL, 1947, Pl.CIV,No.1
Museum: Castle Mus., Colchester

11. HOD HILL (Claudian)
Blade with pronounced waist; midrib defined by grooves; flat tang, handle not inlaid; ?rivets through shoulders (similar to Nos.46 & 47)
Provenance: stray find
Length overall: 33.6cm
References: BRAILSFORD, 1962, fig.12:B2; MANNING, 1985, No.V 7
MUSEUM: B[ritish] M[useum], Durden Colln. 92.9.-1.1210

12. HOD HILL (Claudian) (Fig.12)
Blade with marked waist; midrib defined by grooves; flat tang, handle broken off, not inlaid; 2 rivets through shoulders
Provenance: stray find
Length of blade: 23.9cm
Length of handle, extant: 9.1cm
Width of blade: 5.2cm
Reference: MANNING, 1985, No.V 8
Museum: B.M., Durden Colln. 92.9.-1.1211

13. HOD HILL (Claudian)
Incomplete blade; simple midrib; flat tang; 2 rivets through shoulders
Provenance: stray find
Length overall, extant: 15.1cm
Reference: MANNING, 1985, No.V 12
Museum: B.M., Durden Colln. 92.9.-1.1213

14. MAINZ (Claudian or later)
Slim blade; ?midrib; flat tang; no rivets through shoulders
Provenance & Associations: from a site with a ditch which produced an Augustan or Tiberian coin, and a Claudian coin; ditch overlain by a later feature. Exact location of dagger unclear.
Length of blade: 20cm
Width of blade: 4cm
Reference: BEHRENS & BRENNER, 1911, 114, & Fig.28, No.El; p.67, for context.

15. USK (Neronian) (Fig.2,65)

Slim blade corroded in sheath, form unclear; ?midrib; rod tang, handle intact, not inlaid; no rivets through shoulders
Provenance & Associations: from a Neronian pit, which contained this dagger and its sheath (App.2,No.65), and a second sheath plate (App.2,No.45)
Length overall, with sheath: c.31.5cm
Width of blade: under 4cm (from X-ray)
Unpublished, report forthcoming
(*)

16. KINGSHOLM ? (pre-Flavian)
Slim blade; midrib defined by slight grooves; rod tang; no rivets through shoulders
Provenance: ?
Length overall: 35.2cm
Reference: MANNING, 1985, No.V 10
Museum: B.M., Lysons Colln. 1819.2-10

17. KINGSHOLM (pre-Flavian)
Broad blade; slight upstanding midrib; ?tang; ?rivets through shoulders.
Provenance: ?
Length of blade, extant: 24cm
Reference: MANNING, 1985, No.V 11
Museum: B.M., Lysons Colln. 1810.2-10

18. NIJMEGEN (Mid-Flavian)
Broad blade; simple midrib, piled core; flat tang; 2 rivet holes through shoulders; handle not inlaid
Provenance: "Grube 370": contained late 1st century pottery; with Type B sheath (App.2,No.39)
Length overall, extant: 27.4cm
Length of blade: c.26.5cm
Width of blade: 5.2cm (BOGAERS & YPEY, 4.9cm)
Reference: BOGAERS & YPEY, 1962-1963
Museum: Rijksmuseum G.M.Kam, Nijmegen, 1960/370

19. RIßTISSEN (Vespasianic or later)
Broad blade; midrib defined by grooves; rod tang; 2 rivets through through shoulders
Provenance & Associations: "Kastell 2", found with Type A sheath (App.2,No.32)
Length overall, extant: 24cm
Length of blade: c.23.5cm
Width of blade: c.6.5cm
Reference: ULBERT, 1970, No.259, & Taf.16

20. STRAUBING (Flavian or later)
Much corroded blade; ?flat cross-section; ?flat tang, handle not inlaid; ?2 rivets through shoulders

Provenance: stray find
Length overall: c.30cm
Length of blade: c.21cm
Reference: WALKE, 1965, 152 & Pl.106, No.5

21. CAERLEON (Flavian or later)
Slim blade, corroded; midrib defined by grooves, no piled core; rod tang, part of composite handle in situ, not inlaid; no rivets shoulders
Provenance: ?
Length, extant: 12cm
Width of blade, extant: 4cm
Unpublished, report forthcoming
(*)

22. CAERLEON (Flavian or later)
Tip of dagger with piled core
Provenance: ?
Length, extant: 3.2cm
Unpublished, report forthcoming
(*)

23. CHESTER (Flavian or later)
[It is not possible to give details of this dagger at the time of publication]
Monograph, Grosvenor Museum, forthcoming
(*)

24. VINDONISSA (?Flavian)
Broad blade; grooves down centre of blade; tang missing; no rivets through shoulders
Provenance: ?
Length of blade: 25.9cm
Width of blade: 5.6cm
Museum: V.M.

25. VINDONISSA (?Flavian)
Slim blade, little waisting; midrib defined by grooves; rod tang; no rivets through shoulders
Provenance: ?
Length overall: 31.2cm
Length of blade: 20.6cm
Width of blade: 4.4cm
Museum: V.M. 28:16

26. VINDONISSA (?Flavian)
Blade with little waist; 4 parallel grooves down centre; rod tang; no rivets through shoulders
Provenance: ?
Length overall: c.29cm

Length of blade: c.21cm
Width of blade: c.4cm
Museum: V.M.

27. VINDONISSA (?Flavian)
Slim blade with marked waist; mid rib defined by grooves, and flanked by a pair of grooves; rod tang; no rivets through shoulders
Provenance: ?
Length overall: 27.9cm
Length of blade: 19.9cm
Width of blade: 4cm
Museum: V.M. 10742

28. VINDONISSA (?Flavian)
Slim blade, with waist; 3 grooves down centre of blade, no midrib; rod tang; no rivets through shoulders
Provenance: ?
Length overall: 28.1cm
Length of blade: 19.1cm
Width of blade: 4cm
Museum: V.M. 28:31[13]

29. VINDONISSA (?Flavian)
Slim blade, with waist, badly corroded; ?midrib; rod tang; no rivets through shoulders
Provenance: ?
Length overall, extant: 21.7cm
Length of blade: 19.9cm
Width of blade: 3.9cm
Museum: V.M. 62[...

30. VINDONISSA (?Flavian)
Slim blade, with waist; grooves down centre of blade; stump of ?flat tang; no rivets through shoulders
Provenance: ?
Length overall, extant: 21.7cm
Length of blade: 20.9cm
Width of blade: 3.8cm
Museum: V.M. 2206k

31. VINDONISSA (?Flavian)
Slim blade, little waist; midrib defined by grooves, and flanked by a pair of grooves; no rivets through shoulders
Provenance: ?
Length overall, extant: 22.9cm
Length of blade: 19.7cm
Width of blade: 3.4cm
Museum: V.M. 13.903

32. VINDONISSA (?Flavian) (Fig.1)
Very slim blade, scarcely any waist; slight midrib defined by grooves; ?rod tang, with secondary wooden grip and bronze guard; no rivets through shoulders
Provenance: ?
Length overall: 26.5cm
Length of blade: 18cm
Width of blade: 3.1cm
Reference: FELLMANN, 1966, 219 & Abb.4.1
Museum: V.M. 3302

33. VINDONISSA (?Flavian)
Fragment from point of asymmetrical blade; irregular double grooving in centre of blade
Provenance: ?
Length extant: 14.9cm
Width: 3.2cm
Museum: V.M. 28:3057

34. VINDONISSA (? Flavian)
Fragment from point, badly corroded; grooves down centre
Provenance: ?
Length extant: 14.9cm
Width extant: 3.2cm
Museum: V.M. 15:170

35. VINDONISSA (?Flavian)
Fragment from point, badly corroded; grooves down centre
Provenance: ?
Length extant: 15.4cm
Width extant: 3.3cm
Museum: V.M. 10754

36. GELLIGAER (Trajanic or later) (Fig.1)
Slim blade; midrib defined by slight grooves; rod tang; no rivets through shoulders
Provenance: ?
Length overall, extant: 17.8cm
Length of blade, extant: 15.2cm
Width of blade: c.3.5cm
Museum: Nat. Mus. of Wales "Gelligaer" 02 127

B. Undated Daggers

37. SISEK
Blade with slight waist; midrib outlined by grooves, ?possibly a piled core; rod tang; no rivets through shoulders
Provenance: R. Kulpa
Length overall, extant: 28.5cm

Length of blade: 23cm
Width of blade: 4.6cm
Reference: HOFFILLER, 1912, fig.47

38. SISEK
Blade with marked waist; simple midrib, with fine grooving outlining it; flat tang, handle largely intact, not inlaid; 2 rivets through shoulders
Provenance: R. Kulpa;
Length overall, extant: 30cm
Length of blade: 22cm
Width of blade: 5cm
Reference: HOFFILLER, 1912, fig.46

39. DUNAFÖLDVÁR
Similar to No. 37
Provenance & Associations: R. Danube; with Type A sheath (App.2,No.7)
Length overall: 32.6cm
Reference: THOMAS, 1969

40. NORDENDORF
Broad blade; simple midrib; ?flat tang; ?rivets through shoulders, handle in situ, not inlaid
Provenance: "Die alemannischen Gräber von Nordendorf"
Length overall: c.32cm
Reference: LINDENSCHMIDT, AuhV 4, 1900, Taf.11,2

41. ALLÉRIOT
Broad blade with marked waist; simple midrib; ?flat tang; ?rivets through shoulders, inlaid handle in situ
Provenance & Associations: R. Sâone; with Type A sheath (App.2,No.22)
Length overall: 35.5cm
Length of blade: 23.8cm
Width of blade: 6cm
Reference: BONNAMOUR & FERROUX, 1969

42. MAINZ
Broad blade of flat cross-section, no waist; slight midrib defined by grooves; rod tang; no rivets through shoulders; (may be a late form, and therefore irrelevant here)
Provenance: "bei der Rheinbrucke"
Length overall: 36cm
Length of blade: 25cm
Museum: M[ittelrheinisches] L[andesmuseum], Mainz

43. MAINZ
Blade with simple midrib; handle in situ
Provenance & Associations: R. Rhine; Type B sheath

(App.2,No.38)
Length overall: c.32cm
Reference: LINDENSCHMIDT, AuhV 3, Heft 2, 1881, Taf.3.2

44. MAINZ
Broad dagger with marked waist; simple midrib; handle in situ secured by rivets; decorative rivets in handle (cf. No.48)
Provenance & Associations: ?; with undecorated sheath of Type A (App.2,No.71)
Length overall: 29cm
Length of blade: c.20cm
Width of blade: 5.3cm
Museum: M.L.,Mainz R4001

45. MAINZ
Broad blade; simple midrib; handle in situ, heavily encrusted, not inlaid
Provenance & Associations: R. Rhine; with a Type A sheath (App.2,No.13)
Length overall: 36.2cm
Length of blade: c.26cm
Width of blade: c.7.5cm
Reference: M[ainzer] Z[eitschrift] 12/13, 1917/1918, Abb.6.6

46. MAINZ
Blade with distinct waist, and long point; midrib defined by grooves, piled core; tang missing, ?handle not inlaid; 2 rivets through shoulders (cf. Nos.47, 49, 53, etc.)
Provenance: R.Rhine (Gustavsburg)
Length of blade: 24.6cm
Width of blade: 5.7cm
Reference: LINDENSCHMIDT, AuhV 4,1900, Taf.52,3
Museum: M.L., Mainz

47. MAINZ
Blade similar to No.46
Provenance & Associations: R. Rhine; with Type A sheath (App.2,No.30)
Length overall, extant: c.25cm
Length of blade: c.24.5cm
Width of blade: c.5cm
Reference: M.Z. 12/13, 1917/1918, Abb.6.4

48. WEISENAU (Fig.1)
Similar to No.44, but larger; handle not inlaid
Provenance: ?R. Rhine
Length overall: 35.6cm
Length of blade: c.25cm
Width of blade: 6.9cm
Museum: M.L., Mainz 62/148

49. "UNKNOWN"
Blade similar to Nos.46 and 47, but lacking shoulders and tang
Provenance: unknown
Length, extant: 24.4cm
Museum: Wiesbaden

50. RÖSEBECK
Large dagger, blade corroded; simple midrib; flat tang, handle not inlaid; 2 rivets through shoulders
Provenance & Associations: stray find; with Type A sheath (App.2,No.15)
Length overall: 38.2cm
Length of blade: 28cm
Width of blade, extant: 5.9cm
Museum: Germ.Nat.Museum, Nurnberg R. 381

51. NIJMEGEN?
Blade with simple midrib; stump of flat tang; 2 rivets through shoulders
Provenance & Associations: ?; with a Type A sheath (App.2,No.16)
Length overall, extant: 24.8cm
Width of blade: 5.1cm
Reference: YPEY, 1960-1961, 352f & fig.9

52. NIJMEGEN
Broad blade, lacking shouders, midrib defined by grooves (may be a late form, and therefore irrelevant here)
Provenance: ?
Length overall, extant: 22.8cm
Width of blade: 5.0cm
Museum: Rijksmuseum G.M.Kam, Nijmegen

53. LEEUWEN
Similar to Nos.46, 47 & 54; inlaid handle
Provenance & Associations: R. Waal; Type B sheath (App.2,No.40)
Length overall, extant: 28.2cm
Length of blade: 26.2cm
Width of blade, extant: c.5cm
Reference: YPEY, 1960-1961, 353ff & figs.12 & 13

54. LEEUWEN?
Similar to preceding example; handle is complete, but construction is obscured by covering of bronze sheet
Provenance & Associations: ?R. Waal at Leeuwen; Sheath of bronze (App.2,No.72)
Length overall: 33cm

Length of blade: 22.4cm
Width of blade: 5.4cm
Museum: R[ijksmuseum] v[an] O[udheden], Leiden e1931 /2. 21

55. VECHTEN?
Broad blade, rounded outline; simple midrib; flattang, handle intact and inlaid; 2 rivets through shoulders
Provenance & Associations: ?; Type A sheath (App.2,No.27)
Length overall: 33.6cm
Length of blade: 22.4cm
Width of blade: 6cm
Reference: YPEY, 1960-1961, 347ff & Abb.5

56. VECHTEN
Slim blade with protruding shoulders; slight midrib defined by grooves; flat tang, broken; 2 rivet holes through shoulders
Provenance: ?
Length overall: 26.3cm
Length of blade: 21.5cm
Width of shoulders: 4.7cm
Museum: R.v.O., Leiden VF*. 1054 (I.36)

57. VECHTEN
Poorly preserved blade of diamond cross-section; flat tang; handle not inlaid; ?rivets through shoulders
Provenance: ?
Length overall: 30.3cm
Length of blade: 20.4cm
Width of blade: 5.4cm
Museum: R.v.O., Leiden VF. 537. I.21

58. VECHTEN
Broad blade with simple midrib; flat tang, handle not inlaid; 4 rivets through shoulders
Provenance: ?
Length overall, extant: 30.3cm
Length of blade: 21.4cm
Width of blade: 5.4cm
Museum: R.v.O., Leiden VF. 533. I.22

59. COLCHESTER
Similar to Nos. 46, 47, etc., but poorly preserved; flat tang, handle not inlaid; 2 rivet holes through shoulders
Provenance: "Colchester 1938"
Length overall: 32.7cm
Width of blade, extant: 4.8cm
Reference: MANNING, 1985, No. V 9
Museum: B.M.

60. LONDON (Fig.1)
Slim blade with waist and long tapering point; slight midrib defined by grooves; rod tang, with turned wooden handle, that is not original; no rivets
Provenance: Site of building of National Safe Deposit Co.
Length overall: 33.1cm
Length of blade: 23.6cm
Width of blade: 4.8cm
Reference: PULLESTON & PRICE, 1873, 70 & Pl.VI, No.6
Museum: Mus. of London (formerly Guildhall Mus.) 3506

61. "UNPROVENANCED" (Fig.1)
Blade with waist; midrib defined by grooves; flat tang, handle complete, not inlaid; 4 rivets through shoulders
Provenance: ?
Length overall: 33.4cm
Length of blade: c.24cm
Width of blade: 6.4cm
Reference: MANNING, 1985, No. V 6

APPENDIX 2

CATALOGUE OF SHEATHS AND DECORATIVE MOTIFS

Note: Those sheaths, which I have only been able to study from photographs published without detailed descriptions, are marked (P) in the catalogue

Part I - Bronze sheath bindings

1. TITELBERG (c.30 B.C. to c.10 B.C.)
   2 cross braces, at mouth and across mid point; latter has small panel of decoration. Dagger handle not inlaid (see Appendix 1, No.1)
   METZLER & WEILLER, 1977, Abb.31

2. HALTERN (c.10 B.C. to c.A.D. 9)
   Bronze binding, no cross bracing extant (may not have had any?). (See Appendix 1, No.4)

Part II - Type A sheaths

3. OBERAMMERGAU
   Silver inlay and fittings. Dagger, with inlaid handle, in sheath. Inscription on hilt: "C.ANTONIUS.FECIT.[..."
   ULBERT, 1962a, and 1971

A.

Type A sheaths with brass, or yellow metal, and enamel inlays. Where the metal has been analysed it is brass (copper and zinc alloy), but there are a few slightly ambiguous references to gold inlay in the older literature; more analyses are needed. The daggers associated with these sheaths do not have inlaid handles. Principal motifs found on these sheaths include:

i) Segmented rosettes with alternate triangular petals of brass and enamel often set within a laurel wreath;
ii) laurel wreaths usually of enamel and between concentric circles of thin brass inlay
iii) L-shaped filets in the corners of decorative panels, usually in brass
iv) "palm fronds" and Inverted chevrons are the alternative motifs more often than not filling the fourth, triangular, zone of decoration. They are made up of alternate leaves, or chevrons of brass and enamel

v) borders of laurel leaves similar to the wreaths
vi) peltas usually in brass

4. CONCORDIA (P)
Dagger in sheath
BRUSIN & ZOVATTO, 1960, 79 & fig.111

5. SISEK, from R. Kulpa
Dagger in sheath
HOFFILLER, 1912, fig.48 left

6. SISEK from R. Kulpa
Dagger in sheath
HOFFILLER, 1912, fig.48 right

7. DUNAFÖLDVÁR from R. Danube
see Appendix 1, No.39
THOMAS, 1969

8. AUERBERG (P) (c.A.D. 10/15 to c.A.D. 40/45)
Dagger in sheath
EXNER, 1940, No.1 & Taf.8.3; ULBERT, 1975, Abb.17.1

9. AUERBERG (P) (c.A.D. 10/15 to c.A.D. 40/45)
?Brass. Dagger in sheath
EXNER, 1940, No.2 & Taf.8.2; ULBERT, 1975, Abb.17.3

10. LADINER TAL, Südtirol
?Brass. Dagger in sheath
MERCKLIN, 1928, 462, Abb.172; THOMAS, 1969, Abb.7.2

11. MAINZ, "Umgebung"
?Brass. Dagger in sheath
EXNER, 1940, No.12 & Taf.8.1; THOMAS, 1969, Abb.8.2

12. MAINZ from R. Rhine
Dagger in sheath
M[ainzer] Z[eitschrift] 12/13, 1917/1918, 176-7 & Abb.13a

13. MAINZ from R.Rhine
See Appendix 1, No.45
M.Z. 12/13, 1917/1918, 177 & Abb.13b

14. KÖLN from R.Rhine
Dagger in sheath
LINDENSCHMIDT, AuhV 4, 1900, Taf.52.1;
BUSHE-FOX, 1949, Pl.XXXIII, No.74A (incorrectly assigned to Rösebeck)

15. RÖSEBECK, Kr.Brilon

see Appendix 1, No.50
LINDENSCHMIDT, AuhV 4, 1900, Taf.52.2

16. NIJMEGEN?
    See Appendix 1, No.51
    YPEY, 1960-1961, 352f & Abb.9

B.

Type A sheaths with brass, or yellow metal, and enamel inlays; decoration in a narrow panel. Principal motifs found on these sheaths include:

i) Small segmented rosettes with enamel and brass petals.
ii) Inverted chevrons, alternately in brass and enamel
iii) Broad brass border surrounding decorative panel

17. MOERS-ASBERG (?Claudian)
    Dagger in sheath
    BECHERT, 1974, Abb.66a

18. HOD HILL (Claudian)
    No associated dagger
    MANNING, 1985, No. V 17; RICHMOND, 1968, 137-8 & Pl.40

Possibly of this group

19. WADDON HILL (Claudian)
    No enamel inlay, lacks rosettes and chevrons, but has decoration in narrow panel, with broad border. "Propeller" motifs. No dagger
    WEBSTER, 1961, 104-5 & photograph

C.

Type A sheaths with silver, or white metal, and enamel inlays. In the few cases, where analysis has been done, the white metal is silver. The daggers associated with these sheaths have inlaid handles. The motifs, found on these sheaths are similar to those found on sheaths inlaid with brass, the exceptions listed:

i) Hatched fields, that is, whole areas covered with fine hatched lines inlaid with silver
ii) decoration of the triangular fourth zone consists of a complex of motifs that cannot readily be described in words: incorporates a shape like hanging drapery, or like a tear drop, and often a rosette or roundel

20. RIßTISSEN (Claudian or later)
Silver inlay, but no hatched fields unlike sheaths listed below. No dagger
ULBERT, 1970, No.257, 16-7 & Taf.14

21. VINDONISSA
No associated dagger
Jahresb. der Gesellschaft pro Vindonissa 1968, 1969, 81 & photograph

22. ALLÉRIOT from R.Saone
See Appendix 1, No.41
BONNAMOUR & FERROUX, 1969

23. COLCHESTER (A.D. 43 to 49)
Dagger in sheath
DUNNET, 1971, 24, 28-30 & fig.12

The following sheaths may be of this group

24. MAINZ (Claudio-Neronian)
?Silver or brass. Dagger in sheath, handle ?not inlaid
BAATZ, 1962, 44-5 & Taf.19.10

25. MAINZ (Claudio-Neronian)
No extant decoration on sheath; inlaid handle
BAATZ, 1962, 44-5 & Taf.19.9

26. MAINZ (Claudio-Neronian)
No extant decoration on sheath; ?inlaid handle
BAATZ, 1962, 44-5 & Taf.19.11

D.

Type A sheaths with some Type B features.

27. ?VECHTEN
Silver and brass inlay, zig zag borders, silver headed nails, and hatched fields. Main decorative elements are most unusual, not paralleled on any sheath known to me. Suspension loops fixed by sets of 4 rivets. (see Appendix 1, No.55)
YPEY, 1960-1961, 347ff & Abb.5

28. HOD HILL (Claudian)
Brass, and red and orange enamel inlays. Has rosette, with enamel inlaid diamond shaped petals, of the kind found on some Type B sheaths (see below Nos.33ff). No associated

dagger
MANNING, 1985, No. V 16; BRAILSFORD, 1962, B5, p.5 & Pl.IV

E.

Type A sheaths with decorative motifs of late Type B sheaths. Their decoration is inlaid with brass and silver exclusively, no enamels have been found. Dagger handles ? without inlay. Principal motifs found on these daggers:

i) "Temple" motif consistng of... supported on columns
ii) "palmettes"; stylised, in the form of a fan of large leaves
iii) diamonds filling the centre of decorative panels
iv) single decorative panel, not subdividedlike the more common 4 zone decorative scheme

29. MAGDALENSBERG
Four sets 5 rivets at suspension points. No associated dagger
Carinthia I 145, 1955, 27 & Abb.21

30. MAINZ, ?R.Rhine
Four sets of 4 rivets at suspension points. See Appendix 1, No.47
M.Z. 12/13, 1917/18, 177 & Abb.14

31. LINCOLN (Flavian) (Fig.2)
No associated dagger
MANN, 1981
(*)

32. RIßTISSEN (Vespasianic or later)
Single decorative panel. See Appendix 1, No.19
ULBERT, 1970, No.258, 16-17 & Taf.15

Part III - Type B sheaths

Type B sheaths differ from Type A sheaths not only in their construction,but also in the materials, and in the motifs used in their decoration. With one exception (No.65 below) brass is never used. Few of the motifs found on Type A sheaths occur on Type B, or vice versa, but, as the sheaths listed above and below show, there is some overlap. One feature common to all Type B sheaths is the use of hatched fields.

F.

Type B sheaths with Type A features. In those cases where the metal of the inlay has been analysed it has proved to be

silver. Type A motifs found on these sheaths:
i) Enamel, usually red
ii) laurel wreaths and laurel borders
iii) segmented rosettes, with and without enamel

Other features characteristic of these sheaths:
i) Rosettes with diamond shaped petals, outlined in silver, the petals inlaid with enamel
ii) rosettes with a silver nail at their centre
iii) silver headed nails, singly within the decoration, and in lines as a border around the inlaid plate
iv) circles and triangles, or diamonds, in horizontal rows. The circles actually linked to the triangles, or diamonds, in a single motif.

Not all of the above features are found on each sheath; in the list below I have noted the motif, or motifs, that identify any particular sheath with this group.

33. KEMPTEN (Tiberio-Claudian)
Laurel border, no enamel
No associated dagger
KRÄMER, 1957, 119-120 & Taf.A

34. AUERBERG (c.A.D. 10/15 to c.A.D. 40/45)
Enamel, and rosettes. Dagger handle not inlaid (see Appendix 1, No.8)
ULBERT, 1975, Abb.17.2b; EXNER, 1940, No.3, Taf.9.2

35. VINDONISSA (?Flavian)
Enamel; circles & triangles. Silver lost
No dagger
ULBERT, 1962, No.1, Abb.1

36. VINDONISSA (?Flavian)
No enamel. Silver headed nails
No associated dagger
DRACK, 1946

37. VINDONISSA (?Flavian)
Segmented rosette. Associated with inlaid dagger handle fragment
<u>Jahresb. der Gesellschaft pro Vindonissa</u> 1973, 1974, 65-6 & fig.1

38. MAINZ R. Rhine
?Enamel; rosettes
Dagger handle not inlaid (see Appendix 1, No.43)
LINDENSCHMIDT, AuhV 3, Heft 2, 1881, Taf.3.2

39. NIJMEGEN (Mid-Flavian)
Laurel wreath, rosette, triangles & circles and silver nails
Dagger handle not inlaid (see Appendix 1, No.18)
BOGAERS & YPEY, 1962-1963

40. LEEUWEN R. Waal
Laurel wreaths, and rosettes
Dagger has inlaid handle (see Appendix 1, No.53)
YPEY, 1960-1961, 353ff & fig.12

41. VELSEN (A.D. 15 to A.D. 30)
Laurel wreath, rosette, and silver nails
No dagger handle (see Appendix 1, No.9)
SCHIMMER, 1979, 111-4 & fig.

42. RICHBOROUGH (?Claudian)
No enamel, but circles & diamonds. (Not certainly Type B sheath)
No dagger
BUSHE-FOX, 1949, No.74, 123-4 & Pl.XXXIII

43. HOD HILL
Enamel, rosettes, and silver nails
No dagger
MANNING, 1985, No. V 18

Possibly in the above group

44. VINDONISSA (?Flavian)
Silver nails
No dagger
ULBERT, 1962, No.3, Abb.3

45. USK (Neronian) (Fig.2)
Rosettes
No dagger
Unpublished, report forthcoming
(*)

46. CHESTER
? Rosettes
No dagger
Unpublished excavation (CHE/HW '80, small find no.705)
Monograph, Grosvenor Museum, forthcoming
(*)

G.

Type B sheaths with "palmettes" and "temples". Where

analysis has been done the inlay metal is silver. Principal motifs found on these sheaths include:

i) "Palmettes", formed of large leaves

ii) "temples". Triangular pediment supported on 3 or more columns.

Also found:

iii) Cross-hatched diamonds, filling a whole panel

47. RIßTISSEN (Vespasianic)
"Temple" and "Palmette"
No dagger
ULBERT, 1959, p.71 & Abb.14

48. VINDONISSA (?Flavian)
"Temple" and "Palmettes"
No dagger
ULBERT, 1962, No.4, Abb.4

49. VINDONISSA (?Flavian)
"Temple" and hatched diamonds
No dagger
ULBERT, 1962, No.5, Abb.5

50. VINDONISSA (?Flavian)
"Palmettes"
No dagger
Unpublished, Vindonissa Museum

51. MAINZ R.Rhine
"Palmette"
Dagger in sheath, rod tang, no handle
Inscription on sheath: "LEG. XXII PRIMI"
LINDENSCHMIDT, AuhV 4, 1900, Taf.11.3

52. ZWAMMERDAM
"Temple" and hatched diamonds
No dagger
HAALEBOS, 1981, 114-5 & fig.3

53. COLCHESTER (c.A.D. 75 to c.A.D. 125)
"Temple" and "Palmettes"
No dagger
CRUMMY, 1983, No.4229, 134-5 & fig.154

54. GLOUCESTER (c.A.D. 70 to c.A.D. 90)
"Temple"
No dagger
HASSALL & RHODES, 1975, 79 & Pl.VIc

55. CAERLEON (Fig.2)

"Palmette"
No dagger
Exacavations of Glamorgan-Gwent Arch. Trust
Unpublished, report forthcoming
(*)

56. CHESTER
"Palmettes".
No dagger
Crook St., 1973-4 excavations, small find no.1308
Monograph, Grosvenor Museum, forthcoming
(*)

57. CHESTER
"Palmettes".
No dagger.
Old Market Hall excavations, 1967-9
Monograph, Grosvenor Museum, forthcoming
(*)

H.

Type B sheaths with abstract designs. Where analysis has been carried out the inlay material has proved to be silver. No daggers have been found in direct association with any dagger of this type. The decoration of these daggers still displays a vestigial 4 zone division but the only recognisable motif is the cross-hatched diamond.

58. VINDONISSA (?Flavian)
ULBERT, 1962, No.6, Abb.6

59. VINDONISSA (?Flavian)
ULBERT, 1962, No.7, Abb.7

60. NEATH
Similar to No.61
Exacavations of Glamorgam-Gwent Arch. Trust
Unpublished
(*)

61. LOUGHOR (Fig.2)
Similar to No.60
Excavations of Glamorgan-Gwent Arch. Trust
Unpublished
(*)

62. CHESTER
Crook St., 1973-4 excavations, small find no.458

Monograph, Grosvenor Museum, forthcoming
(*)

63. CHESTER
Goss St., 1968-70 excavations, small find no.278
Monograph, Grosvenor Museum, forthcoming
(*)

64. CHESTER
Deanery Field, 1924-6 excavations
Monograph, Grosvenor Museum, forthcoming
(*)

I.

Miscellaneous inlaid Type B sheaths.

65. USK (Neronian) (Fig.2)
One panel decorative scheme, with stylised floral motif.
Much of sheath survives.
Dagger in sheath (see Appendix 1, No.15)
Unpublished, report forthcoming
(*)

66. CHESTER
Fragment, possibly as No.63.
No dagger
Crook St., 1973-4 excavations, small find no.893
Monograph, Grosvenor Museum, forthcoming
(*)

67. VINDONISSA (?Flavian)
Bird motifs - Cranes ? - within borders of cross hatching.
No dagger
ULBERT, 1962, No.2, Abb.2

68. "NORDFRANKREICH"
2 roundels inlaid with naturalistic motifs in brass, one the bust of a man, or deity, the other an eagle with wings spread
No dagger
M.Z. 30, 1935, 68 & Taf.6.3

PART IV - Sheaths without inlay

There are a small number of sheaths that conform, in general terms, to our two forms of sheath, but which differ in detail and have no inlaid decoration.

69. CARNUNTUM
?undecorated iron sheath of Type A
Carnuntum Museum

70. MAINZ R. Rhine
Undecorated iron sheath, with dagger, handle not inlaid
LINDENSCHMIDT, AuhV 4, 1900, Taf.11.1

71. MAINZ
Undecorated bronze, or brass, sheath, with dagger (App.1,No.44)
M.L., Mainz 4183

72. LEEUWEN
Bronze, or brass, sheath of Type A, decorated with small amount of filigree-like decoration; matching decoration on guard of associated dagger (App.1,No.54)

73. XANTEN
Undecorated bronze, or brass, plate from Type B sheath
Bonner Jahrbücher 176, 1976, 422-4

74. NIJMEGEN
Thin bronze, or brass, plates from a ?Type B sheath. Decorated with patterns of parallel, fine incised lines on both faces. One face more highly decorated than the other.
Rijksmuseum G.M.Kam, Nijmegen 1932: 18 .10 .32 .3

There are six inlaid sheaths, which were listed by Professor Ulbert in 1970, that I have not seen. I give them here with their numbers in Ulbert's list, which will give further references:

Mainz (Nos.7 & 8);
Holzmühlheim (No.14);
Klein-Hettingen (No.16);
Ilischken (No.19);
and Auerberg (No.51)

CAPTIONS TO FIGURES

Fig. 1: 5: Lorenzberg (after ULBERT, 1965); 12: Hod Hill; 32: Vindonissa (after FELLMANN, 1966); 36: Gelligaer; 48: Weisenau; 53 Leeuwen (after YPEY, 1960-61); 60: London; 61: "unprovenanced". (Nos. refer to Appendix 1). Scale 1:3.

Fig. 2: 31: Lincoln; 45: Usk; 55: Caerleon; 61: Loughor; 65: Usk. (Nos. refer to Appendix 2). Scale 1:2.

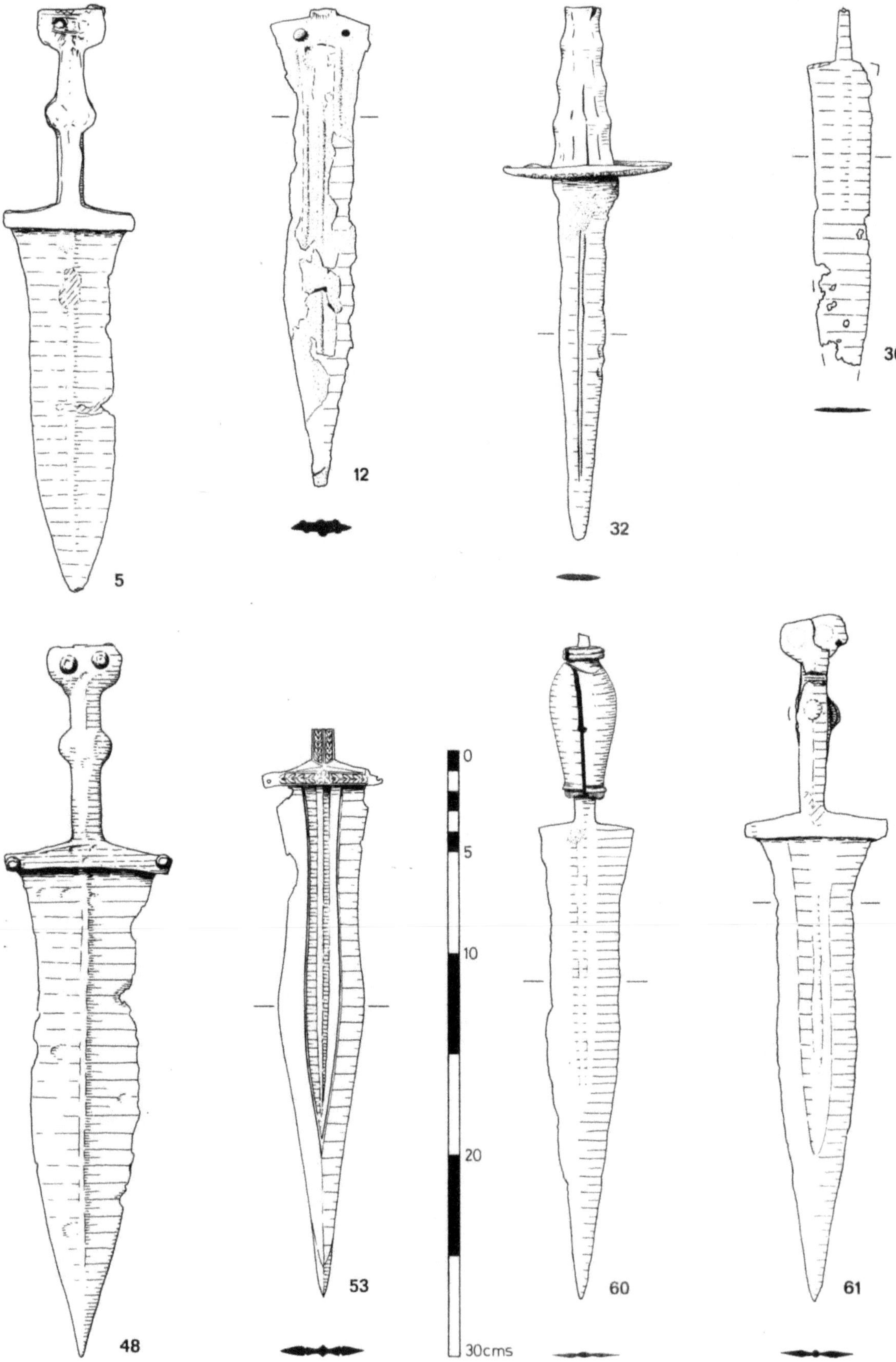

Fig. 1

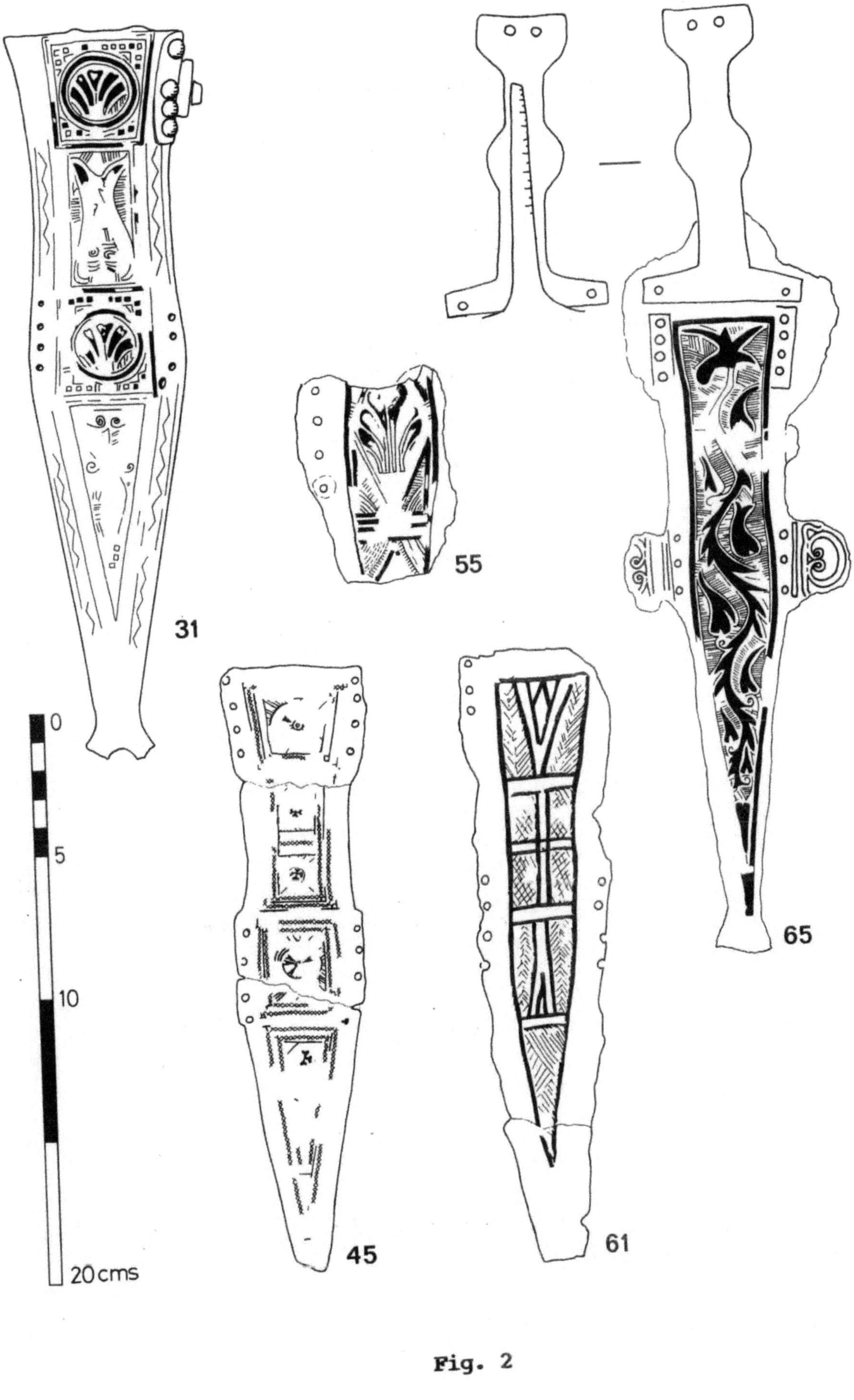

Fig. 2

BIBLIOGRAPHY

ALFÖLDY 1974: G. Alföldy, Noricum, (London 1974)

BAATZ 1962: D. Baatz, Mogontiacum, Limesforschungen, Bd.4, (Berlin 1962)

BECHERT 1974: T. Bechert, Asciburgium - Ausgrabungen in einem Römischen Kastell am Niederrhein, Duisburger Forschungen, Bd.20, (Duisburg 1974)

BEHRENS & BRENNER 1911: G. Behrens & E. Brenner, 'Ausgrabungen im Legionskastell zu Mainz während des Jahres 1910', Mainzer Zeitschrift, 6, 1911, 53-120

BOGAERS & YPEY 1962-63: J.E. Bogaers, & J. Ypey, 'Ein neuer römischer Dolch mit silber-tauschierter and emailverzierter Scheide aus dem Legionslager Nijmegen', Bericht van de Rijksdienst voor het Oudheidkundig Bodemonderzoek, 12/13, 1962-1963, 87-98

BONNAMOUR & FERNOUX 1969: L. Bonnamour, & H. Fernoux, 'Un poignard romain trouvé dans la Sâone', Gallia, 27 1969, 178-85

BRAILSFORD 1962: J. Brailsford, Hod Hill. Volume One, (London 1962)

BREEZE et al. 1976: D.J. Breeze, J. Close-Brooks, J.N.G. Ritchie, 'Soldiers' burials at Camelon, Stirlingshire, 1922 and 1975', Britannia, VII, 1976, 73-95

BUSHE-FOX 1949: J.P. Bushe-Fox, Fourth Report on the Excavation of the Roman Fort at Richborough, Kent, Soc. Antiq. Report No.XVI, (London 1949)

BRUSIN & ZOVATTO 1960: G. Brusin, & O.L. Zovatto, Monumenti Romani e Christiani di Julia Concordia (Pordenone 1960)

CHILVER 1941: G.E.F. Chilver, Cisalpine Gaul, (Oxford 1941)

CICHORIUS 1896 & 1900: C. Cichorius, Die Reliefs der Traianssäule, (Berlin 1896 & 1900)

CRUMMY 1983: N. Crummy, Colchester Archaeological Report 2: The Roman Small Finds from Excavations in Colchester 1971-9, (Colchester 1983)

DRACK 1946: W. Drack, 'Fragment einer römischen silbertauschierten Dolchscheiden', Zeitschr. für Schweizer

Archäologie und Kunstgeschichte, 8, 1946, 113-15

DUNNETT 1971: R. Dunnett, 'Excavations at Colchester, 1964-68', Trans. Essex Archaeological Soc., 3, 1971, 1-130

ÉSPÉRANDIEU 1907-: E. Éspérandieu, Récueil Général des Bas-Reliefs, Statues et Bustes de la Gaule Romaine, 10 vols., (Paris 1907 -)

EXNER 1940: K. Exner, 'Römische Dolscheiden mit Tauschierung und Emailverzierung', Germania, 24, 1940, 22-8

FELLMANN 1966: R. Fellmann, 'Hölzerne Schwertgriffe aus dem Schutthügel von Vindonissa', Helvetica Antiqua. Festschrift für Emil Vogts, (Zürich 1966), 183-208

FINGERLIN 1972: G. Fingerlin, 'Dangstetten, ein augusteisches Legions lager am Hochrhein', Ber. Röm.-Germ. Kom., 51/52, 1970-71 (1972), 197-232

FLORESCU 1965: F.B. Florescu, Das Siegesdenkmal von Adamklissi Tropaeum Traiani, (Bucharest & Bonn, 1965)

FORRER 1927: R. Forrer, Strasbourg-Argentorate, Vol.II, (Strasbourg 1927)

GILLIAM 1967: J.F. Gilliam, 'The Deposita of an Auxiliary soldier (P. Columbia inv. 325)', Bonner Jahrbücher, 167, 1967, 233-43

GILLIAM 1981: J.F. Gilliam, 'Notes on a new Latin Text: P. Vindob. L 135', Zeitschr. für Papyrologie und Epigrafik, 41, 1981, 277-80

HAALEBOS 1981: J.K. Haalebos, 'Door een bezoeker opgeraapt', Westerheem, 30, 3/4, 1981, 111-18

HARRAUER & SEIDER 1977: H. Harrauer & R. Seider, 'Ein neuer lateinischer Schuldschein: P. Vindob. L 135", Zeitschr. für Papyrologie und Epigrafik, 36, 1977, 109-20

HASSALL & RHODES: M. Hassall & J. Rhodes, 'Excavations at the new Market Hall, Gloucester, 1966-7', Trans. Bristol & Glos. Archaeol Soc., 93, 1975, 15-100

HAWKES & HULL 1947: C.F.C. Hawkes & M.R. Hull, Camulodunum, Soc. Antiq. Report, No.XIV, (London 1947)

HOFFILLER 1912: V. Hoffiller, 'Oprema rimskoga vojnika u prvo doba carstva', Vjesnik, 12, 1912, 16-123

HOLDER 1980: P. Holder, Studies in the Auxilia of the Roman Army from Augustus to Trajan, B.A.R. Int. Ser., No.70, (Oxford 1980)

HÜBENER 1973: W. Hubener, Die römische Metallfunde von Augsburg-Oberhausen, Materialh. z. Bayerischen Vorgeschichte, Heft 28, (Kallmünz 1973)

JOHNSON 1983: A. Johnson, Roman Forts, (London 1983)

JONES 1974: A.H.M. Jones, 'Inflation under the Roman Empire' in A.H.M. Jones, The Roman Economy, ed. P.A.Brunt, (Oxford, 1974), 187-227.

KEPPIE 1984: L.J.F. Keppie, The Making of the Roman Army, (London 1984)

KRÄMER 1957: W. Kramer, Cambodunumforschungen 1953. 1, Materialh. z. Bayerischen Vorgeschichte, Heft 9, (Kallmünz 1957)

LINDENSCHMIT 1881 & 1900: L. Lindenschmidt, Alterthümer unserer heidnischen Vorzeit, Bd.3 (Mainz 1881); Bd.4 (Mainz 1900)

MANN 1981: Jenny Mann, 'Small Finds: An ornamented dagger-sheath', Lincoln Archaeol. Trust, 9th Annual Report 1980-81, (Lincoln 1981), 25-6

MANN 1974: John C. Mann, 'The frontiers of the Principate" in Aufstieg und Niedergang der romischen Welt, II.1, 1974, 508-33

MANNING 1985: W.H. Manning, Catalogue of the Romano-British Tools, Implements and Weapons in the British Museum, (1985 forthcoming)

METZLER & WEILLER 1977: J. Metzler, & R. Weiller, 'Beiträge zur Archäologie und Numismatik des Titelberges', Publications de la Section Historique (Musée de Luxembourg), XCI, 1977

PULESTON & PRICE 1873: J.H. Puleston & J.E. Price, Roman Antiquities recently discovered on the Site of the National Safe Deposit Company's Premises, Mansion House, London, (London 1873)

RICHMOND 1968: I.A. Richmond, Hod Hill. Volume Two, (London 1968)

SCHIMMER 1979: E.L. Schimmer, 'De Romeinse waterputten te Velsen', Westerheem, 28, 3, 1979, 109-19

SZILAGYI 1956: J. Szilagyi, Aquincum, (Budapest 1956)

THOMAS 1969: Edit B. Thomas, 'Römischer Legionarsdolch von Dunaföldvár', _Folia Archaeologica_, 20, 1969, 25-46

ULBERT 1959: G. Ulbert, _Die römischen Donau-Kastelle Aislingen und Burghöfe_, Limesforschungen, Bd.1, (Berlin 1959)

ULBERT 1962: G. Ulbert, 'Silbertauschierte Dolchscheiden aus Vindonissa', _Jahresber. der Gesellschaft pro Vindonissa_, 1961/2 (1962), 5-18

ULBERT 1962a: G. Ulbert, 'Der Legionarsdolch von Oberammergau', _Aus Bayerns Frühzeit. Friederich Wagner zum 75. Geburtstag_, Schriftenreihe z. Bayer. Landesgeschichte, Bd. 62, (Munich 1962), 175ff

ULBERT 1965: G. Ulbert, _Die Lorenzberg bei Epfach_, Münchner Beiträge z. Vor- u. Frühgeschichte, Bd.9, (Munich 1965)

ULBERT 1969: G. Ulbert, 'Gladii aus Pompeji: Vorarbeiten zu einem Corpus römischer Gladii', _Germania_, 47, 1969, 97-128

ULBERT 1970: G. Ulbert, _Das römische Donau-Kastell Rißtissen. Teil 1. Die Funde aus Metall, Horn und Knochen_, Urkunden zur Vor- und Frühgeschichte aus Südwurttemberg-Hohenzollern, Heft 4, (Stuttgart 1970)

ULBERT 1971: G. Ulbert, 'Gaius Antonius, der Meister des silbertauschierten Dolches von Oberammergau', _Bayerische Vorgeschichtsblatter_, 36, 1971, 44-9

ULBERT 1975: G. Ulbert, 'Der Auerberg. Vorbericht über die Ausgrabungen von 1968-1974', _Ausgrabungen in Deutschland, 1950-1975_, Monographien, Bd.1.1, Röm.-Germ. Zentralmuseum, (Mainz 1975), 409-33

WALKE 1965: Norbert Walke, _Das römische Donaukastell Straubing-Sorviodurum_, Limesforschungen, Bd,3, (Berlin 1965)

WEBSTER 1961: Graham Webster, 'The discovery of a Roman fort at Waddon Hill, Stoke Abbott, 1956', _Proc. Dorset Nat. Hist. & Archaeol. Soc._, 82, 1960 (1961), 88-108

WELLS 1972: C.M. Wells, 'The supposed Augustan base at Augsburg-Oberhausen: a new look at the evidence', _Saalburg Jahrbuch_, 27, 1970, 63-72

YPEY 1960-61: J. Ypey, 'Drei römische Dolche mit tauschierten Scheiden aus niederländischen Sammlungen', _Berichten van de Rijksdienst voor het Oudheidkundig Bodemonderzoek_, 10/11, 1960-1961, 347-362

DECORATED DAGGER SCABBARDS FOUND IN BRITAIN

Graham Webster

The discovery in recent years of ten finely decorated dagger scabbards, mainly on legionary fortresses, has prompted the idea that it may be possible to isolate patterns peculiar to each legion or at least examine the varieties of style. The standard issue of daggers had a plain scabbard, as seen in the hoard from Künzing.[1] It is reasonably clear that individual soldiers added decorations to their armour[2] and that there was never any concept of uniformity, except generally in the type of armour and equipment peculiar to each unit, or type of unit.

The ten British finds are:

Site	Number	Legion
Colchester	1	Legio XX[3]
Gloucester	1	Legio II Aug[4]
Lincoln	1	Legio IX or possibly II Aug[5]
Usk	2	Legio XX[6]
Chester	1	Legio II Ad or possibly XX[7]
Hod Hill	2	Legio II Aug[8]
Waddon Hill	1	Legio II Aug[9]
Richborough	1	any invasion legion[10]
Loughor	1	possibly Legio II Aug[11]

There are four examples which may be associated with II Aug and possibly another two with XX. In the auxiliary forts at Hod Hill and Waddon Hill, it has been established that legionaries were present.[12] Finds from the other frontiers listed with those from Britain by Dr Günter Ulbert[13] appear to demonstrate that there are very few sites not associated with legionaries.

On examining the decoration as a whole one is aware that some examples are very much finer in the detail work than most of the others. Two published by J. Ypey from Utrecht and Leeuwen[14] are quite outstanding, whereas some of these from Britain are much poorer in quality, two in particular from Chester and Loughor. There could be several reasons for these differences. The scabbards are very difficult to clean and need much time and patience, as a result some rather sketchy designs may be due to the lack of detailed attention. Or it could be argued that the quality deteriorated with time and that the fine pieces may be pre-Claudian and those from Chester and Loughor of Flavian date.

But while there may be some substance in these points, the main factor was probably the amount soldiers were prepared to pay for their scabbards. Centurions would, one imagines require theirs to be superior to those of any miles and as they served until incapacitated by age or illness, their daggers would survive for a much longer period.

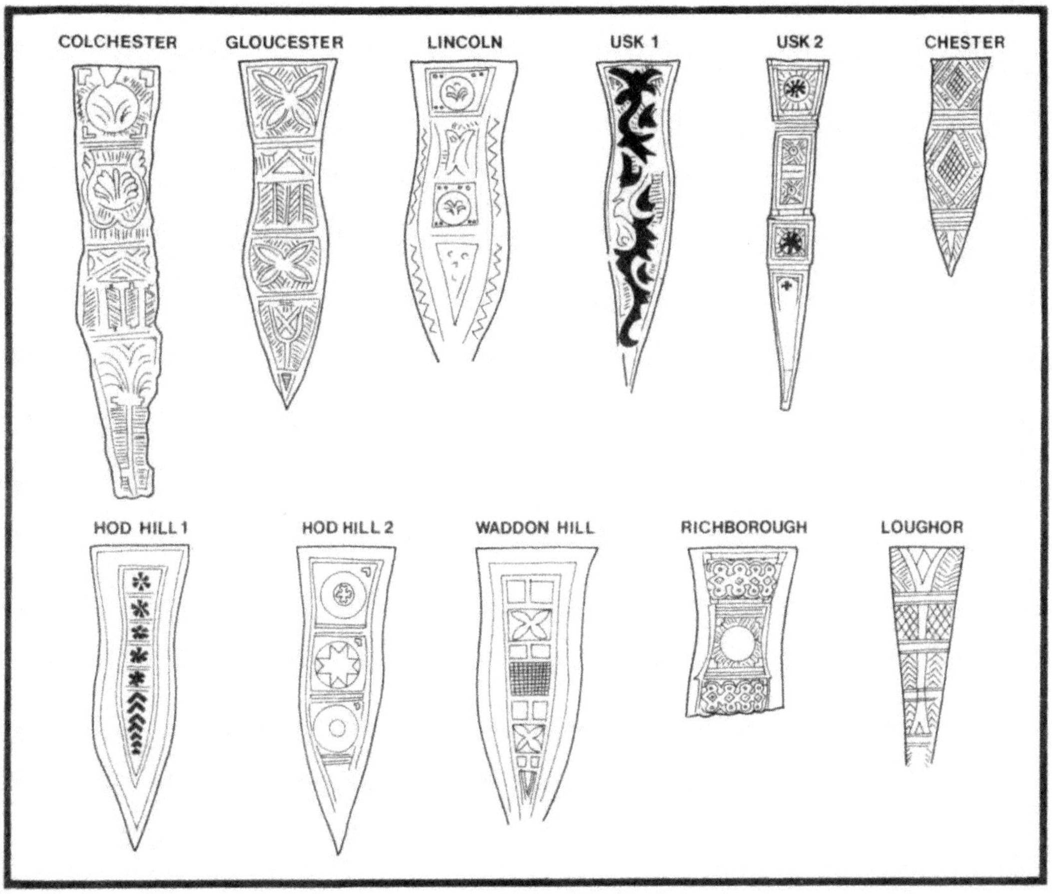

Fig. 1: Sketches of the British dagger scabbards (not to scale).

THE DECORATIVE SCHEMES ANALYSED

The shape of the scabbard determined the basis of the main decorative scheme which was to divide the triangular shape into square and rectangular panels with a long pointed one at the base. Usk 1 however, provides a notable exception to this rule with a free style floral scroll occupying the full length of the panel, and this at present remains unique and so beyond comparison. It could well have been the work of a Celtic craftsman and it is reminiscent of the scroll pattern on the Cerrig-y-Drudian bowl.[15] All the other patterns are highly formalised and, one might argue, the very antithesis of the Celtic style, yet a bronze from a hoard found at Comachio in the territory of the Lingones has precisely this panelled form with diamond shapes,[16] a common feature on the scabbards. Of the British examples, Chester and Loughor are purely geometric and their crude simplicity could be indicative of a later date than the remainder. There is a direct link between Colchester and Gloucester, as both have the temple facade motif which also appears at Vindonissa[17] and Velsen.[18] Colchester has elaborate floral sprays and a tree at the base, while Gloucester has a formalised floral quatrefoil which is also on Waddon Hill. The two Hod Hill examples are quite different, one is a simple pattern of rosettes and chevrons, but the other is a much finer piece - sadly much of the detail has been lost. The same applies to Lincoln but the cleaning and conservation work is not yet complete. It could be a very sophisticated pattern analogous to to those from Utrecht[19] and Leeuwen.[20] The fragment from Richborough shows this scabbard to be a finely worked piece and it is most unfortunate that the details of the roundel have been lost.

It will be seen that the variety in design and execution of the British examples is considerable, but there are no discernible relationships between those of the same legion. The conclusion can only be that 1) the earlier pieces tend to be richer in design and workmanship than those of a later period, 2) the main differences are most probably due to the amount a soldier or centurion was prepared to pay a craftsman for the work.

NOTES

1. SCHÖNBERGER & HERMANN, 1971, Abb.12.

2. In the civil war of A.D.69 legionaries are said to have donated to Vitellius their belts and silver ornaments on their armour (Tacitus *Hist* i,57).

3. CRUMMY, 1981, 134-5 and Fig.154.

4. HASSALL & RHODES, 1974, 79 and Pl.VIc.

5. Information kindly supplied by M.J. Jones, Director of the Lincoln Archaeological Trust.

6. MANNING, 1976, 152-3 and Fig.260 and information kindly supplied by Dr W.H. Manning.

7. Grosvenor Museum Newsletter, June 1983, 3 No.2.

8. BRAILSFORD, 1962, 5 and Pls.III and IV; RICHMOND, 1968, 137-8 and Pl.40.

9. WEBSTER, 1960a, 104 and Pl.3; *Id*. 1960b, 92 No.217 and Pl.XII.

10. BUSHE-FOX, 1949, 123-4 No.74 and Pl.XXXIII.

11. Popular Archaeology, 4 No.6 (Dec. 1982) 7.

12. RICHMOND, 1968, 121-3; WEBSTER, 1981, 55-6.

13. ULBERT, 1970, 16-19.

14. YPEY, 1961, 347-62.

15. GRIMES, 1939, Fig.38 and 197, No.461.

16. Illustrated by JACOBSTHAL, 1969, No.401k and the pattern No.214, Pl.268.

17. Jahresbericht der Gesellschaft Pro Vindonissa 1962 Abb.4 and 5.

18. SCHIMMER, 1979, 109-16.

19. Fn.14, Abb.5 F, p.351.

20. *Ibid*. Abb.12.

BIBLIOGRAPHY

BRAILSFORD, 1962: J.W. Brailsford, Hod Hill I; Antiquities from Hod Hill in the Durden Collection, (London 1962)

BUSHE-FOX, 1949: J.P. Bushe-Fox, Fourth Report on the Excavations of the Roman Fort at Richborough, Kent, (Rep. Res. Comm. Antiqs. No.16) (Oxford 1949)

CRUMMY, 1981: N. Crummy, The Roman Small Finds from Excavations in Colchester 1971-9, (Colchester Archaeological Report 2) (Colchester 1981)

GRIMES, 1939: W.F. Grimes, Guide to the Collection Illustrating the Prehistory of Wales (Cardiff 1939)

HASSALL & RHODES, 1974: M. Hassall and J. Rhodes, Excavations at the new Market Hall, Gloucester, 1966-7, Bristol and Glos Archaeol Soc Trans 93, 1974, 15-100

JACOBSTHAL, 1969: P. Jacobsthal, Early Celtic Art (Oxford 1969)

MANNING, 1976: W.H. Manning, Blacksmithing, in STRONG & BROWN, (1976), 143-53

RICHMOND, 1968: I.A. Richmond, Hod Hill II; Excavatons Carried Out Between 1951 and 1958 (London 1968)

SCHIMMER, 1979: E.L. Schimmer, De romeinse waterputten te Velsen, Westerheem 28, 1979, 109-16

SCHÖNBERGER & HERMANN, 1971: H. Schönberger and F.R. Herrmann, Das Römerkastell Künzing - Quintana, Jahresbericht der bayerischen Bodendenkmalpflege 8/9, 1967/8 (1971), 60-85

STRONG & BROWN, 1976: D. Strong and D. Brown (eds.), Roman Crafts (London 1976)

ULBERT, 1970: G. Ulbert, Das römische Donau-Kastell Rißtissen Teil 1, Urkunden zur Vor- and Frühgeschichte aus Südwürttemberg - Hohenzollern Heft 4, (Stuttgart 1970)

WEBSTER, 1960a: G. Webster, The discovery of a Roman fort at Waddon Hill, Stoke Abbot, 1959, Proc Dorset Nat Hist and Archaeol Soc 82, 1960, 88-108

WEBSTER, 1960b: G. Webster, The Roman military advance under Ostorius Scapula, Arch. J. 105, 1958 (1960), 49-98

WEBSTER, 1981: G. Webster, Final report on the excavations of the

Roman fort at Waddon Hill, Stoke Abbott, 1963-69, Proc Dorset Nat Hist and Archaeol Soc 101, 1981, 51-90

YPEY, 1961: J. Ypey, Drei römische Dolche mit tauschierten Scheiden aus niederländischen Sammlungen, Berichten van Rijksdienst voor het Oudheidkundig Bodemonderzoek 10/11, 1960-1, 347-62

ROMAN ARCHERY EQUIPMENT

J.C. Coulston, B.A., M.Phil.

INTRODUCTION

In the past Roman archers have been studied by students of the army concerned with prosopography and recruitment. Little attention has been paid by them to the equipment and skills which lay behind unusual recruitment patterns and the deployment of archers in the field or along the limites. On the other hand archer-antiquaries have examined the archery equipment with little understanding of Roman military contexts. The present study attempts to fill a gap recognised by Richmond[1] combining a detailed investigation of all the weapons and accessories in use in the Roman period with a study of their military and developmental setting.

A problem of nomenclature arises because types of bows, quivers and bow-cases were not 'Roman' per se but belonged to Levantine or Central Asiatic traditions influencing troops in Roman service. Many items of 'Roman' arms and armour may be traced back to Gallic, Celt-Iberian or other origins so that 'Roman' must be understood as a political rather than a cultural or strictly descriptive term.

The existence and geographical extent of the Roman Empire was important in spreading the use of composite bows but not in their typological development. Syria, Arabia and Armenia within the Eastern provinces provided Levantine archers for the Roman forces but they were part of a wider tradition embracing a region stretching across Mesopotamia and Persia to India. The Roman Empire and this eastern region were directly affected by Central Asiatic nomads who were both enemies of, and auxiliaries in, the Roman and Sassanid forces. Thus archaeological finds outside the Roman sphere must be studied in conjunction with those from inside the Empire.

Composite archery became important to Roman armies with their first Late Republican oriental contacts and became increasingly so through the Imperial period. The employment of archers in Roman armies has been studied in detail elsewhere making an historical rehearsal unnecessary here.[2] Following Jones the Roman period has been understood as extending up to the early 7th century A.D. Regiments with traditional titles were in existence in Syria and Egypt at least until the reign of

Justinian and in the latter area probably lasted until the Sassanid invasion during the reign of Heraclius.[3] Hunnic and Avar material is of direct relevance because of the influence exerted by these peoples in the 4th to 6th centuries A.D. on Roman forces.

Some technical aspects of archery are relevant to all historical periods, especially with regard to composite bows (Theory, below). One particularly rich source of comparative evidence is Islamic archery literature. Several treatises have been translated and edited by linguists collaborating with archer-antiquaries. The most useful are Taybughā's Ghunyah, dating to the second half of the 14th century A.D.; an anonymous Moroccan treatise written c. A.D.1500; a manual embodying traditional and contemporary Ottoman Turkish practices written by Mustapha Kani in the early 19th century A.D.[4] These all provide invaluable information on constructional techniques, the processing of bow stave materials and on archery training exercises. A 'control' for Roman comparisons is provided by remarkably similar constructional information recorded in the 1940s A.D. with regard to workshops still working in Ch'engtu, Sichuan Province, China.[5]

The archaeological and pictorial evidence examined in the present study points to the use of composite bows in Roman military contexts. This does not preclude the employment of self-bows in training or hunting in some regions but the main concern here is with war-bows. The types of bow discussed below may have been in predominant use at various times but contemporary multiplicity of variants may be assumed.

I. BOWS

1. THE EVIDENCE AND TERMINOLOGY

The principal problem in the study of 'Roman' bows is the paucity of contemporary evidence which is restricted to one class of archaeological finds, the 'laths', to a very small number of reliable pictorial representations and to scattered literary references.

No bows have survived intact from within the Roman Empire. Only the 'ear' of one with bone laths, and horn and sinews attached to a wooden core, remains preserved by arid conditions in Egypt (Fig.32). This is not surprising because of all the organic constituents making up a composite bow-stave, only the bone or antler laths are generally imperishable. The horn material employed is taken from the outer, keratin sheath, the least resilient part of the animal's horn.[1] The Parthian bow from Yrzi, discussed in detail below (Fig.2), is thus important because so much of it survives, again in desert conditions. It is also contemporary with the Early-Middle Roman Empire and it has ear-laths identical to most of those found in Roman contexts. However, it did come from outside the Empire and in many respects it may be atypical of bows in Roman service as suggested by the Roman pictorial evidence. Comparative material from Central Asiatic contexts is just as, if not more, helpful in providing direct analogies to the Roman weapons.

When first discovered the identification of the role of laths eluded scholars. Their "curving, sabre-blade" shape was emphasised and a 'rib-knife' explanation was offered.[2] Nash-Williams in referring to a lath from Silchester, in a supposedly civilian context, commented that it was "suggestive of a connection with primary industry like weaving or netting".[3] Von Groller with reference to the Carnuntum material, and MacDonald and Park with regard to the Bar Hill examples, admitted themselves baffled.[4] An assuredly correct explanation of the problem was provided in the 1920s and 1930s by the discovery of very similar bone laths in the Eastern European graves of Asiatic nomad cultures. In particular the work of Rau, Rykov, Sebestyén and Werner proved from the positions of the laths in situ, from their association with arrow-heads and quiver-fittings, and with the use of ethnographic parallels, that the laths in Roman contexts were attached to composite bows.[5]

The pictorial evidence can be unreliable as Rausing and Maenchen-Helfen warned,[6] particularly in general discussions of the structure of ancient bow-staves. Very often a binding around the stave obscured all useful details and shape alone can present interpretational pitfalls. This problem is not so debilitating in the Roman sphere because most, if not all, of the bows in military service were of composite construction. Different problems arise from the degree of stylisation or misrepresentation of the artist and these must be recognised so that 'reliable' representations may be distinguished. Moreover, features exhibited by a stave's profile at rest may change or disappear at full draw. Ears which are acutely angles forward when the stave is unstrung may follow the limb's curve at rest or at full draw.

Most ancient authors, like the artists, were not skilled archers or bowyers and their descriptions of bows, arrows and releasing techniques tend to be more enigmatic than helpful. They are useful for the actual deployment of archers rather than for the elucidation of bow-types in use. The partial exceptions to this include the accounts of Ammianus Marcellinus and Procopius, and an anonymous 6th century Roman treatise on archery.[7]

The laths are usually referred to by archaeological commentators as 'bow-stiffeners'. This only indicates part of their function and 'bow-levers' might be more descriptive of their mechanical action. The present writer prefers the term 'lath' used generally in toxophilogical literature[8] and it will be used throughout this study.

For general discussion of the bow several specific terms must be employed (Fig.1). When a bow is held in the archer's left hand, his left arm outstretched, the face towards the target is the 'back', that towards him is the 'belly'. The left and right sides of the stave are the 'sides' forming the 'side view' or 'profile'. The archer holds the bow by its 'grip' or 'handle', which may be set back, and the parts above and below are known in general as the 'limbs'. The flexing sections of the limbs are <u>dustārs</u> (Arabic), the extremities are the 'ears' or <u>siyāt</u> joined at the 'necks' or 'knees'. The notches on the ears which take the bow-string are the 'nocks', a term also used for the notch at the feathered end of the arrow into which the bow-string slips. The feathers are 'fletchings', the arrow-shaft is the 'stele' (Fig.8). When the bow-string is removed the stave assumes an 'unstrung' or 'reflexed' position (Fig.7). When it is strung it is said to be 'at rest' or 'braced', and when the string is drawn to its furthest extent the bow is 'at full draw'. Bows are discussed with respect to the force (weight) needed to draw them so they may be light or heavy, weak or strong, and their length or 'side measurement' is taken along the line of the stave, not along the string between the ears. Thus laths may be referred to according to their position on the stave, as 'ear laths' or 'grip laths'. In describing ear laths in detail the end nearest to the nock is the 'upper' end, the other is the 'lower'. No inference is made from this as to position on upper or lower limbs, this is purely a convention employed for clarity. The two faces of the lath are the 'convex' (smooth and polished sloping towards the 'back' and with a steep 'belly' edge), and the 'flat' (usually with scoring and saw-marks). On the convex face at the lower end sometimes occurs an area sawn away towards the end, referred to as the 'laterally sawn section'.

2. CATALOGUE OF ROMAN LATHS

This catalogue is thought to include all the British laths but it is unlikely to be complete with respect to the rest of the Empire. The scatter of finds in the Germanies and Raetia suggests that the Danubian picture is far from comprehensive and many Roman laths must exist unrecognised or unpublished in museum collections. The British material is important because the garrison of the province is tolerably well-known. On the Continent finds from Late Roman contexts may have been left by Roman troops influenced by steppe practices, or by Alans or Huns in Roman service, or by independent barbarian forces. Laths from Carnuntum and Intercisa for example have been included because they are associated with Roman military sites, whereas the Wien-Simmering burial material, although sited within the Empire, definitely belongs to one of the non-Roman classes.[9]

The laths are listed in roughly geographical order with dating where known. Types of garrison are given with names of known sagittarii purely for reference. No relationship between small-finds and type of garrison is necessarily suggested.

1. Bar Hill, Strathclyde (Britannia): 6 antler ear lath sections found in the principia well (1), in Refuse Pit 1 (1) and in the fort ditches (4).[10] They make up five distinct laths which, judging from the back zone scoring, represent three dexter and two sinister laths if positioned on the upper ear. Two are curving, two are straight. None are paired except the longest which is a special case. This lath, 27cm long, 2cm wide, tapering to 0.9cm (Figs.9 & 10) has back zone scoring on the convex face up to the nock, but not above it. On the belly edge of the convex face a zone 45cm long at the lower end of the lath has also been scored. On the flat face scoring overlies saw-marks. Unusually this lath has a double globular copper-alloy pin through it, 1.5cm long, at the upper end above the nock. The upper end in fact effectively has two convex faces, one only 3cm long terminating in a horizontal break below which the surface is a normal flat face with scoring. This lath originally formed a one-piece 'hairpin', sawn and scored up the middle, into which the

wooden bow core would have been inserted. The space formed by the removal of a sliver of antler would have been only a couple of millimetres wide and the distribution of the scoring on the flat face demonstrates that the uppermost part was inaccessible to the scoring tool. The lower tip probably ended in a sharp point. The nock is round in profile and unworn. The rivet passes through solid antler so is clearly not structural.

A second lath 8.5cm long, 1.5cm wide has a rounded upper tip and an iron pin through it. It is one of a pair with a wholly flat face and undamaged upper tip. At the lower end on the convex face occurs a laterally sawn section suggesting that only a little of this end is missing. This sawn section appears on a third lath, 6.5cm long, 1.5cm wide, which also has a very wide zone of scoring on the convex face. A fourth lath, 13cm long, has faint saw marks overlain by scoring in three directions on the flat face. The fifth lath, 6.1cm long, tapers to a point which seems to be an alternative lower end termination to the lateral cut-off. In addition to the usual scoring scored 'dashes' appear on the convex face, unaligned with the back zone scoring. The lath curves both towards the back and out from the convex face.

A minimum of 3 bows are implied here (Theory below).

Antonine. Auxiliary fort.
<u>coh. I Hamiorum sag.</u>; <u>coh. I Baetasiorum c.R.</u>

2. <u>South Shields, Tyne and Wear (Britannia)</u>: 2 broken bone ear laths.[11] One is 5.1cm long, 1.7cm long, and is broken off irregularly 1.8cm below the proportionally large nock. The upper tip is rounded and the flat face has diagonal saw marks. No score-lines are present but scored 'dashes' appear. On the convex face the back zone is diagonally scored above the nock.

The second lath is 8.5cm long, 1.8cm wide, with a rounded upper tip, 'U' shaped nock and light back zone scoring. It is broken off at the lower end.

Two other bone pieces from the site are not laths although published as such.

Undated. Auxiliary fort and supply base.
Non-archer auxiliaries.

3. <u>Corbridge, Northumberland (Britannia)</u>: 5 bone and 2 antler ear lath fragments making 6 distinct laths.[12] All were found in excavations of the 1970s. Maximum lengths are 7.1-16.8cm,

widths 1.6-2.3cm. All are incomplete.

One antler lath has an almost right-angled cut-off upper tip; the other exhibits a rounded tip. One bone piece appears to have a little back scoring above the nock and a second has scored 'dashes' in addition to scored diagonal lines all over the flat face. One bone lath has a diagonally cut-off upper tip and all the others have rounded tips.

A bone object with a very coarse cellular structure on its flat side exhibits a typical lath cross-section and was clearly originally worked towards use as an ear lath.[13] Perhaps because of its coarseness the craftsman changed his mind and started to cut it up for use as a latch-lifter (?). However, one deep knife cut and the general shape suggests that even this was abandoned as a bad job. There is no scoring as seen on all the other laths.

A minimum of 4 bows are implied here.

1st-4th century A.D. Auxiliary fort and supply base.
Non-archer auxiliaries and legionary personnel.

4. Chesters, Northumberland (Britannia): 2 bone ear laths, both broken.[14] Lengths 12.5cm and 9.3cm. Both have rounded upper tips and back zone scoring on the convex sides.

   2nd-4th century A.D. Auxiliary fort.
   Non-archer auxiliaries.

5. Chesterholm, Northumberland (Britannia): 1 antler ear lath, 13.1cm long, 2.2cm wide.[15] It is broken at its lower end, rounded at its upper with a slight point. The edges are damaged. On the flat face longitudinal scoring is present below the nock, lateral lines above it. Four longitudinal planes exhibit saw-marks on the convex face. Scoring along the back zone is less distinct than usual and some random scoring appears above the nock. The latter exhibits signs of wear.

   Flavian II. Auxiliary fort.
   Non-archer auxiliaries.

6. Colchester, Essex (Britannia): 1 bone ear lath, 2.15cm long, with a broken lower end, a 'U' shaped nock and a rounded upper tip.[16] It was found in destruction debris on the Balkerne Gate site.

   c.A.D.100/125-c.150. Civitas capital/early provincial capital.

7. London (Britannia): 3 sections of antler ear laths.[17] The first, 20.8cm long, 13cm wide, has lost its lower end but this very likely pointed to judge from the narrowing profile. The convex face has a 0.5cm wide scored back zone and the flat face is heavily scored on several alignments. The upper tip is rounded and there is no scoring above the nock which is too broken for comment. Two lines of scored 'dashes' appear on the convex face of the lower end. This piece was found at the Bank of England site, unstratified.

The second piece, 32.5cm long, 0.75-2cm wide, is badly split and eroded on the back edge for all but the uppermost 3.01cm. The lower end is broken off but cannot have contributed a great deal more to the length. the upper tip is rounded and the nock is 'U' shaped in profile. The flat face is coarsely sawn and is horizontally scored lightly right at its lower end and over its upper half with several alignments. On the convex face a wide scored zone is overlain above and below the nock by some randomly scored lines. The unscored zone is reduced to 0.5cm in width. Found in the Walbrook, unstratified.

The third, 25cm long, 1.4cm wide, is broken at both ends and has coarsely sawn flat face with no additional scoring. The scored back zone on the convex side is up to 0.8cm wide and the back edge is eroded all along its length. The belly edge is very sharply angled away from the rest of the convex side. Found on the Bucklesbury House site, c. 1st-2nd century A.D.

The laths curve to the observer's right, left and left respectively, viewed from the convex face. The second and third laths do not form a pair, therefore at least two bows are implied here.

c. 1st-2nd century A.D. Provincial capital.
Mixed military presence.

8. Silchester, Hampshire (Britannia): 1 ear lath, 10.4cm long, broken off at the lower end.[18] It has a 'U' shaped nock and a rounded upper tip and a back zone of convex face scoring. Found at the West Gate. Boon dated it to the 3rd century on analogy with the Caerleon material.

Undated. Civitas capital.
Mixed military presence.

9. Caerleon, Monmouthshire (Britannia): Of 215 lath fragments approximately 127 represent 'middle' ear sections, broken at

both ends, with or without a back zone of scoring on the convex side, and a varying degree of curve. These average c. 1.75cm wide, some tapering in profile. A small number can be pieced together, fewer joined to ear or lower tip sections, and overall only 4 complete, or near complete, laths can be assembled (Figs.11 & 12).[19]

The 38 fragmentary lower tip sections are mostly pointed, some sharply, but one has a square-ended tip and two are rounded. 16 curve to the observer's left, 22 to the right when viewed from the convex face.

37 upper parts are either rounded or horizontally cut-off in profile. Only one example is pointed. Nocks are semi-circular, rounded or triangular, some being very crudely cut with knife-marks suggesting unfinished work. One has two nocks of different sizes on the same edge. Another fragment has a very small, unfinished nock. 15 have nocks to the left, 22 to the right when seen from the convex face.

Only one lath survives unbroken and intact, 30cm long, 1.7cm wide.[20] It has a rounded upper tip and a gentle curve overall. Four others are complete or near complete, but in pieces. The longest, in two pieces, is 37cm long, 1.9cm wide, with a rounded upper tip and a pronounced overall curve. the lower tip is pointed and there is a wide zone of back scoring on the convex face. a second lath, 35cm long, 1.8cm wide, is incomplete at its lower end to which c. 5cm might be added in length. Overall it is very straight, curving only very slightly at its lower end. A less complete third lath, 26.5cm long, 1.4cm wide, has a square-ended upper tip and curves in two planes. Its lower end is lost but curiously the zone of concave face scoring is on the belly edge, not the back. Lastly, a complete lath in two pieces has a rounded lower end and a square-ended upper tip. It is very thin and without any scoring.

Many laths curve out from the concave face and some even do so then curve inwards again. In this respect they follow the irregular curve of the parent bone, usually rib in these cases. Some appear slightly twisted.

Another group of laths display distinctly different features to those above (Figs.13 & 14). The pieces are characterised by a 'waisted' profile, two square ends and two laterally sawn sections at the ends on the convex face. They are much narrower than the normal ear laths, c. 1.2-c. 1.8cm in width in the middle and are 12.4-16.5cm long. They are slightly curved or flat when viewed from the long edge. The second longest has scoring all over its convex face and some of the others are characterised by a cross-section with sharply angled sides which are likewise scored. 12 fragments make up 6 complete and 2 fragmentary laths.

96 fragments of bone and antler debris or wasters were also found. One substantial piece of antler, in course of shaping, 17.5cm long, 2.5cm wide, had been roughly cut in outline, preparatory to sawing off a lath. These fragments indicate a constructional context for the lath group which is the largest from any Roman site. Other bone objects such as the scabbard-chapes found in the same building may also have been manufactured there. The crudely cut nocks and very roughly cut upper tips of the ear laths suggest that many were unfinished. Moreover, the irregularly curved ribs employed, the extremely coarse and brittle cellular structures of some bone pieces and the example with two nocks do not inspire admiration of the workmanship exercised. Several laths have knife marks across the convex face which in at least one case has caused a subsequent break. Only two ear laths make up a pair (the longest lath and one other) and in such a constructional deposit the conclusion must be drawn that most, if not all, of the laths were never actually applied to bows. Whilst none of the laths are quite as badly tooled as the Corbridge example, it is possible that some of them were 'failed' pieces and consequently discarded. As a bone-working shop the site need not necessarily have been a fabrica producing bows. Composite staves may have been constructed elsewhere in the fortress. It is very difficult to distinguish between bone and antler materials except where cellular structure makes it obvious. Ox-ribs were found on site with varying degrees of tooling. The majority of laths are most likely made of bone but a large proportion of antler is present.

The laths were found in a rampart-back building in the Prysg Field with arrow-heads, spear-heads, helmet-fittings, scabbard-fittings, ballista bolt-heads, pilum-heads, caltrops and mail. The building was constructed in c. A.D.200 and was in use until the end of the 3rd century. The finds from this workshop/weapon-store seem to date from the later part of the period.[21]

A minimum of 19 bows are implied by the fragments with nocks if they were ever applied to bows.

Later 3rd century A.D. Legionary fortress.

10. Waddon Hill, Dorset (Britannia): 1 ear lath, apparently of bone, length 11.6cm, width 1.8cm.[22] It narrows to a horizontally cut off upper end. The convex side has a wide

zone of back scoring above and below the 'U' shaped nock.

Claudian-Neronian. Invasion period fort.
Mixed garrison.

11. Velsen (Germania Inferior): A group of unpublished ear lath fragments from the harbour area.[23]

Auxiliary fort.

12. Oberaden (Free Germany): 1 ear lath, 8.5cm long, 1.5cm wide.[24] It curves to the observer's right when viewed from the convex face. Up to half of the width of the latter has back zone scoring. On the lower half of its length this is split away. How much of the overall lath length is lost is difficult to estimate. The piece comes from a well and is one of the earliest datable pieces yet identified.

c. 10-8 B.C. Lippe legionary base.

13. Mainz (Germania Superior): At least 3 ear laths have been found outside the legionary fortress together with evidence for an antler-working industry in the canabae.[25] One straight lath is possibly complete, 14cm long, 1.4cm wide, and has a sharp lower tip, a rounded upper end and a 'U' shaped nock (undated). A second from Obere Zahlbacher Strasse, 17.6cm long, 1.8cm wide, curves gently, has a rounded upper end and a very wide zone of convex face scoring. Saw-marks are very evident on the flat face and the lower end is broken off. Lastly, an incomplete lath, 7.9cm long, 1.7cm wide, has a large nock and a rounded upper tip. This piece comes from the canabae outside the south-east side of the fortress, and dates to the 3rd century A.D. A fourth ear lath, exhibiting a diagonal channel, or 'string guide', cut across the convex face from the nock, is most likely Early Medieval in date.[26]

At least 2 Roman bows are implied if the laths were ever applied.

3rd century A.D. Legionary fortress, canabae and nearby auxiliary fort.
Pre c. A.D.70 coh. I Ituraeorum sag. and ala Parthorum et Araborum.

14. Zugmantel (Germania Superior): 1 ear lath, apparently of bone.[27] Possibly unfinished with a very small nock similar to one Caerleon example. Undated.

Later 1st century to pre c. A.D.260. Auxiliary fort.

Units of non-*sagittarii*.

15. *Heddernheim (Germania Superior)*: 1 ear lath of bone, length 11.2cm, width 1.7cm.[28] Rounded upper ends above 'U' shaped nock. Tapers slightly towards the broken lower end. Back zone scoring on the convex face . Found in a barrack area.

2nd century A.D. Auxiliary fort.
Units of non-*sagittarii*.

16. *Stockstadt (Germania Superior)*: 2 ear laths, length 33.7cm, width 1.7cm and length 10.6cm and width 1.7cm.[29] The longer example is broken off approximately level with the nock and the upper tip is lost. When complete it must have been *c.* 35cm long. It has a scored back zone on the convex face and the lath tapers to a diagonally cut off lower end. The shorter lath has a diagonal cut-off or break and a downward inclined nock. Both laths were found in Mithraeum II which went out of use after A.D.210.

Date unclear. Auxiliary fort.
Units of non-*sagittarii*.

17. *Osterburken (Germania Superior)*: upper end of 1 ear lath, apparently bone.[30]

Pre *c.* A.D.260. Auxiliary fort.
Unknown garrison.

18. *Selz, Alsace (Germania Superior)*: 1 ear lath.[31]

Undated. Auxiliary fort?
Unknown garrison.

19. *Windisch (Germania Superior)*: 10 ear laths found between 1904 and 1928 in the *Schutthügel*, lengths 8.5-22.5cm, widths 1.5-2cm.[32] Scored flat faces. All are square-ended except two with rounded upper tips. The 2 largest laths, which seem to be complete, represent one of each tip variant (length 22cm and 22.5cm) so do not form a pair. Of the others four are straight (one with rounded tip) but none of these form pairs either. 4 have the nock on the observer's right, 5 on the left when viewed from the concave face. The usual profiles and back zones scoring are present. At least 5 bows are implied here.

c. A.D.45-100. Legionary fortress.

20. *Dangstetten (later Germania Inferior)*: 1 ear lath.[33]

Pre c. 9 B.C. Legionary base.

21. Rißtissen (Raetia): 4 fragmentary bone ear laths.[34] Two have semi-circular nocks, and one is square-ended, the other has a slightly rounded upper end. Length 7.2cm, width 1.4cm and length 6.3cm, width 1.4cm respectively. The other 2 pieces are broken at both ends and taper, lengths 3.5 and 4.7cm. Nocks correspond but these do not form a pair. At least 2 bows are implied.

    c. Claudian-Domitianic. Fort.
    Mixed garrison.

22. Buch (Raetia): 2 ear laths, apparently bone.[35] Rounded upper tips and 'U' shaped nocks. One lath possibly complete. Not a pair although the nocks are on corresponding back edges.

    Pre c. A.D.260. Auxiliary fort.
    Non-archer auxiliaries.

23. Straubing (Raetia): 7 ear laths, apparently bone.[36] The longest and only complete example is 28.5cm long, the others are 9.4-17cm long. 4 curve to the observer's right and 2 to the left when viewed from the convex face. The complete lath is widely scored on the convex face along the back zone with a little scoring above the nock. All have 'U' shaped or semi-circular nocks. The seventh lath has been repeatedly cut into horizontally and was never used on a bow. At least 2 bows are implied here.

    Pre c. A.D.260. Auxiliary fort.
    Hadrian onwards coh. I Flavia Canathenorum mill. sag. (eq.?).

24. Carnuntum (Pannonia Superior): 32 bone and antler ear laths were discovered by von Groller in Building 6 by the west wall.[37] This was probably a weapon-store in use in the 4th century A.D. according to coin evidence. All are broken, ranging in length 6-34.5cm, in width 1.5-2cm. The lower tips are either blunt and rounded or taper to a point. The upper ends are either square-ended or are slightly rounded. The longest example curves to the observer's left when viewed from the convex face and is very similar in appearance to the longest pieces from Bar Hill and London. Nocks are semi-circular, 'U' shaped or rounded triangular, some exhibiting signs of wear. The usual scoring appears on flat and convex faces. Apparently, many form pairs. One has a metal pin through it above the nock, as seen on two of the Bar Hill pieces. There is some question about the date of the context because the deposit was disturbed and scholars have confusingly suggested an Asiatic cultural explanation

rather than a Roman one.

Three more lath fragments were found in the 1968-74 excavations, one in association with 1st to 2nd century A.D. glass.[38] This piece has a rounded, slightly pointed upper tip and an asymmetrical rounded nock. A second fragment is very sharply pointed so appears to be from the lower tip of an ear lath.

1st-2nd century A.D. Legionary fortress.

25. Intercisa (Pannonia Superior): 17 fragments of ear laths found outside the fort to the north and west and inside associated with Building 6.[39] They range from a near complete lath 32.5cm long to a fragmentary lower tip, 4.6cm long. Six pieces have nocks and all surviving upper ends are rounded. Most fragments have some convex face scoring and two pointed lower tips are represented. Some are clearly unfinished.

The presence of the laths within the fort suggests bow-manufacture on site. Intercisa is rich in Late Roman material but the presence of a Hunnic cauldron and the insistence by some commentators that all Danubian lath finds are associated with foederatae or Attila-period Huns complicates the dating problem. Units of sagittarii were present throughout the Roman period and there is nothing distinguishing the Intercisa laths from either 'Hunnic' finds or pieces found in Early Imperial contexts within the Empire.

At least 3 bows are implied if the laths were ever applied.

4th-5th century A.D. (?). Auxiliary fort.
equites sagittarii (Notitia Dignitatum, Oc., XXXIII,38).

26. Dura-Europos (Syria): 4 bone ear lath fragments, 2 with nocks, were found by the Joint Expeditions excavating sites in and around the town but none were published in the interim reports.[40] They will be fully described in the forthcoming Final Report on the arms and armour.

27. Belmesa (Aegyptus): The complete ear of a composite bow was presented by Flinders Petrie in 1897 to Henry Balfour for the Pitt Rivers Museum collection.[41] It had been preserved in arid conditions. This is the most completely preserved fragment of a composite bow found within the Roman Empire to date and it is immensely important for its constructional details (Figs.15-18).

The wooden core is enclosed by a pair of polished bone ear laths which are yellowed and ivory-like in appearance, overall length 15.5cm, width 2.2cm. The laths butt together above and around the nock. They touch at the back edge down to the point where the ear curves, then they diverge. On the belly they diverge exposing the core to view. The upper ends are gently rounded and there is no back zone of scoring.

Black remains of glued sinew appear on the back of the laths where they diverge. This is the most vulnerable constituent and intact fibres do not survive. However, more black material adheres to one edge of the horn backing suggesting that the ear at this point was completely enclosed in sinew leaving only the horn exposed. Diagonal score-marks on both laths across the convex faces where the curve starts suggest that the sinew was trimmed with a knife and that it did not extend up onto the straight section of the ear. The horn strip on the belly is damaged at its upper end but a horizontal knife-cut on the core, extending onto one of the laths, strongly implies that little horn has actually been lost. The line corresponds with the lath scoring indicating a similar termination level of sinew and horn on the ear. A black substance higher on the back is probably glue which had oozed out from between the laths and does not represent sinew. A thick layer of hard glue is visible between horn and wooden core and filling a place exposed to the lath fracture at the lower back. The horn is dark brown, slightly translucent and highly polished. It overlies the belly edges of the laths. There is no evidence for any overall binding but a sinew whipping, covered with black leather, binds the laths together above the nock where there is no core between them. Light scoring on the bone is visible where the binding has fallen away. The grain of the core-wood follows the curve of the ear.

Undated.

3. ROMAN BOW REPRESENTATIONS

The sculptural monuments of the Roman capital were, by their very nature, removed from the detailed reality of the frontier events they purported to depict. Nevertheless, in a few instances credible attempts were made to represent bows carried by Roman troops.

On Trajan's Column Roman archers are depicted on five occasions.[42] In Scene LXVI four archers shoot from a wood in support of a body of auxiliaries but their bows have not survived. In Scene LXX six sagittarii fulfill a similar role. Their bows have a uniform 'segmental' curve, and the ears curl over, presumably without ear laths (i.e. they are 'whip-ended'). In Scene XXIV an archer in normal auxiliary dress shoots such a bow (Fig.19). The Daci use a similar weapon (e.g. XXIV, XXXII) as do the Sarmatae (XXVIII). Its size is comparable to bows seen in Scythian art but the Scythian bow had a set-back handle, not a segmental curve.[43] Another type of bow appears in Scenes CVIII and CXV. In the former a marching sagittarius holds a large, strung composite bow with recurved limbs. Curiously the stave does not appear to be bound because lines on its side indicate the laminated construction (Figs.22 & 23).[44] The archers in Scene CXV hold bows at full draw with strongly recurved limbs and set back handles. One bow displays curled-over ears but another does not (Fig.20).

The curled-over ears are difficult to reconcile with the lath evidence. They are a common feature of Roman deity depictions and may belong to the stylising Hellenistic element of which Robinson was so cautious.[45] Whilst curled ears do appear in the contemporary art of the Sarmaticised Crimea,[46] this feature does not occur in the Column's pedestal reliefs. Here strung, barbarian bows are represented with gently curving limbs and set back handles (Fig.21). Perhaps the curled ears and staves of Scenes CVIII and CXV are a feature of that patchy accuracy which marks the depiction of objects on the Column. Helmet details, shield-sizes and lorica segmentata fittings, for example, vary greatly in their correspondence to archaeological finds. Some are approximately correct, others are wildly stylised. The pedestal reliefs by contrast are something of a 'still-life' study in stone.

The detail on the Marcus Column is reduced to a minimum for the purposes of clarity. Overall the reliefs are more stylised and even less reliable than those on the earlier column. Archers appear in Scenes XV, XXVIII, XXXIX and LXXVIII, and a horse-archer in Scene LVII.[47] All the bows are very small and most have curled ears over which the strings are looped (Fig.24). The infantrymen appear to be Levantine symmachiarii and the horseman is a regular auxiliarius (Fig.25). It is doubtful if any reliance may be placed on these bows.

The development of stylised depictional shorthand continued to such a degree that none of the Eastern barbarians on the Arch of Septimius Severus (Forum Romanum) hold bows (surviving or in antiquarian drawings). Without the monumental context they are quite indistinguishable from Dacians, Sarmatians or other adversaries. However, an interesting contemporary (?) siege relief in S. Paulo fuori le mura depicts an archer shooting a bow with set-back handle (the ears are lost). On the Arch of

Constantine archers of curious aspect appear shooting very short bows reminiscent of those seen on the Marcus Column. These soldiers have been identified as Moors so it is difficult to know what to make of these simplistically rendered weapons.[48] The pedestal reliefs of the Arcadius Column in Constantinople apparently depicted many composite bows with set back handles but such details may merely be noted considering their non-survival and transmission through antiquarian sketches.[49]

In the West a number of figured Roman military tombstones depict auxiliary sagittarii. Unfortunately these are not as useful in supplying details of equipment as this genre generally proves to be. The 1st century A.D. stelei of Hyperanor and Abdes of the cohors I sagittariorum from Kreuznach (W. Germany) depict bows but the surviving details are insufficient for any firm conclusions to be drawn.[50] They would be weakened anyway because the overall style and other equipment details suggest that the same sculptor produced the Daverzus stele, thus casting doubt on the detailed individuality.[51] On the other hand the half-figure relief of Monimus, of cohors I Ituraeorum at Mainz is more promising. The deceased is depicted holding a bow by the grip, with the upper limb, string, and a little of the lower limb visible (Fig.28). The stave is clearly double convex with a set back handle and without prominent ears. It appears to be slackly strung because the string touches Monimus' hand and the upper ear is not bent very far back. Rather similar is the bow on the stele of Dagnas of cohors II Cyrrestarum from Salaria (Yugoslavia). In the lower dexter panel of this 'door tombstone' is a pair of sharply barbed arrows. In the sinister panel a bow is depicted strung, with a slightly canted, set back handle but again without prominent ears.[52]

A full-figure stele from Housesteads (Northumberland) represents an archer holding a bill-hook in his right hand and the upper ear of a well-depicted bow in his left (Figs.26 & 27).[53] The short ears of the stave angle forward at he necks, the limbs are well recurved and the handle is considerably set back. The marked foreshortening of the man's legs affects any calculations of the bow's side length proportional to the man's height. However, it may be noted that the upper limb is longer than the lower (23cm to 20cm). Although the bow is braced no string is depicted. The strung state is obscured by the sculptor's desire to fit the bow into a restricted space. The ears should project back onto the frame of the niche whereas the actual positions would make the string-line cross the handle obliquely rather than pass it as it should on its belly side with room to spare. The upper ear is also further back in the belly direction than the lower ear so that it might be said that the upper limb is strung and the lower is unstrung! The handle is slightly angled forward or 'canted' as appears to be the case with Sassanid bows, but it is doubtful if the accuracy of the carving may be taken so far. The sculptor was certainly familiar with composite bows or was working to the specifications of a knowledgeable customer. If this is indeed a military tombstone it may depict a member of cohors I Hamiorum sag., stationed at Carvoran, another fort on Hadrian's Wall. It cannot be pre-Hadrianic and the style of depiction may be closely compared with early 3rd century figures of Mars Thincsus from Housesteads.[54]

There are some five cavalry tombstones depicting horse-archers. Whilst these constitute important evidence for the other equipment of the equites sagittarii they are rather crude and add little to the foregoing material. A Tiberian horse-archer from Mainz, belonging to the ala Parthorum et Araborum, draws a short bow the upper ear of which seems to be stiffened but the lower limb is unrealistically rendered (Fig.31).[55] The modelling of the figures is not to the best contemporary standards and the deceased appears to be nocking three arrows at the same time! A second stele at Mainz, dating to the second half of the 1st century A.D. (?), belongs to an eques singularis Augusti, and the figure is even less well-proportioned (Fig.29).[56] Both ears of a small bow curve towards the back, suggesting a reflexed bow, but it is too crude to support further comment. The man has drawn his bow in such a way that the stave and string encircle the horse's neck! Clearly the sculptors of these two stones were not familiar with their subject. Two very similar tombstones of equites from ala I Augusta Ituraeorum sag., dating to the first half of the 2nd century A.D., have been found at Győr (Hungary)[57] and Tipasa (Algeria)[58] (Figs.30 & 32). They both depict a very small bow with set back handle. The similarity between the two horsemen is striking but they are too crudely executed for detailed discussion. Lastly, a 1st century A.D. stele from Walbersdorf (Austria) of an eques of ala Scubulorum depicts a horse-archer charging a kneeling adversary.[59] The archer draws a slightly longer bow than that seen on the preceeding two examples. All that may be said about the bows on these five stelei is that they are short, as may be expected of bows to be used on horseback.

Palmyrene sculpture hardly ever depicts bows being drawn and when associated with stylised cuirassed gods the strung bows are small, asymmetrical and of dubious value. Rider-gods only appear with unstrung bows in bow-cases though occasionally the stave profile strongly suggests the application of ear laths. Most useful are the pages which appear associated with funerary banquet reliefs.[60] These figures often carry hunting equipment, including strung bows at rest very similar to those with set back handles on Trajan's Column (Fig.34). No details of laths or stave construction are provided but the staves are medium-sized (compared with the archer), have recurved limbs and ears which are not angled forward. The best examples appear to be asymmetrical with a longer upper limb and a canted handle.

Several Roman deities were normally depicted bearing a bow, notably Hercules, Diana, Atys, Cupid and Apollo. These attributes are almost always a stylised version of the small Scythicus arcus and consequently of no value to the present discussion. Bows carried by Mithras and his attendants, with the exception of the Dura mithraeum frescoes, exhibit no useful features.[61] A signal exception is an oolite statue of Atys from London.[62] The deity is depicted with a bow cradled on his left forearm which is unusual in being seen from the back of the stave (Fig.35). The sculptor has scaled it down ridiculously but preserved the aspect of the working limbs which rapidly widen out above and below the grip. They taper towards the ears and here the sculptor has compromised, perhaps to clarify his subject, because the ears are turned over to reveal their profiles and nocks. They are narrow and bulbous at the tips, the upper being turned over to the observer's right, the lower to the left. Ignoring the overall diminution this is a good rendition of a composite bow corroborated by the Yrzi example (Fig.2) and more modern bows.[63]

Rather unexpectedly Late Roman mosaics from Syria prove to be of some value for their inclusion of archers in hunting scenes. The 'Worcester Hunt' mosaic from Daphne has a vigorous horse-archer holding a large bow at full draw (Fig.36). The handle is very long and stiff, the ears have been damaged but the overall aspect is very realistic.[64] The 'Trichinos' hunt mosaic from Apamea Syriae depicts a pair of archers with bows also at full draw (Fig.37).[65] These staves are strongly recurved with stiff handles and it takes very little imagination to see in the mosaic details the delineation of ear laths and horn belly strips. Unfortunately the 6th century Great Palace horse-archer in Constantinople has lost most of his bow (Fig.38).[66] A strung bow also appears on a 4th century AD silver dish displaying very short working limbs, a well set back handle and long, stiff ears, angled at the knees.[67] Its proportions stongly suggest Sassanid influence.

4. COMPARATIVE EVIDENCE

The Roman evidence discussed above is rather scant. Many of the artistic depictions are unreliable. Apart from the Belmesa ear only the bone or antler laths from Roman bows survive archaeologically. It is inevitable, therefore, that comparative evidence be sought from the areas adjacent to the Empire and from other, more or less contemporary, cultures.

It is true to say that the Roman period is the only time in which composite archery employing the hand-bow was widely employed in Western Europe. This was a function of the Roman Empire's spheres of interaction with the Near East and Asia.[68] There existed what might be termed the 'Near Eastern' or 'Levantine Tradition' and the 'Steppe Tradition'. These were not mutually exclusive because of the areas of joint influence, such as India, Persia and the Roman Empire itself, and because the Parthians who constituted the greatest Mesopotamian influence on Rome originally came from Central Asia bringing new bow-types with them. Something of a synthesis was brought about in the areas of the Early Islamic conquests but by the Late Omayyad period the Steppe Tradition was dominant and remained so into modern times.[69]

Mountainous areas where hunting was at long range and where closing with an enemy was difficult were particularly favourable to the development of archery skills. The Cretans and Armenians amply demonstrated this but even greater skills were exercised from the back of a horse in favourable countryside.[70] The plains of Mesopotamia and Syria, with access to suitable horse breeds and bow materials, was a region which saw from the Late Hellenistic period onwards the dominance of horse-archers in warfare and the concomitant developments in armour for protection against their arrows.[71]

A near complete bow was discovered within the Parthian sphere at Yrzi, near Baghouz on the Euphrates, dating between the 1st century B.C. and the 3rd century A.D. (Fig.2).[72] The wooden grip and one limb are intact. Only a short section of horn belly survives in the grip area. The sinew backing remains around part of the grip and approximately one-third of the length of the surviving limb. Some overall sinew binding appears also around the grip and on the belly of the limb. One pair of laths is 22.5cm long, the other is 19cm. There are no grip laths. The wooden core consists of two long limb section (wood unidentifiable) scafed together with two oak and elm grip sections making for a very long, stiff handle. The profile of the stave when strung and drawn is difficult to determine without a working reconstruction (Fig.7). Even then differences in proportions of the constituents could have affected the profile considerably but it is likely to have been a slightly flattened segmental curve. Assuredly the bow was without a set back handle or sharply forward angled ears but had asymmetrical limbs. The side measurement was c. 147cm. The wood, horn and bone laths were scored to increase glue adhesion.[73] The ear laths are identical to those found on military sites within the Roman Empire, having rounded tips, 'U' shaped nocks and back zone scoring on the convex sides. Mr Edward McEwen kindly informs the writer that his working replica, in course of construction, has a draw-weight of 60-70lbs. Rausing saw the whole bow as typical of those in Parthian use.[74] The bows employed by the Roman auxilia, numeri and irregular symmachiarii he judged to be of one type which he termed the 'Yrzi Bow'. The presence of ear laths distinguished it from his 'Scythian' type and the absence of grip laths (and a set back handle) from his 'Qum-Darya' type.[75]

A similar bow in Parthian use is seen on a terracotta plaque in the Staatliche Museum, Berlin (Fig.39).[76] A galloping horseman holds his bow at full draw with only the upper limb visible. The curve of the stave strongly suggests the lever action of a stiff ear. The handle does not seem to be set back though the profile at rest may have exhibited this feature. A beaker discussed by Rostovtzeff has a bow drawn back so far that the handle profile may not be gauged.[77] It is likely that the use of ear laths was introduced by the Parni in the mid 3rd century B.C. because no laths appear in the Achaemenid contexts.[78]

The present writer objects to the 'Yrzi' classification on the grounds that the bow found near Baghouz was not necessarily typical of those in wide 'Roman' employment. The few reliable Roman pictorial representations suggest that a set back handle was a common feature. Contrary to Rausing's assertion there is evidence of grip laths in Roman use (see below).[79] Where care is taken in the depiction of bow profiles in Dura graffiti (to which 'Roman', 'Palmyrene', 'Parthian' or 'Sassanid' labels are equally applicable) they are virtually all with set back handles and strongly recurved working limbs. The assumption that the single surviving bow from Yrzi is a representative sample is, therefore, unwarranted.

The Dura evidence points the way towards another bow type conveniently termed 'the Sassanid Bow'.[80] This is best observed on Sassanid silver dishes where mounted kings are depicted hunting animals with bows (Fig.44).[81] The latter exhibit well set back handles, proportionally very short, curved working limbs and long stiff ears. The upper limb is often longer than the lower when both are visible and the handle is generally canted forward. It must be assumed that the ears are stiffened with laths, the upper pair longer than the lower it seems. In some cases the constituent materials of the limbs may be indicated, and this is best seen on a 4th century A.D. Kidarite bowl in the British Museum.[82] Here the stylisation of the relative ear to dustār proportions marking the Sassanid dish representation is less-marked. Lines on the dustārs may indicate horn and sinew construction. The ears are sharply angled at the knees. The Kidarite link with Central Asia of course raises the problem of how far the Sassanids were influenced directly by Asiatic practises. Similar bows appear on a 3rd to 2nd century B.C. Greco-Bactrian bowl suggesting on early development of the type. On the Sassanid dishes bows are often shown over-drawn with too great an angle between string and ear. Paterson reconstructed these bows with 18-20cm long ears, reducing their proportions to the dustārs considerably, calculating an overall side length of 130cm for the stave (Fig.7).[83] The hunting scenes at Tāq-i-Bustān contain less stylised bows seen both in braced and in full draw positions (Fig.40). The lever action of the ears is clear but the angle of the knee is far less acute and at rest the ears lie approximately parallel with the string.[84] The bow's profile at full draw is very similar to that suggested for the Yrzi bow, the set back handle disappears, and perhaps this represents a third stave variant.

Whether or not the profile of the Yrzi bow is representative of bows in use by Roman forces it is vitally important for its information on constructional materials and techniques. The Belmesa ear is of little help in discussing stave proportions but again the relationships of its constituents are very important. It may be the case that the Yrzi bow represents the earlier Levantine tradition with the specifically Parthian addition of ear laths. This might suggest that the set back handle was also a Parthian period innovation influenced by steppe practices. Whatever the evolutionary relationships it must be noted that bow types can never be exclusively applied so that 'Yrzi' and 'Sassanid' bows may be seen as contemporaneous variants.

The Steppe Tradition impinged directly on the Roman frontiers along the Danube and around the Black Sea. Each wave of nomads following the 'steppe gradient' bore modified bows differing in size, construction and proportions from those of their Asiatic predecessors. The Scythian bow was very small, c. 75-100cm side length, and was whip-ended. No laths appear in Scythian cemeteries and bows were represented in contemporary art with curled ears.[85] These went into the classical repertoire and appeared in the Roman period born by deities. Before Hunnic contacts, laths are lacking in Sarmatian contexts and this explains the whip-ended bows depicted on the Crimean tomb reliefs and tomb frescoes.[86] the Greek cities of this area were heavily Sarmaticised, particularly in weaponry, and it is likely that a Scythicus arcus form was used by Arrian's cohors Bosporanum.[87] The Thracian alae and cohors sagittariorum probably also carried this weapon given the long Thracian history of mounted skills and Scythian contacts.[88]

Thraci and Bosporani apart, the Steppe Tradition was far subordinate to the Levantine before the Late Imperial period. Sarmatian contacts were influential in the use of the contus and of increasingly heavy forms of cavalry armour, but there is no evidence for the Roman adoption of Sarmatian bows.[89] On the contrary, Levantine sagittarii were greatly favoured for use as a counter to Sarmatian and the latter's archery may have been inferior.[90]

This situation changed dramatically with the appearance of the Huns in the West. This Turkic horde brought a bow radically different from the Scythicus arcus. Rausing termed it the 'Qum-Darya Bow' from the type-site at the mouth of that river.[91] The alternative, 'Hunnic', begs questions of nomenclature for the Central Asiatic peoples so will only be applied here with reference to the forces of Attila, Ruga etc. The type-site was in fact not 'Hunnic' but Han Chinese.

The bow itself was found in Mass Grave 1 and was preserved almost intact because of arid conditions. The cemetery was associated with the Chinese frontier post of Lou-lan and the deposit dated by analogy to c. 1st century B.C. to 3rd century A.D.[92] One photograph exists of the bow complete, from which a measurement of 140-150cm, ear to ear, was calculated, but the stave was broken up on a camel-back journey before it could be examined in detail (Fig.42).[93] Only the upper half of one limb (Piece A) and the ear of the other (Piece B) now survive (Fig.4). It was, incidentally, very similar in profile to the bow in Scene CVIII of Trajan's Column (Fig.23). The ears curve without a change of angle at the knees. Piece A, 339cm long has a surviving wooden core, a horn belly strip and a sinew backing on the back and sides up to just below the nock. The ear laths have a rounded lower end, a horizontally cut-off upper end and a 'U' shaped nock. They are 25.5cm long, 1.5cm wide. Piece B, 31.5cm long, has its wooden core and a horn belly consisting of two overlapping strips. The sinew backing survives for much of its length as does a sinew binding over one section. The ear laths have been lost except for the end of one (7.5cm length remaining) which tapers to a point. An overall lath length of 32cm was rather dubiously calculated. A grip lath is identifiable from other surviving fragments, apparently with a wedge-shaped end.[94]

Fittings from this type of bow appear right across Asia from Korea to the Crimea.[95] Alanic graves in the Volga region dating to the 3rd to 4th century A.D. signal the adoption of the Qum-Darya type by Sarmatian peoples from Hunnic groups advancing from the East. It was these Alanic deposits which led to the original identification of the role of ear laths, not only for Eastern European contexts but also for the Roman material.[96]

Hunnic finds per se appear along the Volga, in the Crimea, and in Western Europe on sites such as Blučina and Wien-Simmering.[97] This has led to a good deal of confusion between 'Hunnic' and 'Roman' contexts, notably at the latter site. Researchers keen to extend the body of Hunnic finds have gone so far as to assume that all laths are Alanic or Hunnic representing Asiatic troops in Roman service.[98] The Carnuntum group (Catalogue No. 24) has been interpreted in this way and indeed the disturbed stratigraphy does not make a Roman Waffenmagazin context assured.[99] The finds in Britain, however, are of course securely Roman, despite Maenchen-Helfen's fictions.[100] One lath from Mainz with a string-guide is likewise insecurely dated. The guide appears on other, definitely Hunnic pieces.[101] There are also problems with the Intercisa laths.

In general, Hunnic/Qum-Darya bows had two pairs of ear laths identical in every respect to those found on Roman limes sites.[102] The only difference is that there are proportionally a greater number of longer laths (like those Roman examples from Bar Hill and London). In addition the grip of the bow was stiffened by three laths. On the sides were glued a pair of trapezoidal laths with their longest edges towards the back. On the belly was glued a third lath, varying in shape but often narrow with parallel sides and splayed ends.[103] Therefore, each bow possessed seven grip and ear laths, compared with none on the Scythian and Sarmatian bows and four (ear) laths on the Yrzi bow. The bow may be reconstructed with a set back handle and asymmetrical limbs, c. 130-60cm long overall to judge from lath positions in situ.

A rather more spectacular Hunnic practice was to sheath model bows, or cover only the ears of working bows, in gold sheeting. Two-thirds of a model bow's sheathing was found at Jacuszowicze (Poland) and stippled decoration carefully outlined the shape of the ear laths as they would have appeared on a real bow. Golden ears from working bows appeared at Novogrigorjevka, Borovoje (Russia) and Pécsüszög (Hungary). The latter exhibits a rivet-hole above the nock passing right through the ear. Both ear sheaths were recovered and the holed piece exhibited a tapering upper end whilst the other piece was square-ended.[104]

Hunnic skills in horse-archery were profoundly influential on the Roman Army of the 5th to 6th century AD.[105] It is most likely that Levantine bows in Roman use were superceded by the Hunnic type for at least the best quality archers. Both Avitus and Aetius are recorded as having been skilled archers, one surpassing the Huns (panegyric!), the other taught archery and horsemanship by them whilst in captivity.[106] Stilicho employed Hunnic troops as did Belisarius whose personal archery skills Procopius mentions.[107] Roman cavalry in the campaigns described by Procopius, with their Hunnic allies, were almost all bow-armed and devastatingly effective against Germanic adversaries.[108] There is no evidence as to what type of bows the psiloi used.

Unfortunately, because of changing burial practices the laths do not appear in Late Roman funerary contexts, although they do occur in some Frankish and Alamannic graves.[109] Bivar saw Procopius' description of Roman and Sassanid archery at Callinicum as a confrontation between 'Hunnic' and 'Sassanid' bows, the advantage being with the former.[110]

The next wave of nomadic invaders were the Avars in the later 6th century AD. Maurikios' Strategikon demonstrates the influence this people had on Roman military equipment. Avar lamellar armour, saddles,gorgets, stirrups, horse-armour, lance-pennons, belts, cloaks and, one may add. bows were adopted.[111] The numerous laths from Avar graves reveal a number of bow modifications demonstrating that the Qum-Darya bow was superceded by an 'Avar' type.[112] This differed radically in the number and shape of the laths. The grip laths stayed essentially the same except that a fourth piece was sometimes glued to the back of the handle enclosing it with bone on all four faces. The belly lath was often parallel-sided with splayed ends.[113] The ear laths became much wider in profile above the nock and less rounded, giving a bulbous aspect. The nock was often further away from the upper end than on Qum-Darya type examples.[114] Additional laths were usually added to the belly and back of the ear thus enclosing both ends of the stave on four faces. This made a total of up to 12 laths on an asymmetrical bow with stiff, set back handle.[115] Examples measured in situ suggest bow lengths of 120-40cm. When unstrung the ears reversed sharply forward at an angle of 50-60°. Some ear laths, for example a pair from Üllő, have a rivet-hole near the upper end and several pieces from Feherto B and Gátér (Hungary) actually had bronze rivets surviving in place.[116] Working reconstructions of Avar and later Magyar bows have been made with great success by Dr G. Fabian.[117]

A clear picture emerges of the Hunnic and Avar bows in Late Roman use, each type incorporating modifications in design and, presumably, improvements in performance. Apart from the original identification of laths as bow fittings, two features of these steppe laths add to the understanding of Levantine bows in Roman use. Firstly the Hunnic and Avar belly grip laths are identical in size and shape to the group of eight laths from Caerleon (Figs.13 & 14; Catalogue No. 9). Without reservation the latter may be identified as grip laths and a bow with five laths must have been in Roman use, at least in Britain. The number of finds from the Roman East or Mesopotamia is small and it is not at all surprising that no others have appeared on Western sites but with the present publication of the Caerleon pieces it is hoped that this class of artifact will be recognised elsewhere. Secondly, rivets were put through the ears of Hunnic and Avar bows in a similar fashion to the rivets in the Bar Hill and Carnuntum examples (Catalogue No. 1, 24). These are clearly not structural, as proved by the Bar Hill one-piece pair (Figs.9 & 10). Some other function must, therefore, be sought. For the Bar Hill examples Macdonald and Park suggested a suspension role without knowing the bow-attachment of the laths.[118] This seems to be the only reasonable explanation which presents itself. When stored unstrung English longbows were hung by a ribbon tied around the upper ear. Any good composite bow must be unstrung when not in use to retain the elastic qualities of the sinew backing. To keep out the moisture the Ch'engtu bowyers kept bows stored in a drying cabinet over a charcoal brazier. On the other hand Lt-Cdr Paterson sees no need for the rivet, and suggests that bows would have been hung up by the nock or in a bow-case from a strap.[119]

5. THE COMPOSITE BOW: THEORY

Rausing defined the composite bow as "any weapon where the constituent layers of the stave have been joined with any kind of adhesive, and where the materials employed have been selected so as to make the back stretch around an incompressible belly, and where at least one ingredient, generally wood, gives the necessary dimensional stability".[120] In the examples from Belmesa, Yrzi and Qum-Darya the stave was bellied with horn and backed with sinew. This is the case with surviving Mongolian, Manchu, Indian, Persian and Turkish bows and, it must be concluded, these materials were in general Roman use.

The thin wooden core provides adhesive strength and the general shape but plays a minor part in the bow's physical actions. When a stave is drawn the horn belly is pulled into a compressed curve and the back sinew is stretched. The bow is constructed to elicit the maximum distance of curve and stretch by making it in a 'reflexed' shape so that it reverses itself when unstrung (Figs.2 & 7). The extra distance from the reversed to strung positions gives a greater potential energy storage than with a straight self-bow stave. Small Turkish flight-bows reverse into a 'C' shape with the ears almost touching. Longer bows, such as that from Qum-Darya, assume a squarer 'C' or a very shallow 'V' shape. Thus the materials are used to create an artificial elasticity whereby on release the belly springs back and the backing pulls the stave to a rest position.[121]

The superior properties of horn and sinew allow the stave to be much shorter than a comparable wooden construction, and for the limbs to be drawn through a greater arc.[122] Scythian and Sarmatian bows demonstrate this shortness which is convenient for horseback use and the crudely depicted bows on Roman horse-archer tombstones may reflect this (Representations above). However, "it is unlikely that an archer could shoot as strong a bow on horseback as he could on foot. Though the main power comes from the shoulders and the back muscles, with the arms playing a relatively minor part, when shooting a powerful bow an archer normally needs a firmer footing and, if the ground is slippery, he may have difficulty in controlling his bow. Some loss of such support would naturally result from shooting from the saddle so that the mounted archer would probably have used a slightly lighter bow on horseback than he would have used on foot".[123] For this reason units of infantry-archers, for example those in Roman use on the Parthian front, would have been of great tactical importance in confrontation with horse-archers because of their ability to outrange them and to cast heavier and thus more penetrative arrows.[124]

It is not always the case that bows used on horseback were short, however, and the trend in central Asia was for an increase in length over time, but a composite design still made for a weapon mechanically superior to a wooden stave. One advantage is the possible increase in draw length. This depends initially upon the length of the archer's left arm and the width of his chest. In addition he may draw the string to various points: to chin, ear or right shoulder. If the handle of the bow is set back, as on the Qum-Darya or Housesteads bows, this draw distance may be slightly increased. The greater possible curve of the composite limbs allow for a proportionally longer draw and the addition of long, stiff ears increases this by allowing a greater angle between string and ear. The angle of the 'V' of the string at the hand depends upon the variant of the 'lock' or 'release' employed (below, Thumb-Rings and Bracers).

The bow's handle must be rigid and not bend with the limbs during the draw because if it does the bow will 'kick' when the stave returns to rest. The handle of the Yrzi bow has a scarfed construction designed to stop it from curving. The Qum-Darya bow and at least some in Roman use, as evidenced by the Caerleon finds, had a lath attached to the grip for this purpose. Hunnic and Avar bows went further with the addition of laths to three or even four faces of the grip. If the bow kicks it can be uncomfortable for the archer's hand and impart inaccuracy to the arrow at the last moment of leaving the string.[125]

Ear laths performed a specifically mechanical function only partially intimated by the term 'bow-stiffeners'. The stored energy of a bow, and from it the force of impact and distance of cast, depends upon the archer's strength and length of draw, thus the weight of the bow he can use. The weight and draw-length of two bows with different constituents, but the same weight and draw-length, can be plotted on a graph as a 'force-draw curve'.[126] This demonstrates that when an English longbow, for example, is drawn the weight experienced by the archer builds up uniformly over the draw but increases markedly, or 'stacks', towards the end. However, with a stiff-eared composite the weight builds up quickly for the first half of the draw but in the second half the rate of increase slows appreciably. This is because the stiff ears act as levers pulling the flexible dustārs back mechanically.[127] Paterson comments that "When shooting this bow it feels, as the shoulder and back muscles come into exerting their power, that the pull on the bowstring reduces, though this is not, in fact the case".[128] For a given exertion the composite stores more energy and with the addition of stiff ears avoids 'stacking'. The latter makes it difficult to hold the string at full draw with the hand tending to 'creep' forward or 'shake' under the strain. It is crucial that the last part of the draw be perfectly executed in order to hit the target with consistent accuracy. The Housesteads Archer's bow and some weapons in Sassanid use seem to have had ears angled forward at rest (Figs.27 & 40). The main advantage of this during release is that at the moment the string hits the knees of the stave it is effectively shortened and the arrow is given a final 'push' before leaving the string.[129] Avar bows reconstructed by Fabian working from the angle of ear laths in relation to grip laths as found in undisturbed graves, also have such forward ears.[130] It might be suggested that the additional ear laths were designed to protect the ear from the impact of the string which could adversely affect the cohesion of the components. The addition of ear laths to wooden bows would be unnecessary and might even lead to damage of the limbs by the lever action.[131] Thus the incidence of ear laths in Roman contexts is in itself an indication of composite bow use.

The occurrence of pairs of ear laths of differing lengths on the Yrzi and Qum-Darya bows, and from Hunnic and Avar weapons, leads to the conclusion that such bows had asymmetrical limbs. This is clearly seen in artistic representations discussed above, notably of Scythian and Sassanid bows. Surviving Middle Eastern bows have symmetrical limbs and could be used either way up in the heat of battle but Taybugha's Mamluk Syrian bow had an upper limb and ear slightly longer than the lower as did the bows recommended in the 15th century Moroccan treatise.[132] This is logical because the arrow passes above the hand on the grip, not along the grip's mid-point. The comparative properties of the limbs had to be carefully adjusted to compensate but this is not a problem as the Japanese longbow most clearly demonstrates. This was held with two-thirds of its length above the hand to enable it to be shot from horseback (not done with the English longbow) or whilst the archer was kneeling.[133]

The longest Roman laths and those with rivets for suspension may thus be attributed to upper limbs. However, Alfoldi positioned the longer ear laths on the lower ear of his Hunnic bow reconstructions.[134] He reasoned that the string was permanently affixed to the lower ear and when the bow was strung it was slipped over the upper. He thought that the upper limb was the one pulled back during stringing so it had to be more flexible than the lower which was stiffened for a greater length by his longer lath positioning. Brown followed this reasoning for the Yrzi bow.[135] Against this it must be said that there is no need to permanently tie the string to one ear. On the contrary it would be far more convenient to remove it altogether when the bow is not in use. The risk to the string from damp would be obviated by storing it in a pouch with spares. Although Alfoldi was quite correct in supposing that the string was slipped into the lower nock first, especially on horseback, performance after release was the governing factor in limb-design, not ease of bracing.

The very longest Roman ear laths all have rounded upper ends as do both sets of the Yrzi bow laths. The ends of the longer Qum-Darya laths are missing but the shorter ones are square-ended and it might be suggested that the lower ears of some bows were so treated. The shorter delineated Jacuszowicze golden 'laths' accord with this. However, there is no consistency amongst Roman (or Hunnic) laths, short examples with rounded ends surviving complete in large numbers. Moreover, it is obvious from the variety of Roman lath lengths that there was a great variety of bow lengths and proportions in use. This is unsurprising considering the temporal and geographical distribution of these finds and the variety seen in artistic depictions.

The positioning of the longest laths on the longer, upper limb of the bows used by Roman forces aids in the estimation of the minimum number of bows represented by a given assemblage (Lath Catalogue). Two non-matching laths could come from two different bows or just two ears of the same bow. Ten non-matching laths constitute a minimum of five bows, and so on.

In Hunnic and Avar funerary contexts very few complete weapons were deposited, defective weapons being substituted.[136] A custom-made bow could, with care, last an archer for his entire life and indeed 18th century Ottoman bows may still be shot today. Neglect resulting in damage from damp and insects considerably shortens the working life.

6. THE COMPOSITE BOW: CONSTRUCTION

The construction of a really good composite bow demanded the utmost skill and craftsmanship as the major sources of comparative information stress throughout. The basic properties of the materials are unchanging, as are the best methods of treating them.[137] Consequently the constructional processes described in the surviving 14th to 19th centuries AD archery manuals are remarkably similar. Likewise methods and tools applied in the Ch'engtu bow industry in China are closely comparable. The design, proportions and choices of materials for the stave vary with availability, geography, climate and function (for war, hunting, target shooting or flight shooting). Climate may have been very important judging from the Moroccan treatise which specified adjustments in limb-widths, core thickness and amount of sinew applied according to local conditions.[138] Weight, draw-length, amount of recurve, presence of a set back handle and inclining forward of the ears are other variables. The exact choices of materials made by Roman arcuarii cannot in many cases be proven but the recommendations of more modern experts and the conclusions drawn from modern reconstruction work may be taken into account, subject to availability to Roman craftsmen.

According to Taybughā "the fashioning of a bow calls for more competence than anything else if a truly good weapon is to be obtained. Its manufacture also calls for more patience, since it cannot be properly completed in less than a full year. Autumn must be devoted to the carving and preparation of the wooden core on the one hand, and to the sawing and fitting of the horn on the other. Winter is the season for binding and reflexing, and the at the beginning of the spring the sinew is applied. Next, in summer, the bow, as yet unfinished, is strung and rounded to the curvature required. It is then veneered and painted".[139] The four seasons are used to obtain optimum conditions for glue-setting.

For a good bow a full year was the minimum time period. Inferior staves could be produced over a shorter time dependent on the glue drying sufficiently. The physical work of fashioning and glueing the components may take as little as eight full days.[140] Luschan quoted a period of five to ten years for Turkish bows and those produced at Ch'engtu each took three years' work.[141] These longer periods were partially dictated by careful gathering of materials at the right seasons but treatises often recorded practices in periods of marked decline in customer demand.[142] Paradoxically high levels of craftsmanship and pure virtuoso work were being exercised as a result. It is true that Taybughā was writing in a period when archery and bow-construction had been developed to unequal levels of skill but he was an expert horse-archer himself writing a practical manual for warriors. Thus, his full year is a good framework on which to base a discussion of construction, given the undoubted contemporary demand for bows and the logical use of seasons for glue-drying. Taybughā stated that the best bows were made in Syria, so his remarks on climate and materials must be seen in this context.[143]

The first task in composite bow manufacture is the construction of the wooded core (Fig.5.1). Maple, cornus and mulberry woods were preferred in the Levant.[144] The handle should be oval in section, the limbs elliptical, rectangular or 'D'-shaped, a few millimetres thick, and the ears triangular or 'D'-shaped. The belly is usually flat or slightly convex (Figs.2 & 4).[145] The choice of pieces demands great working experience if irregularities of grain patterns are to be accounted for. The actual component make-up can vary widely from region to region. The most complex forms have a hardwood grip section, two dustārs,, two ears and two ear inserts with nocks, making a total of seven sections.[146] Fewer are seen on Medieval-Islamic bows with the whole siyah in one piece. These are all joined by 'V'-form splices and glued.[147] The Yrzi bow has four sections, two for the limbs and two laterally applied grip sections scarfed together (Fig.2). Only the wood of the latter, oak and elm, was identified by Brown.[148] The Qum-Darya bow was not examined before the grip area was lost. Both ears, however, seemed to have additional wooden strips attached to the back of the core. From published drawings (Fig.4) it is not clear how these pieces were related to each other but they may have served to deepen the ear profile ('stacking' them).[149] The resultant rectangular section, quite different from that of the dustārs, contributed to the lever function of the ears. The evidence indicates that the cores of the bows used by Roman forces might have had continuous cores from ear to grip with a separate piece(s) for a stiff handle. Fish-tail splicing was employed by Islamic craftsmen to achieve the necessary core curvature in preference to shaping with heat.[150] A combination of spliced grip section and heat-curved limbs may have been the Roman period practice judging from the occurrence of Roman grip laths. These would have been made redundant by the Yrzi scarfed method. The wood must absorb glue well and the core must be very carefully assembled and shaped to prevent lateral twisting and fracture during the draw.[151]

Glueing is the method used throughout the construction for attaching the components to the core. These would have been bound together during the setting process but neither binding nor other methods such as nailing play a part in the final stave cohesion. The glue itself is produced from tendons simmered in rainwater. The liquid is strained off and then evaporated to a viscous solution then cooled and gelled so it may be stored indefinitely. In this viscous state it is clearly visible between the laths on the back on the Belmesa ear (Fig.16). It only has to be heated through again for use. Variants of fish glue were preferred in the Ch'engtu workshop and for slower setting Turkish bowyers favoured glue made from the roof of the mouth of the Danubian sturgeon. Heating at all stages of construction is done over charcoal braziers.[152]

Judging from the Belmesa ear and the longest Bar Hill lath the tips of the core should be extremely fine and triangular. Von Groller calculated that the Carnuntum lath with a nail formed a pair enclosing something no more than 3mm in width and the longest Bar Hill lath had an even finer tip inserted into it.[153] As already remarked, the rivets had no constructional role.

Both bone and antler materials were employed for ear laths but antler is much the tougher of the two. It is less likely to break under sudden loads and less vulnerable to weakening by small notches and nicks cut into it or accidentally incurred in use. Several Caerleon bone laths broke along lines made by non-functional knife-cuts. There is evidence to suggest a preference for antler over bone in some Roman contexts for these reasons.[154] However, it is often very difficult to distinguish the two materials in Roman laths. Most reports assume bone, but both were used at Caerleon and Carnuntum. The Bar Hill and London examples are all antler and their waterlogged contexts have given them a characteristic staining not present in the Caerleon deposit. Only when the cellular structure is visible can definite conclusions be drawn because bone laths may be cut from the thick, outer material of long bones avoiding the weak, springy inner cells. In the finished bow the choice of material might not have been very significant but if it was dropped or given a hard nock the laths might be more easily chipped or cracked if made of bone.

It is likely that the ear laths were applied to the ears of the core before the other materials. The Belmesa laths are overlain by the belly-horn and the sinew backing. Most unusually they do not have a back zone of scoring up to the nock but with nearly all the other Roman laths this feature is probably designed to give a better purchase to the sinew material. The zone of belly scoring near the lower end of the longest Bar Hill lath and one of the Caerleon pieces strongly suggests they were overlain by horn. The laterally sawn sections on the convex faces of the other Bar Hill laths may have been intended simply to present an unstepped surface for the sinew on the sides of the stave. The position of the grip laths in the constructional sequence is less clear. Those with splayed ends may have been applied directly to the core with the laterally sawn sections overlaid by the belly-horn, the sawn sections making a less broken surface for the overall binding (Fig.1). The steep long edges of some of these laths appears to be scored so the sinew backing on the sides of the stave probably abutted against them. Thus the grip lath would be glued on after the horn but before the sinew. The Yrzi bow's grip had horn along its length (Fig.2) as do more modern bows.[155] Most of the Roman ear and grip laths, like those in position on the Yrzi bow, are heavily scored on the flat face in order to improve glue adhesion.

Filing and smoothing of the core in addition to heat-shaping is carried out before the next constructional stage. The cooler weather and higher humidity of autumn in Taybughā's scheme allows for a slower glue setting rate, imparting superior adhesion to that possible in hot, dry weather.[156] During this process the horn for the belly must be prepared. It must only curve on one plane, not twist and must be long enough to serve most of, if not the entire length of one limb. Experimentation has shown that horn from Western domestic animals tends to be too friable and splits into thin layers when subjected to stress.[157] In Roman contexts the Celtic shorthorn (Bos longifrons) would have been quite unsuitable. In the Levant the long-horned caraboa or Indian Buffalo (Arni, Bos bubalis) was available together with the Armenian Wild Sheep, or Asiatic Mouflon (Ovis orientalis typica) and wild or domestic goats.[158] Two strips from each animal horn, on the outer and inner curves could be used. It is unclear how the horn was laid on the Yrzi bow because so little survives. On fragment B of the Qum-Darya bow (Fig.4) two strips of horn are present, the piece on the ear underlying that on the dustār. Perhaps longer horn was not available to the bowyer. The horn on the Yrzi bow is only 2.5 to 3mm thick. The outer keratin layer is employed so the horn is perishable, prone to insect damage and semi-translucent.[159] Horn on the Belmesa ear is a light brown in colour. On Turkish and Sino-Tatar bows it is black or dark grey.[160] The short dustārs associated with bows on Sassanid silver dishes may indicate the use of Mouflon horn which is shorter than Buffalo.[161]

For the horn to adhere to the core both surfaces are heavily scored as seen on the Yrzi bow and which may be assumed for all bows in Roman use. Scoring is visible on more modern bows and was also practised in the Ch'engtu workshop.[162] When the horn is applied the core is strung in a reversed, squared 'C' shape (Fig.5.2). Turkish bows with siyāt had 'false nocks' cut in the belly edge of the ears for this reversed stringing. It is not impossible that these nocks were employed on bone ear laths, especially as the ends were butting bone without other materials between them. It is unlikely that this was the case, however, because of the absence of false nocks on the Caerleon laths which were in all probability never applied to bows. One piece, assembled by Nash-Williams, appeared to have a nock on the belly but this may either have been wrongly put together or may have been another feature of the low level of competence exhibited in the assemblage.[163] This piece is in any case shorter than would be expected because later in the constructional process the end with the false nock was cut away and a 'true' nock fashioned

(Fig.5.4).[164]

When the stave is increasingly reflexed for the sinew application the horn strips butting at the grip are inevitably pulled apart. The resultant gap must be plugged to prevent harmful shearing stresses when the bow is later strung and drawn. A bone, ivory or hardwood piece is inserted (Fig.1) on Turkish and other Oriental bows (Arabic ibranjak; Turkish çelik)[165] and if made of wood it would not survive in Roman archaeological contexts. It may also contribute to the stiffness of the handle. The Chinese bowyers bound the horn to the core for a day and a night then removed the binding and left the glue to set for four to five months before polishing and rubbing the horn surface.[166] They insisted on June for the binding and a setting period extending to November which roughly corresponds with Taybughā's autumn stipulation.

These two sources are at variance, however, on the season for sinew application. For Taybughā, working on a one-year timetable, the sinew was best applied in early spring but the Chinese chose winter with an obligatory commencement in October. They spread their glueing over three years using the first and second autumns' for glue-setting, the last winter and spring for sinew laying and setting.[167] The sinew application is the most important stage of the whole process, this substance determining the final recurve. Sinew is also the material most affected by temperature and humidity variations. Defects in the wood and horn could be rectified to some extent at this stage. Spring was chosen by Taybughā for its warmer conditions, inferior elasticity resulting from application on a cold day.[168]

Practical experience demonstrates that the best sinews come from cattle or deer leg tendons.[169] The Ch'engtu craftsmen used cattle back tendons deeming it important to remove them just after slaughter to facilitate the separation of flesh before it cools.[170] The sinews are dried, combed out, pounded with a wooden mallet into separate flax-like fibres, and graded by length. At this stage they are hard, translucent and stiff. They are then soaked in warm glue. The bow is heated and the sinew is applied onto the back of the wooden core by hand and combed into a homogeneous layer. The core must be deeply scored for adhesion beforehand. When the first layer is dry a second and even a third is laid on. These are applied to the back of the dustārs, the back and sides of the grip and more or less 'stacked' on the ears. With each layer the stave is increasingly recurved in order to elicit the maximum stretch when the bow is strung. Only with the final layer is the full-reflex achieved (Fig.5.3).[171] With small weapons, such as the Scythian bow, the ears would probably touch and even overlap, as with a Turkish flight bow in the Pitt Rivers Museum, Oxford.[172]

The back zone of scoring on Roman ear laths may be intended for the adhesion of sinew. The Qum-Darya fragment A has sinew on the back and sides of the ear up to approximately 3cm from the upper end (Fig.4). The Belmesa ear has traces of oozed glue on the back of the laths but the sinew only extended up as far as the point where the laths straighten out, level with the termination of the horn. On both laths knife scorings appear on the convex face (Fig.3) suggesting a trimming or tidying up of the sinew at that level. This explains the lack of back edge scoring. Traces of sinew overlaying the edges of the horn on the belly and at the lower end of the laths confirm that sinew was applied to the sides as well as the back of the ear.

After two to six months or more the sinew is hard enough to be filed and the exuded glue removed.[173] Shaping continues during the 'tillering' process as the bow is gradually bent, using heat to increase flexibility, and eventually the stave is strung (Fig.5.4). Careful study and adjustment accompanies the slow drawing of the bow and experimentation with the comparative physical performance of the limbs. Once the stave has been strung the true 'C' of the constructional reflex is lost as the sinew partially adjusts to the stretch. The bow is drawn on a 'tiller', a wooden beam which holds the handle at one end and takes the string in notches at intervals along its length. Filing and adjustment commences after the bow has been on the tiller for a few hours and after it has been sun-dried or brazier-warmed for hours to expel moisture.[174] Thus the stave can be monitored at measured stages and the draw-weight is estimated. This process may be carried out in consultation with the archer if the bow is custom-made.

Work on the ears would be completed before stringing. False nocks would be cut away from the tips. If the ears of a bow are designed to angle forward at rest, as was the case with Avar weapons, 'string bridges' are necessary to prevent the bow 'capsizing'.[175] The 'string guide' on the Mainz lath and other Hunnic examples might also be cut at this stage. Roman, Parthian and Sassanid bows do not seem for the most part to have had such ears, a point suggested by the Belmesa ear which has a leather-covered sinew binding above the nock (Fig.3). This was probably intended to bind the ear assemblage more firmly together preventing it from coming apart with the string's impact. Where the binding has come away scoring for additional purchase on the polished bone is visible. A Sassanid silver dish may depict a bow similarly bound below the nock (Fig.44). On Medieval bows this binding was termed the 'aqab' and a leather-covered example survives on an 18th century Bashkir bow.[176]

Islamic and Chinese sources recommend strings of silk or sinew for cold and humid climates, intestine strings only for hot conditions because they stretch when damp. For all weathers the best are hide strings, especially from a young emaciated camel, followed by wild ass and deer. Goat hide is only good in warm climates because it also stretches. Vegetable fibres could also be employed. A thick string increases accuracy and on a powerful bow it must be strong enough to withstand the tremendous strain of checking the forward movement of the limbs. A wide safety margin was allowed for war bows.[177]

Next, the stave is bound. Back edge scoring on Roman laths (and the scoring all over the convex face of one of the Caerleon grip laths) may in fact be for the adhesion of binding rather than exclusively for the sinew backing. The Yrzi bow has a sinew binding surviving particularly around the grip (Fig.2). Neck tendon could be used for this and it is also used as additional siyah stacking on more modern bows.[178] Sinew binding appears on fragment B of the Qum-Darya bow and it may have retained the surviving ear lath fragment (Fig.4). Over this, or instead of it, a binding of glued birch bark was universally popular in the Levant and China. This substance does not stretch during the draw and is laid spirally, taking glue as well.[179] Sinew and bark may bind the whole surface as with Persian and Indian bows, or the horn belly may be left exposed as with Mamluk, Sino-Tartar and Ottoman flight bows.[180] Sassanid bows are depicted as covered up to the knees but not it seems on the laths (Fig.44). The bow in Scene CVIII of Trajan's Column curiously does not appear to be bound at all, nor perhaps are those seen in the Syrian hunting mosaics (Figs.23, 36, & 37).

Lastly, several coats or sandarac of lacquer were applied to the binding, perhaps with painted designs to taste. For what little it is worth the bow held by both attendants in the S. M. Capua Veteres (Italy) mithraeum tauroctony fresco has a convincing profile and is browny-yellow in colour. This suggests an unpainted binding.[181] The onager hunting fresco from Dura-Europos has an unpainted bow as do the Kizil and Pendzhikent Early Medieval paintings.[182] Apart from the aesthetic reasons, the binding and varnishing is carried out to protect the stave from moisture damage. Western European climatic conditions may have caused problems in this respect.

The tools involved in the construction process were very similar in widely different contexts. Turkish implements described by Kani bear close comparison with those employed in Ch'engtu.[183] The core work demanded a saw for cutting components, a knife for shaping them, a glue pot and brush, a small adze for smoothing surfaces, a wood file, a scoring tool and a brazier for drying and shaping. A horn working tool for polishing, a rasp, a scraping knife, a binding tightening tool and a pressing horse were needed for the belly application. For the sinew a mortar and pestle, a sinew comb and a glueing knife for scraping were necessary. For the general shaping of the limbs curved 'formers' may be bound to the dustārs. These also prevent the stave from twisting during the tillering process. The Turkish tepelik was very similar to the Chinese 'movement tool'.[184] Likewise, the slotted destagh frame, used in conjunction with the tepelik for curving the limbs, performed the same function as the Chinese 'big board bench' which was a work-bench with a slotted top also generally used for sawing, cutting and filing the stave at various stages (Fig.6.E).[185] The Japanese yumidame was of a similar design.[186] Drying frames, drying boxes, storage cabinets and charcoal braziers were used in common.[187]

In the Ch'engtu shop the horn binding demanded the hands of four people: a heater of glue, bow, and horn; a glue-spreader; a binder to apply the binding-cord; and a press-horse operator.[188] Application of the sinew would perhaps require three: a heater, an applier and a spreader to comb and mould. The workshop's staff of four consisted of owner-manager, the skilled artisan, the apprentice and the labourer. The owner helped during important stages such as horn-binding and the labourer performed unskilled tasks such as tending braziers and peeling birch bark. The shop itself (Fig.6) was 4.94m by 3.88m fronting onto a street with a shuttered front. The workbench (E) extended out the front with plenty of room to work. A storage cabinet (A) had a brazier burning in it in all weathers so that components could be dried out and valuable, completed bows be safely stored. A work table (B) was used for the general cutting and shaping of components. Tools and materials hung on the walls, materials and partly completed bows hung from the rafters, and materials were also piled on the floor in empty spaces (F).[189]

A workshop of course produced more than one bow per one, three, five or ten years. The materials could be stored and the physical work was intensive but not very time consuming, so bows could be made in large, staggered batches. With a staff of four the Ch'engtu shop was producing bows in batches of 50 during the 1940s. In the past, presumably with a larger staff, it had produced 500-600 bows in a batch.[190] Even when, latterly, there were very few customers the shop was busy every day with the preparation of materials. Klopsteg commented that if a bowyer newly started up worked on 100 bows in one year he would need 500 bows in preparation before he could do business.[191] This represented a large investment of materials and time, a high cost product and, in urban centres at least, a stable industry. What this meant in terms of nomadic steppe peoples moving around in wagons and living in felt tents can only be guessed at, although the itinerant nature of such societies can be over-emphasised.

A distinction must always be drawn between bowyer and archer beyond that of producer and customer. Whilst the bowyer might be skilled in shooting, particularly for stave-testing, most archers were technicians in the use of the bow, rather than its construction. The archer was capable of minor repairs and maintenance with regard to the string, the stave's binding and producing and repairing arrow stele and fletchings.

Before the establishment of large, centrally organised fabricae three levels of Roman equipment production might be rather simplistically postulated for the East. The production of large items such as bladed weapons and armour would have been done in areas of 'civilian' industry, particularly in the cities of Syria. These products would then be moved through negotiatores to the legionary fortresses on the second level where large fabricae would make supplementary pieces and artillery, plus shields, shafted weapons, missiles etc. in bulk. These would be distributed to the smaller forts on the third production level, the fabricae of which might be capable of making large items from scratch but would specialise in smaller scale manufacture (arrow-heads, spear-heads, caltrops, mail rings, scales etc.) and running repairs. On this level the scrupulous recycling of materials from broken pieces would be an important feature. In the West the fortresses were the first level though civilian manufacture may have played a part as urban development increased. Obviously these levels blur together, particularly with troops stationed within Eastern cities and once a unit had been raised and equipped further production would consist of repairs and topping up the equipment pool. The vici also had a role in manufacture and most likely in the decoration of pieces to individual order.[192]

The secure identification of bow manufacturing sites is difficult. Nothing of the Ch'engtu workshop would survive in the archaeological record, except an anonymous shop groundplan and perhaps some iron tools. The latter were not particularly diagnostic and could be mistaken for textile industry or wood-cutting implements. In the case of Sino-Tartar bows bone laths were not used so nothing of the workshop products would give a clue.

The rather ineptly worked and often unfinished Roman laths from Caerleon suggest a manufacturing context rather than just a weapon-store. The finds associated with other legionary fortresses suggest by analogy that such sites were generally producing bows. By the time of Commodus, and probably much earlier, the legionary immunes included arcuarii and sagittarii, bow- and arrow-workers.[193] Vegetius included bows and arrows in the production activities presided over by the legionary _praefectus fabrorum_.[194] Unfortunately none of the Corbridge laths were found in association with the legionary workshops of the Severan campaign supply base. One piece was in course of manufacture, however, when rejected. For campaign purposes some construction of spare bows would have been necessary though bow workshops might have had stocks of completed weapons already. The mass-production of arrows would have been far more important and this is exactly what the Corbridge _fabricae_ were engaged in.[195]

The construction of bows was a specialised task but the possibility of general legionary production plus the state of the Caerleon laths excites the suspicion that the level of skill and competence involved here was not of the highest order. The situation may have been very different in the context of specialist _alae_, _cohortes_ and _numeri sagittariorum_ (Calibre below). The construction and maintenance of _c_. 480 or _c_. 800 composite bows would have been a major undertaking. It is reasonable to assume the existence of specialised _fabricae_ centred round master-bowyers, with their apprentices and assistants, in the forts housing such archer-units. The evidence for bow construction at Intercisa is important but the late antler-working industry and possible Hunnic associations makes a link with a series of known Roman units of _sagittarii_ problematical.[196] The competent excavation of Carvoran (Northumberland) on Hadrian's Wall, repeatedly occupied by _cohors I Hamiorum sag_., might be informative in this respect. It is not impossible that when such units were originally raised in the Levant the _tirones_ brought their personal weapons with them because eastern recruitment in itself supposed prior skill and a cultural background of archery.

The general commercial and manufacturing industries in the Walbrook area of Roman London might suggest that the laths here were associated with on-site bow construction, though the pieces themselves exhibit every sign of completion.[197] The Mainz laths may partly be associated with the Mainz-Weisenau fort garrisons but the 3rd century _canabae_ antler-working industry suggests either civilian bow construction or the extension of production work outside the fortress to gain additional space.[198]

The areas most active in bow construction in the East would have been Syria, Arabia, Palestine and the client states. Bows were a major hunting weapon throughout the region to judge from the pictorial record (Roman Representations and Comparative Evidence, above). Developments in tactics, armour and fortifications were all dominated by archery. Cities such as Palmyra and Hatra were more than capable of producing bows for mural defence and the equipping of the horse-archers.[199] Within the Parthian Empire bow workshops have been identified at Merv (Ferghana)[200] and Toprak Kala (Transoxiana).[201] As in China and in the Medieval Levant the bow manufacturing industry was urban-based.[202]

In the Western provinces bow construction is likely to have been carried out in the legionary fortresses for general training and mural defence for legionary and auxiliary troops (Bow-Armed Units, below). The fabricae of specialist sagittarii would have met the needs of such units except, perhaps, the mass supply of arrows for campaign purposes. All the materials for composite bow staves would have been available except the belly horn, there being perhaps no suitable supply from wild or domestic Western animals. This reason, plus the climatic conditions was put forward by Paterson to explain the failure of medieval Western peoples to adopt composite archery when faced with its effective use by invading steppe peoples.[203] Composite bows do appear in Frankish, Alammanic and Lombard graves and there is evidence for Carolingian use but no suggestion that these peoples either manufactured their own composite bows or that they adopted them widely from the Huns, Avars and Magyars.[204] This is in contrast with the effects of contemporary Asiatic archery on Byzantine and Islamic states having direct access to horn materials. It is true that composite cross-bows were in Western use from the 9th to 14th centuries A.D. and that some suitable horn was found in the Baltic region, but the most common belly material was whale-bone acting, it may be said, as a horn substitute.[205]

The establishment of the centralised fabricae, itemised in the Notitia Dignitatum, may have been an extra production level superimposed upon the earlier situation rather than replacing it. These would have supplied the needs of mobile field-army units, especially for bulk orders and the larger classes or armour. None of the eastern fabricae are specifically designated as bow and arrow producers but in the West Ticinum (Pavia) is labelled 'arcuaria' and both Concordia (Concordia) and Matisco (Maçon) are 'sagittaria'.[206] Possibly Ticinum acted as an inlet for horn materials from the Levant and the two arrow-producers supplied arrows in bulk for the field-army and built up campaign reserves. The appearance of three such fabricae in the West but none in the East must be entirely due to the long Levantine cultural tradition of military and hunting archery. Significantly large elements of the fabricenses of the Ticinum and Concordia workshops were of oriental origin.[207]

7. ROMAN CROSS-BOWS

The bows so far discussed are 'hand-bows' as distinguished from 'foot-bows' which are more commonly referred to as 'cross-bows'. The foot-bow is so called because the stave is held down on the ground by the archer's two feet (Medieval Latin arbalista ad duos pedes; Arabic qaws ar-rijl) or one (arbilista ad unum pedem; qaws ar-rikāb) whilst he pulls back the string using his arm and back muscles. The former type is more primitive involving placing the stave actually on the ground whilst the latter employs a stirrup. The stirrup method was a later medieval development with advantages of avoiding stave contact with potentially damp ground and of allowing use in the saddle. The stave is lashed onto and set into the end of a wooden stock or 'tiller'. The string is pulled back along the tiller and locked on a trigger nut mechanism.[208]

The direct evidence for the Roman use of cross-bows amounts to two clear sculptural representations of 3rd century AD date, references by Vegetius and possible mentions in Byzantine military manuals. Comparative evidence from later periods to some extent illuminates this material.

A Gallo-Roman relief from Salignac-sur-Loire, in the museum at Puy, Haute Loire, depicts a hunting dog, a knife, a quiver and a clearly represented cross-bow hanging up.[209] The stave is unstrung with reversed limbs strongly suggesting a composite structure. The tiller is grooved with an eliptical stock at one end and possibly a delineated trigger nut. It is likely that the tiller is foreshortened to fit into the restricted panel space so no conclusions may be drawn with relation to the type or length of projectile used with the weapon.

A second relief, from a Gallo-Roman villa at Espály, also in the Puy Museum, depicts a hunting party with two pack animals (?), two men and a dog.[210] The sinister man rather awkwardly carries under his arm a long quiver and a cross-bow. The tiller is similar to the Salignac example except that it is of a larger scale and the trigger area is obscured by drapery. Although strung, the recurving of the surviving limb again denotes composite construction. The quiver is approximately the length of the tiller minus the eliptical stock, and this would imply the use of normal projectiles, not of short heavy quarrels.[211]

The gastraphetes of Heron of Alexandria was similarly a very large bow attached to a tiller but with a curved stock.[212] The latter fitted the archer's middle and a 'slider' was employed to push back the string until it could be locked by the trigger mechanism. The operator would prop the end of the slider firmly against a wall or the ground and exert his weight against the stock pushing with his stomach (hence 'belly-bow'). The weapon would have been heavy and unwieldy but would have needed only one operator and no stand. The major drawback inherent in the stave's composite construction would have been the extreme length of the horn belly materials needed. Extra layers of sinew on the back would have given the desired increase in strength. The paucity of evidence for the gastraphetes apart from Heron's description might suggest that light torsion bolt-throwers were preferred to this weapon. The period in which it was used, if ever, is unclear because Heron drew from Hellenistic sources and probably wrote in the Late Roman period.[213]

The small-scale fittings from bolt-shooters found at Orsova and Gornea on the Danube, dating from the late 4th century A.D. have been equated with Heron's cheiroballistra, effectively a torsion artillery version of the gastraphetes.[214] The differences between the two sets of finds suggest a contemporary variety of models in use. Like the belly-bow these demanded a solid surface against which to push back the slider and would be heavy and difficult to handle. Their forte would have been in mural defence with a parapet on which to rest the weapon for aiming and shooting.[215]

Several times Vegetius refers to a 'manuballista' which may be equated with the Danubian finds and with the cheiroballista in scale and name.[216] In a passage discussing the legionary order of battle Vegetius describes a support line of tragularii, grouping together arrow-shooting manuballistae and archuballistae.[217] 'Arcuballista', etymologically similar to 'manuballista', with its bow element may be compared to the Gallo-Roman weapons. A trigger nut was found at Carnuntum (Austria), although a Roman date for it is not secure.[218]

The Late Roman units designated 'ballistariorum' were suggested by Marsden to have consisted of large numbers of carroballistae accompanying mobile field-armies.[219] This view was partly based on the assumption that they were formed by pooling legionary artillery and represented the only available engines, at least in the West. This scenario is unlikely because it would not make for 'mobile' units, moving as they would with carts as slowly as baggage trains. One possibility is that such units were equipped with the whole range of artillery types, the heavier pieces being detached to important bases for mural defence. The light variants would have served in the field. An 'artillery legion' would have been of little use as a block of men and heavy machines on the battle field and would demand parcelling out along the battle-line. A unit of cross-bowmen would have been a different proposition. Julian's oft-quoted journey with an escort of catafracti and ballistarii is readily understandable if the latter were arcuballistarii giving the necessary mobility.[220] A lack of post-conflict pursuit might be put down to tired, slow-moving cataphracts and a desire not to throw forward unsupported missile-troops. With regard to manuballistae, Vegetius' line would have demanded the lightest weapons of the Gornea type and firm ground for slider movement.

The arcuballista would, however, have been most effectively deployed in mural defence, except on the eastern frontiers where enemy cavalry was heavily armoured. Interestingly Taybughā commented of cross-bows that "my own view is that in the manoeuvres of combat, in the desert, and on expeditions the hand-bow is a better and more serviceable weapon whereas in fortresses, sieges and ships greater power and advantage will be derived from the cross-bow".[221] The possibility that not all ballistarii were creamed off from the Legiones to the field-armies, and that some remained is suggested both by Vegetius' order of battle and by the provision of combined legionary artillery units in Danubian bridgehead forts.[222] In the latter static garrisoning by long-range missile troops commanding the areas around river crossing, as specifically recommended by Vegetius, would have made great tactical sense.

Maurikios' Strategikon (later 6th century AD) refers to solenaria which go with small quivers and arrows and have a long range. Similarly, Leo's Taktika and Problemata (end of 10th century AD), and the anonymous Sylloge Taktikorum (10th century AD) mention solenaria with small arrows.[223] These all suggest the field-use of a weapon most suitably interpreted as a cross-bow. As such, the solenarion might represent a continuous use of the arcuballista right through into the 10th century AD.

Whether or not cross-bows had a history of use by the Romans prior to the 3rd century AD cannot be determined on the present evidence but they were not necessarily derived from the gastraphetes. As a hunting weapon the cross-bow had distinct advantages over the hand-bow. It could be shot and cocked (by the archer lying on his back) from a concealed, lying position. It could be held cocked thus ready for use immediately game appeared. It delivered an arrow or quarrel over a greater distance, with heavier impact and with more accuracy (at least without demanding skilled marksmanship from a novice) than a hand-bow and was especially useful for fowling.[224]

In warfare cross-bows were most useful in marine-conflict and in mural defence where advantage could be taken of their greater range. The latter feature would not compensate for their slow rate of shooting in a direct field confrontation with infantry handbow-men but might very well do in exchanges with horse-archers. The penetrative qualities of cross-bow missiles hitting armour would have been most useful in the Parthian and Persian wars.

If a tiller was widely used in bow construction (Construction above) the cross-bow development would not have been a huge innovatory leap of imagination. The main prerequisite was the provision of a trigger mechanism and hints for this may have been provided by contemporary artillery practices or, more likely, such a device represented a new approach. A trigger nut remarkably similar to Medieval European examples had been developed in China by the late 3rd century B.C.[225] The Chinese cross-bow had a composite stave and a variety of strength classifications. It was an exclusively infantry weapon developed for frontier defence as an answer to Asiatic horse-archer adversaries. This is not to suggest any technological link between Roman and Chinese empires, but simply to draw attention to an interestingly parallel development.

From the Gallo-Roman representations it may be inferred that cross-bows in Roman and Byzantine use had composite structures. The necessary stave length was shorter than for hand-bows and the increased power would have been provided by additional sinew backing, say up to five layers. In the West cross-bows appear on 9th and 10th century A.D. Pictish stones and were in continental use by the late 10th century.[226] Subsequent cross-bows were generally of composite structure employing whale-bone on the belly as a horn substitute, at least until the 14th century introduction of steel staves.[227] Wooden staves were also in use, the Pictish examples presumably being of thus material. Taybughā discussed composite cross-bows but recommended yew staves for ship-board use because of the detrimental effects of moisture on glue and sinew.[228]

II. OTHER EQUIPMENT

1. ARROWS

Arrows vary in their dimensions, weight, shape of fletchings and type of head according to the size of the archer, the use of the bow and the vulnerability of the target. The metal heads are the commonest archaeological finds providing some information about these factors. Very occasionally the organic materials making up the stele, nock and fletchings (Fig.8) survive in arid contexts. Artistic representations are usually of little use because of scale or the omission of fletchings which might obscure an archer's face (Fig.44). A few literary mentions provide evidence for types of wood employed.

Detailed examination of arrow-head types in this study is rendered unnecessary by the work of Erdmann and Davies.[1] The British material does repay review because of its geographical distribution with relation to known sagittarii and the incidence of laths. Arrows are easily lost and likely to appear in the archaeological record wherever archery was practised. The arrow-heads which appear in British contexts are all of iron and may be grouped in seven broad categories.

1. Trilobate tanged: Characterised by three vanes, barbed or otherwise, c.3-3.5cm long without the pointed tang. Davies suggested an early Scythian introduction of the type into the eastern Mediterranean region.[2] The employment of oriental sagittarii from an early date ensured its ubiquity in use by archers of whatever origin in the Imperial period. Although Scythian arrow-heads were characteristically of cast bronze and socketed, the tanged trilobate (or quadruple-vaned) iron head was of Central Asiatic origin spreading eastwards into China and southwards into Achaemenid Persian, Parthian and Syrian use. Tanged trilobate heads appear in Sarmatian, Hunnic and Avar contexts.[3] Within the Roman East many examples occur on Palestinian sites associated with the two Jewish Revolts (Fig.46).[4]

   Davies lists British provenances: Ham Hill (Dorset), Hod Hill (Dorset), Kingsholm (Gloucs.), Richborough (Kent), Margidunum (Notts.), Wall (Staffs.), Wilderspool (Cheshire), Corbridge (Northumb.), Turret 25b on Hadrian's Wall (Northumb.), Newstead (Roxs.), Bar Hill (Strathclyde).[5] To these must be added Gloucester (Gloucs.), Maumbury Rings (Dorset), Walbrook (London), Brecon Gaer (Brecon), Chichester (Sussex), Watercrook (Cumbria), Ebchester (Durham), Godmanchester (Hunts.), Carlisle (Cumbria), York (Yorks.), Caernarvon (Caerns.), Burnswark (Dumfries) and several examples from Housesteads (Northumb.).[6] This list has no pretensions of completeness but demonstrates that these heads are a very common find on military sites.

Post-Antonine pieces from dated British contexts are few as seems to be the case on the German-Raetian limes.[7] In the East the type continued in use as the finds from Dura-Europos demonstrate, and it may perhaps be seen in Dura graffiti and on Sassanid dishes.[8]

2. Socketed, vaned: A socketed head with three or four barbed vanes c.10cm overall. This represents a socketed variant of the trilobate, tanged and probably replaced it in Britain by the early 3rd century A.D. In the Severan workshop III at Corbridge three tanged, quadruple-vaned heads were found with five socketed examples.[9] Hearths and tempering tanks suggest production on-site and a total of seventeen heads were found with javelin-heads, pilum-heads and spear-heads. The predominance of the socketed type is suggested by the late 3rd century deposit in the rampart-back building at Caerleon which included twenty-two examples with three or four vanes.[10]

3. Tanged, 'bodkin': A type with a square or triangular cross-section, c.4cm long. Erdmann lists British examples from Newstead (Roxs.), Corbridge (Northumb.), Poltross Burn (Northumb.), Kirkby Thore (Westmoreland) and Richborough (Kent).[11] This type of arrow may have been shot against ox-skull targets found at Corbridge and Chesterholm (Northumb.) but it occurs much more frequently on the Continent (Calibre below).

4. Socketed, 'bodkin': A long, slim head, square or rhomboid in section, c.5cm long. Few occur in Britain, rather more on the German limes. An example comes from Richborough (Kent).[12]

5. Socketed, flat-bladed: Much simpler in construction than the foregoing types, an example from Watercrook (Cumbria), c.5.1cm long, has a flat, triangular blade and was affixed to the stele with a pin. At Dura-Europos similar heads were used alongside the trilobate tanged but the latter were in the majority.[13] Sockets were glued to the stele, not pinned.

6. Tanged, flat-bladed: The blade of this type varies from triangular to leaf-shaped and may be barbed. This category might in many instances best be termed 'improvised'. Some 800 were found in room 12 of the Housesteads (Northumb.) principia in a 4th century context.[14] Some still had wood enclosing their tangs and had been arranged in bundles.

Scrap-metal suggests hammering into shape on the spot and they are very crude compared with the trilobate, tanged heads found elsewhere in the fort. Similar examples occur at Richborough (Kent).[15] The later date, simple design and local manufacture at Housesteads suggest improvisation for mural defence.

Quite different is a group of 44 tanged heads from Bearsden (Dumb.) on the Scottish wall, dating to the Antonine period.[16] They are characterised by a tang, square in section and a flat head with a triangular blade. The only approximate contemporary parallel is a head from Vindolanda (Northumb.).[17] They are reminiscent of the triangular point on the Dura-Europos incendiary ballista-bolt head.[18] The writer can only suggest that the Bearsden group was either locally produced for mural defence or was manufactured simply for hunting purposes. When sharpened this type would accord with heads recommended for soft-skinned targets.

7. Fire-arrows: Five examples were found at Bar Hill (Strathclyde) on the Antonine Wall, 5.2 - 6cm, and one at Wroxeter (Shrops.), 7.6cm long.[19] They consist of a tang joined to a short point by three outwardly curving bars. There is no doubt that these are fire-arrow-heads, of a type described by Ammianus and Vegetius, termed *malleoli*.[20] Davies' suggestion that they were constructed to reduce the time and skill needed for trilobate, tanged heads is mistaken because the bars are not sharpened, would hinder penetration and would buckle on impact.[21] The real problem with these projectiles is the tendency to be extinguished when shot.[22] The British examples are closely related to the Dura incendiary ballista-head in structure.[23]

Shades of local variation in dimensions and proportions are to be expected within these types.

The Elder Pliny remarked upon the use of reed for arrow stele: "The people of the East employ reeds in making war; by means of reeds with a feather added to them they hasten the approach of death, and to reeds they add points which deal wounds with their barbs that cannot be extracted... And if anybody should make a rather careful reckoning on the Ethiopians, Egyptians, Arabs, Indians, Scythians and Bactrians, and the realms of the Parthians, almost half of mankind in the whole world lives subject to the reed."[24]

During the sixth season of work at Dura-Europos three posterior ends of reed or cane arrows were found, 27.5cm, 21.5cm and 21cm in length (Fig.45).[25] The stele are without taper towards either end, the first two being 1cm in diameter, the third 0.95cm. The longest has a nock 0.95cm deep with the tips and sides rounded off. The end was first bound with glue-soaked sinew for 2.5cm, then the nock was cut so that its sides and base were strengthened by this binding against this thrust of the bow-string. Three slightly ballooned vanes, 15cm long, 1.1cm high, fletched the arrow with the cock feather aligned with the string. A sinew whipping extends along the waist to give a glue purchase on the reed for these fletchings. A red line encircles the stele on the waist. On the second stele the fletching are badly damaged but in other respects it is similar to the first, with the addition of black painted lines and red circles on the waist. The vanes are 14.3cm in length. The third piece has lost its fletchings, which were 15cm in length, but has white-rimmed red circles, red dots, and black and red bands. These markings were presumably intended to identify ownership and/or an equally matched set of arrows.

Reed or cane stele presuppose tanged heads. However, for the socketed heads at Dura wooden piles were used (Fig.45). Two identical examples were found with the three stele discussed above, cut from tamarisk shoots.[26] One was intact, 17cm long, 1.1cm in diameter at its widest point. A dowel, 0.5cm long and 0.2cm in diameter acts as a 'tang' to be inserted into the reed stele. The remainder of the length from the stepped stele tapered to a 0.6cm diameter end onto which a socketed arrow-head would have been placed. The last 2.5cm has traces of glue remaining and there is no evidence for pinning.

At Masada (Palestine) Yadin found "hundreds of arrows in such places as the middle terrace of the palace-villa, the western palace and elsewhere, literally in heaps where they had been piled and intentionally set on fire. Only very few weapons for hand-to-hand fighting like the sword and spear, were found at Maṣada; the most effective defence weapon on the walls of the summit was of course the bow and arrow."[27] Roman troops would have removed any metallic weapons that they found, accounting for their absence. Some stele were found made of wood and some heads had very long tangs, presumably to distribute the shock of impact and to increase the purchase within the stele (Fig.46). Shafts from the Second Jewish Revolt have also survived in arid conditions. These exhibit an unusual two-part construction; the half of the stele with the head was wooden, the half with the fletchings was of reed (Fig.8).[28] An arrow from the Cave of Letters has only its wooden half and with it a small, trilobate, tanged head, the vanes of which are flush with the sides of the stele leaving in effect no barbs. The remains of a gut binding are in place at the other end and this would leave attached the reed section with fletchings as seen on better-preserved examples from Nahal Se'elim and Wadi Muraba'at. The wood is of the <u>Tamarix</u> species and was sharpened to a point so as to be driven into the reed. The wooden section was clearly intended to lessen the possibility of the tang splitting the stele on impact. In both Jewish Revolts the Roman targets would have been metal-armoured.

In comparison a number of tamarisk stele, 70cm long, were found in the same cemetery as the Qum-Darya bow. Lighter, reed examples, 56 - 65.5cm long, were also discovered. Both tanged and socketed heads were present and some painted stele decoration.[29] These and the Dura stele indicate the use of thin bow-strings in contrast to some Sarmatian and later Islamic nocks which were more bulbous to accommodate thick string.[30]

In the Roman West stele do not survive so well but probably less use was made of reed than in the East. The Sarmatians used cornel-wood arrows, the Germans employed pine and hazel.[31] Fabian used heat-straightened reed for his Magyar reconstructions and birch was common in Medieval Europe.[32] Ammianus mentioned reed for malleoli.[33] Taybughā recommended vulture feathers as the best material for fletchings, followed by eagle.[34] The Ch'engtu workshop used vulture or eagle feathers, and white poplar for the stele.[35] Head and fletchings were attached to the stele using glue-soaked sinew-whipping.

Wood still adheres to the tangs of some of the Housesteads 'improvised' heads and inside the sockets of some of the Caerleon and Corbridge vaned examples.

The most efficient war arrow was a heavy one which would absorb all of the bow-string's thrust. This arrow would not have a great range, but if the head was of an appropriate type it would impact hard and penetratingly.[36] Hunting arrows had a maximum of sharpened edge for use against soft targets which would be hit with very little shocking power but deep penetration. They did not drop large animals but caused heavy haemorrhaging.[37] The Bearsden arrows would have been effective in this way. Bone heads were apparently employed by the Scythians, Sarmatians, Huns and Germans and these could have inflicted very messy wounds.[38]

War-arrows had quite different heads unless they were shot at unarmoured targets.[39] According to Taybughā "the most reliable and effective arrow-heads are those which are either triangular or square (in section). These are used in battle and for piercing all kinds of armour."[40] The penetrative qualities of various heads were laid out in the Moroccan treatise with reference to helmets, body-armour, shields, unarmoured man and game. Barbed heads were recommended for the last categories.[41] The barbed, trilobate form in Roman use might not be thought of as the best type for armoured targets. The tanged 'bodkin' was an improvement but developmental impetus was not really present in the West where most of the barbarian adversaries were unarmoured.

The Elder Pliny, quoted above, remarked upon the difficulty of extracting eastern barbed heads from flesh. The barbed arrows shot by Parthian horse-archers at Crassus' legionarii were credited by Plutarch as having good armour (presumably mail in this case) penetration and they pinned hands to shields and feet to the ground.[42] Procopius described the removal of a barbed Gothic arrow from a face-wound, it being necessary to cut off the fletchings and pull the shaft out through the exit wound.[43] Procopius also states that Sassanid arrows "hitting a corselet, perhaps, or helmet or shield of a Roman warrior, were broken off and had no power to hurt the man who was hit." Although Persian bows were said to be weak, the Roman bows shot with such force "as to kill whoever stands in the way, shield or corselet alike having no power to check its force."[44] Without experimental reconstructions it would be dangerous to make inferences about arrow-types in use here.

The tang attachment method had the disadvantage of the stele tending to split on hitting a hard target. On the other hand a socket could snap off at the point where the collar ends. The Moroccan treatise was heavily in favour of the tang method.[45] The wooden piles from Dura and the wooden stele from Palestine lessen the incidence of both types of damage. Reed would be especially problematic and sinew-whipping was applied with this in mind. The Islamic sources speak of wooden stele and Taybughā also approved of tang attachment. "It (the head) must be so made that it is perfectly straight from the point to the end of the tang (sīlān), the end of the tang being narrower than its base. The wood of the shaft should be bored right in the centre to a distance that is less than the length of the tang of the arrow-head so that its end can be inserted into unbored wood in its natural state."[46]

Presumably, units of auxiliary sagittarii would have had specialist fabricae producing arrows for their needs. With the exception of the heads this work could have been done on a personal basis to make matching sets of arrows appropriate to an archer's draw-length. Mass-production is most likely within the Roman army, but more personal specification would have been possible in the Levantine 'private sector' of urban workshops. Legionary fortresses produced arrows for training and mural defence and most forts probably had a small supply for these purposes. The imperial fabricae at Concordia and Matisco (Maçon) in the Notitia must have produced them in bulk.[47] Garrisons under prolonged siege would have been constantly manufacturing projectile-heads in addition to using up prepared stocks and returning enemy missiles. Some of the cruder Dura-Europos finds may fall within the hastily-fabricated category.[48] Metallurgical analysis of one of the tanged, trilobate heads from Masada revealed no evidence of quenching or tempering, and manufacture was done from a piece of bloom iron with an uneven carbon content. "This arrow-head, hence, would be effective against animals, or humans protected by leather armour, but ineffective against other iron or bronze armour."[49] In the field the archer might be expected to have his quiver full of arrows made to his specifications. Thereafter he would have relied on mass-produced supplies from the baggage train. Surena's horse-archers pursuing Crassus were constantly resupplied from a train of 1000 arrow-bearing camels.[50]

2. BOW-CASES AND QUIVERS

Whenever a composite bow is out of use it must be unstrung to preserve the elastic qualities of the sinew. Thus when Surena met Crassus he unstrung his bow as a sign of peace. When Parthian horse-archers fraternised with Antonius' troops they pointed to their unstrung bows and when they eventually gave up pursuit they unstrung their bows and left the Romans unmolested.[51] Dampness is extremely detrimental to the glue binding the constituents of the composite stave. The sinew backing is very sensitive to moisture and temperature changes. Likewise the bow-string suffers especially if it is made of leather or gut. In sedentary conditions unstrung bows may have been hung up in a drying cabinet, as at Ch'engtu, or kept in a bow-case. In the field a bow-case of some kind was vitally important.[52]

There are numerous instances of horse-archers being rendered ineffective by sudden downpours. Frontinus recorded that rain demoralised Antiochus' forces facing Publius Scipio in Lydia and made their bow-strings damp and useless. Likewise bows used by Varus' troops were rendered useless by continuous rainfall. In A.D.624 the Eastern Turks were decisively defeated in battle by Li Shih-min (later T'ang T'ai-tsung) because of a rainstorm. In 838 Theophilus escaped capture by the Turks at Dazimon when a sudden shower neutralised his pursuers' bows. There is a tradition that the Ottonian victory over the Magyars at Lechfeld in A.D.955 was partly attributable to rain before the battle.[53]

Similarly, the glue attaching head and fletchings to the arrow stele may be adversely affected by dampness and in any case the fletchings are delicate and easily damaged by mishandling. Many forms of quivers enclosing part of all of the arrows have been developed by various cultures. In Borneo, for example, fletched darts were kept safe from the high rainfall by cases with wooden lids.[54]

Bow-cases and quivers would, therefore, have been very necessary items in the equipment of archers in Roman service, probably joining the numerous leather cases and covers needed for such items as musical instruments, artillery torsion frames, parade helmets, standards and shields.[55] Nothing survives archaeologically of these objects, Roman quivers apparently were lacking the metal fittings of some Avar and Magyar equipment (below). A number of Roman, Parthian and Sassanid pictorial representations of archers allow a reasonable estimate of the types of quivers and bow-cases in use by Roman forces.

The <u>gorytus</u>, a combined bow-case and quiver, was suspended from a waist-belt on the archer's left hip. It was developed by the Scythians, was primarily for cavalry use and took the small unstrung <u>Scythicus arcus</u>.[56] A quiver was attached to the outer face of the bow-case but the withdrawal of arrows with the right hand was very inconvenient because the top was angled backwards. In the later Medieval period Turkish horse-bows and arrows were of similar proportions to the Scythian but in spite of the use of cases for strung bows the Turkish quivers were always separate, hanging from the right hip, with the tops angled backwards.[57]

Judging from the Crimean frescoes and funerary reliefs the Sarmatians probably used the <u>gorytus</u>.[58] Therefore this is likely to have been employed by Bosporan <u>sagittarii</u>, by units of Sarmatian <u>dediticii</u> in the Roman forces and by Thracian archers.

Bow-cases do not appear in Roman military representations, a visible bow making a more interesting subject. All the military quivers are of a cylindrical type with parallel sides and round mouth. Generally this is worn on the archer's back on a <u>balteus</u>, with the mouth projecting above his right shoulder for ease of arrow withdrawal. This is a form useful for infantry, avoiding the danger of entangling quiver and legs possible with hip suspension. Archers in Scenes LXX and CVIII on Trajan's Column display this type of quiver and method of carriage. In Scene LXX conical caps cover the mouths as with the Housesteads archer's quiver. Presumably this is a measure to protect the fletchings from rain. Some form of lid may also be seen on the <u>ala Scubulorum</u> tombstone from Walbersdorf, an unusual instance of <u>balteus</u> suspension on a cavalryman. No quivers appear on the Marcus Column. The lid is commonly shown with deities and personifications with archery attributes.

The usual cavalry suspension-method of the cylindrical quiver was from the right side of the saddle behind the rider. The Mainz <u>eques singularis Augusti</u> has a quiver hanging vertically with no cap and the fletchings entirely visible above the mouth (Fig.29). The ala I Augusta Ituraeorum stone at Győr depicts a quiver, conceivably with a cap, with the mouth angled backwards in the later Medieval manner (Fig.30). This would demand two vertical suspension straps.

Eastern quiver practices went with eastern bow types in Roman employment as suggested by the Mainz ala Parthorum et Araborum tombstone which displays a curious, curved object where the quiver usually hangs (Fig.31). The sculptor's grasp of archery details was not good (Roman Representations above) and the object's profile may have been badly depicted. However, it corresponds most closely to the combined quiver and bow-case used by Crimean, Parthian and Palmyrene horse-archers. This was quite different from the gorytus because it served a longer bow, unstrung for storage, and comprised a cylindrical quiver hung vertically behind the saddle, with a 'sheath' bow-case attached to its side. Numerous Palmyrene caravan god reliefs of 1st to 3rd century A.D. date depict this type with varying degrees of decoration and stylisation (Fig.33). Sometimes the case is curved, sometimes straight, with angled ends for unstrung, angled forward bow ears. It always appears behind the saddle on the rider's right side.[59]

Palmyrene funerary banquet reliefs often depict pages with a cylindrical quiver hanging over one shoulder from a strap held in the hand. This gives the appearance of a quiver removed from the usual saddle attachment, not suspension from a balteus. When a horse is present a quiver is seen behind the saddle.[60] The pages normally bear the hunting equipment of the deceased who would have hunted on horseback with the bow.

Quiver and bow-case combinations appear on late Crimean tombstones, though it is unclear whether they are similar but independent development to the Palmyrene type, or whether there is some direct, possibly Parthian connection between the regions. Some proportionally very long bows are depicted with angled ears.[61] It is possible that late Sarmatian peoples were using this form of quiver-case instead of the gorytus. The Parthian horseman carving at Tang-i-Sarvak seems to depict a type very similar to the Palmyrene combination.[62]

Variants of the quiver and bow-case appear in two Palmyrene reliefs. One may represent a case on the rider's right side, for a strung bow but this is not clear because of the damage.[63] The other is for an unstrung bow but does not seem to have the clearly defined separate bow-sheath.[64] Another variant, most clearly seen on the Mordechai and Esther fresco in the Dura synagogue (mid 3rd century A.D.) incorporates the usual bow-sheath alongside a tapering quiver.[65]

This type points to Sassanid developments. At some point in the 3rd century A.D. a tapering quiver came into use which bunched the arrow-heads at the lower end and perhaps giving the fletchings greater space. This is seen on the early Sassanid rock reliefs at Bishāpūr, Naqsh-i-Rustam and Fīrūzābād suspended on the rider's right side from a hip-belt, not behind him from the saddle.[66] The Dura-Europos graffiti provide numerous additional examples as do the the Sassanid silver dishes.[67] Curiously when the rider and horse on the latter are viewed from the rider's left side no attempt is made to depict a bow-case. This is most regrettable because there is no definite indication as to the exact form of this early separate bow-case which, by later analogy, would have been suspended on the rider's left side from a belt. The separation of bow-case and quiver, and the respective sides for suspension, is an important innovation which continued right through into modern times, despite later reversion to cases for strung bows. Forces on both sides of the Roman-Sassanid frontier are likely to have influenced each other's equipment.

A completely different type of quiver appears on the Tāq-i-Bustān armoured horseman sculpture which probably depicts Chosroes II (AD590-628).[68] It is a waisted 'hour-glass' quiver, so-called because the foot and mouth are usually wider than the body, suspended on the rider's right side by two vertical straps from a hip belt (Fig.41). It hangs diagonally with the open mouth uppermost and pointing forward. The lower end is much wider than the upper where the arrow stele are visible.

This quiver-type was developed in Central Asia, and was in use in China from at least the T'ang period when the Chinese were influenced by Turkish equipment. The 6th-8th century AD Turkish petroglyphs from the Minusinsk region exhibit hourglass quivers and many are seen on frescoes depicting Iranian and Turkish armoured cavalry from Pendzhikent (7th century AD), Kizil (mid 8th century), Chotscho (8th-9th century) and other sites.[69]

When the open quiver is depicted in detail it is clear that the arrow-heads are positioned at the mouth with the fletchings inside the lower end. This situation is reflected in western Avar graves where a concentration of arrow-heads is found in close proximity to the hip of the deceased in association with decorated belt-plates and pendants. The quivers themselves do not seem to have had a metal mouth or 'chape' but may have been decorated with worked bone strips.[70] Metal fittings at mouth, foot and long edges do appear on Magyar quivers, however. An entirely preserved mouth with seven arrow-heads attached to it by corrosion was found at Magyarhomorog (Hungary).[71]

The Sassanids may have adopted the hourglass quiver in the

mid 4th century AD from the neighbouring Chionitae. Late Roman horse-archers almost certainly would have acquired it from the Avars if they did not possess it already. The question arises as to whether the Huns introduced the hourglass quiver to the Roman army. Unfortunately, concentrations of arrows do not occur in Hunnic funerary contexts similar to those in Avar graves. The 6th century AD mosaic horse-archer from the Great Palace at Constantinople although viewed from the man's left, has the top of a quiver appearing in front of him (Fig.38).[72] From its profile and angle of suspension this is very probably of the hourglass type. Regrettably this mosaic is not dated closely enough to distinguish between Hunnic or Avar influence. On the other hand a rather crudely executed horse-archer in the church of St George at Khirbet el-Mekhayyat (Jordan) depicting a quiver slung diagonally with the open upper end forward dates by inscription to the 530s AD.[73] This may point to a Hunnic introduction although little weight can be placed on it. The Syrian hunt mosaics do not depict quivers.

No separate bow-cases appear in Roman art. A 3rd-2nd century BC Bactrian bowl suggests an early date for the sheath type which is depicted both containing a bow and empty.[74] An object suspended on the left side of the Tāq-i-Bustān horseman looks very much like the upper 'ear' of a sheath quiver.[75] This type is most clearly seen in Central Asian frescoes at Pendzhikent and Kizil and in the Jenissei petroglyphs. Again it was adopted in China, suggesting an Asian origin.[76] A sheath quiver occurs on a post-Sassanid silver dish with a curved profile and angled ears for an unstrung bow.[77] Sassanid period dishes do not depict bow-cases. Non-perishable fittings do not appear in Hunnic funerary contexts but Magyar bow-cases had decorated bone plates attached to the mouth of the sheath to enclose the bow's upper ear.[78] With the 'Sassanid' and hourglass types of quivers sheath bow-cases would have been necessary and presumably widely used on both sides of the Roman-Sassanid frontier. Citing Maurikios' Strategikon, Haldon comments that bow-cases of 'the Persian type' were in Roman use, as opposed to steppe sheath type bow-cases.[79] However, this seems to be a false distinction because the Persian type very probably was a sheath bow-case.

A variety of quiver and bow-case types may have been in contemporary use but the evidence discussed above allows the following conclusions about Roman usage. Infantry archers employed cylindrical quivers, with caps in wet conditions at least, suspended on the back and they probably carried a sheath bow-case. There is no evidence for the latter supposition but some such form of bow-case must have been used. Sarmatian and Bosphoran troops probably used the gorytus suspended from the left hip for a short Scythian bow. The majority of oriental horse-archers in the 1st-3rd centuries AD would have had the quiver and bow-case combination suspended from the right side on the rear of the saddle, for longer unstrung bow. This type spread also to the Crimea.

During the first half of the 3rd century AD the quiver and bow-case elements were separated. A tapering quiver with a narrow foot remained on the archer's right but was now suspended vertically from his hip. The necessary bow-case must have been carried on the left side hanging from the same belt as the quiver, probably diagonally with the upper ear rearwards. Again there is no direct evidence for this except for temporally wide analogies.

At an indeterminate point both Roman and Sassanid forces adopted the Central Asiatic hourglass quiver suspended horizontally with the mouth towards the front and the arrow-heads uppermost. Dated representations are scarce from both areas but if the Huns did not introduce it into Roman use, the Avars almost certainly did. The sheath bow-case would have been unaffected by this new quiver type.

3. BRACERS AND THUMB-RINGS

Bracers and thumb-rings are mutually exclusive pieces of equipment associated with bow-string release in shooting. Their use is dictated by the method of release, 'Mediterranean' or 'Mongolian'.

The Mediterranean release is the drawing and loosing of the string using two or more fingers with the back of the hand vertical.[80] This is the most common release used by modern archers. As the hand is opened the string springs away to the archer's left (presuming he is holding the bow in his left hand) and takes a curved course to the position at rest. The acceleration and force of the string presses the arrow rightwards so that with this release the arrow has to be laid on the left side of the stave (the 'inside') or it would fly off uncontrollably.[81] The path of the string runs close to the archer's left forearm and may actually contact it resulting in a painful burn or bruise. A leather band, the bracer, is employed to protect the arm.

The Mongolian release involves the use of the thumb to hold the string 'locked' by the index finger, with the back of the hand horizontal or sloping downwards (Fig.43).[82] When the string is released it springs rightwards bending the arrow around the right of the stave. The arrow is laid on the 'outside' and the string does not go near the left forearm, thus a bracer is unnecessary. To protect the thumb, to make the holding of the string more comfortable and to aid consistent accuracy a leather 'thumb-stall' or a bone 'thumb-ring' is slipped over the thumb.[83] The ring is characterised by a flange positioned on the inside of the thumb to take the string.

The question of the type of release employed by Roman period archers has important ramifications for shooting performance and the use of other equipment. The advantages of the Mongolian over the Mediterranean release are many. Firstly, the string-path and 'out-side' placing of the stele of the Mongolian allows a small shield to be strapped to the left forearm, the straps of which could be caught by the string in the Mediterranean release.[84] A baggy left sleeve must be bound up but a bracer is unnecessary.[85] This is an important advantage for horse-archers who were particularly vulnerable to melee attack by other skirmishing cavalry if they were shieldless. The smaller angle made by the string at the thumb alone, compared with that when held by several fingers, enables an archer to use shorter bow with a greater draw-length in comfort. An archer may draw to the right shoulder with a thumb-ring imparting greater power to the arrow for increased distance or penetration. Alternatively a large bow may be used with a commensurately longer draw.

Robinson suggested that archers in the Roman forces used the Mongolian release, citing a bone ring from Chesters (Northumberland) as a thumb-ring.[86] This object in fact does not have the characteristic flange. To the writer's knowledge no clear examples of bone thumb-rings have been found in securely dated Roman contexts, although leather thumb-stalls would not have survived. One ring was found at Dura-Europos but the context of the find was unrecorded and it could easily be a Late Roman or Medieval stray-find.[87]

Artists might be expected to be lax in portraying such details as fingers on a bow-string but a few representations prove to be of use for the discussion.

Parthian terracottas rarely depict archers at full draw. The Berlin figure is seen from the wrong side (Fig.39), but a plaque on display in the British Museum, London, has such an archer with a hound which may be executing a Mediterranean release.[88] The clearest depiction is a piece of mother-of-pearl box inlay, 2nd-1st century BC in date, from Shami (Iran).[89] The bow is lost but the back of the archer's hand is clearly vertical and the string is held by the two middle fingers. The index and little fingers are extended. Fragments of a terracotta frieze was recovered from the small palace at Khalchayan (Transoxiana) dating to c. 50BC-50AD. One drawing arm and hand survived from a light horse-archer figure. Despite Pugachenkova's reconstruction drawing with Mongolian releases, the back of the hand is vertical and the string appears to be held by the index and second fingers.[90]

The Dura-Europos Parthian-Sassanid horse-archer graffiti are of little help because of their small scale and often crude execution. Sassanian silver dishes depicting kings hunting on horseback clearly depict what might be termed the 'Sassanid release' which is really a continuation of that seen on the Shami piece. This was a Mediterranean variant by which the middle two fingers hold the string and the index and little fingers are extended paralleled with the stele (Fig.44). Only on evidently post-Sassanian dishes does the Mongolian release appear. The Moroccan treatise described a similar 'Slav release' with just the index finger extended.[91] It might be suggested that some of the advantages of the Mongolian thumb-lock might be gained by holding the string with only the two middle fingers. The angle of the string at the hand would be more comfortable with a long draw though a shield could not be worn simultaneously on the left forearm. This may account for Ammianus' description of Sassanid archers drawing to the right side of their chest.[92]

The Scythians and presumably the Sarmatians used the Mediterranean release.[93] On the above evidence eastern archers in Roman service would also have used this release. All the archers on Trajan's Column draw with their fingers and those in Scene LXX are clearly depicted with bracers. There are no shields associated with archers on military tombstones, though these do not show bracers, mainly because of lack of detail.

The Mongolian release was another steppe development, as the name suggests, and is seen in Asiatic frescoes.[94] There are no thumb-rings in Hunnic funerary contexts known to the writer though this does not preclude the use of thumb-stalls. However, some decorated bone thumb-rings do survive in association with Avar material.[95] Late Roman forces would almost certainly have adopted the Mongolian release by the later 6th century A.D. and Bivar suggested that Roman archery at Callinicum was so penetrative because the Hunnic bow and the Mongolian release were being employed. The Sassanid opponents by this token would have used their Mediterranean variant with the smaller Sassanid bow. The silver dishes may of course be stylised and conservative in this respect, and the contemporaneous use of both releases is not impossible. Bivar further implied that the Arabs adopted the Mongolian release from the Romans not from the Sassanids but there is no contemporary evidence to prove it. The Umayyad Khasr el-Hayr el-Gharbī fresco which he cites in support (Fig.43) is too late to be relevant and the archer represented is probably from Khurasan or Transoxiana anyway.[96] The distinction between

'Roman' and 'Persian' shooting in the anonymous 6th century archery treatise may perhaps be interpreted as referring to Mongolian and Sassanid releases.[97]

It would be surprising if the Sassanids were slow in absorbing the Mongolian release, considering their longstanding steppe contacts. The late use of the hourglass quiver demonstrates that Asiatic practices were influential. A Kidarite Hun dish of late 4th century date depicts horse-archers employing thumb-locks (no bow-cases or quivers are represented) proving the steppe use of this method at this time.[98] Another relief at Tāq-i-Bustān, the aquatic hunt scene, depicts Chosroes II in a boat with a bow held at full-draw.[99] The king's right hand seems to be bent downwards, a common artistic convention to denote a thumb-lock (Fig.40). Moreover, with only one exception, all depictions of Parthian, Palmyrene, Sassanid and other oriental horse-archers include no shields. The predictable exception is the armoured Chosroes II at Tāq-i-Bustān which has shield, bow and contus (Fig.41).[100]

There is no evidence that the eastern foot- and horse-archers in Roman service in the auxilia, numeri or irregular bodies of troops employed anything but the Mediterranean release which required a bracer. With the arrival of the Huns in the later 4th century it is likely that the use of Mongolian release, requiring a thumb-ring, spread at least to the best bow-armed Roman cavalry. Other archers may have continued to use the Mediterranean or Sassanid releases. Some evidence supports a Sassanid use of the Mongolian method, again adopted from Hunnic peoples, by the late 6th century A.D. at the latest.

4. GENERAL EQUIPMENT

The soldiers' general requirements of mill-stones, tents, pack-animals etc. would have been the same for all classes of troops but it might be expected that eastern archers exhibited oriental features in their clothing and armour. On the contrary the tombstones of auxiliary sagittarii suggest that they differed not at all from other auxiliarii except in the carrying of archery equipment. The two archers of cohors I sagittariorum are dressed in the normal sagum and tunica as seen on the very similar stone of Annaius Daverzus from cohors IV Delmatarum, a unit equipped with shields and hastae.[101] The gladius, pugio and two cingula are identical. The Mainz archer of cohors I Ituraeorum wears a paenula and a tunica (Fig.28). A tubicen of the same cohors is depicted on another tombstone in Roman civilian dress.[102]

These figures all date to the first half of the 1st century

A.D. whereas the Housesteads archer must be Hadrianic or later in date (Fig.26). The man wears a long-sleeved tunic[103] which may be an oriental feature if the stone is 2nd century in date and the man came from cohors I Hamiorum stationed at Carvoran, two forts along Hadrian's Wall to the west. Alternatively it may be a 3rd century feature. No body armour appears to be present, even considering the weathered state of the stone. A conical projection on the head has been interpreted as a conical helmet or as a normal helmet with crest.[104] A large knife with bird-headed pommel hangs from a cingulum. In his right hand the archer holds a bill-hook, not an axe. This combination of bow and bill-hook is also found on a Gallic-Roman statue of a forest deity from Mont-Saint-Jean (Oise).[105] The funerary nature of the Housesteads relief, rather than a deity depiction, is indicated mainly be the debased rosettes and pine-cones flanking the aedicula.[106]

The horsearcher from the ala Ituraeorum at Tipasa appears to wear knee-breeches and tunica. The Gÿor and Walbersdorf reliefs suggest the same attire with the latter adding a helmet with neck flange and a normal cavalry spatha (Figs.30). Similarly the eques singularis Augusti at Mainz has short, tight breeches and a tunica. He appears to be bare-headed. Lastly, the member of the ala Parthorum et Araborum is badly damaged but provides very important details of clothing (Fig.31). The servant is clad in the usual short-sleeved tunica but the deceased appears to have a long-sleeved tunic with a cuff appearing on his left arm. Most importantly the profile of his eroded leg looks as though he is wearing baggy Parthian anaxyrides. On the back of the damaged head appears a helmet neck-flange.

On Trajan's Column an archer appears in the background of Scene XXIV dressed in the same fashion as other auxiliarii but substituting a bow for shield and shafted weapon (Fig.19). The other archers appear in the well-known ankle-length robes, with conical 'spangenhelme' and either loricae squamatae or hamatae (LXX, CVIII, CXV). The helmets are similar to one variant seen on the Column's pedestal and may denote Sarmatian origin for the men (Fig.20). The archers in Scene LXVI wear helmets identical to those worn by Sarmatian cavalry in Scenes XXXI and XXXVII. The writer would see the archer in XXIV as typical of members of cohortes sagittariorum. The long robes of the other archers probably denote Sarmatian origin rather than Levantine dress (Fig.20 & 22). In Scene C the Sarmatian envoys wear long, asiatic kaftans. The sculptors have misinterpreted their subject as they did elsewhere in depicting horses with scale-armoured legs. No dress in Parthian, Palmyrene or Syrian art suggests that such long robes were common Levantine wear. The Conon fresco at Dura is the closest approximation but these robes are priestly in character.[107] Conceivably the sculptors could have misrepresented the anaxyrides by running the baggy legs together, or the 'sarongs' worn by rider-gods. Definitely the archers in LXVI and possibly those in CXV were intended to be Sarmatian Iazyges who may have provided the Romans with contingents for the Dacian wars. The archers in CVIII may be the same (hamatae are seen on the pedestal reliefs) or badly depicted Levantines (Fig.22). Cichorius favoured such a dual interpretation.[108]

On the Marcus Column the horse-archer in Scene LVII (Fig.25) is dressed like any other mounted auxiliarius with lorica hamata, substituting bow for shield and hasta. The infantry in Scenes XV, XXVII, XXXIX and LXXVIII all wear long-sleeved tunics, tight, long trousers and Phrygian caps (Fig.24). These men are clearly orientals, stylised but not out of place in Levantine art.

From these representations, funerary and monumental, it appears that the alae and cohortes sagittariorum were dressed and armoured in the same manner as other auxiliary units. The ala Parthorum et Araborum may have been an exception to this because of its unusual origins. The Dura Terentius fresco, depicting officers and men of cohors XX Palmyrenorum sag. mill. eq., supports this conclusion. All the figures wear normal 3rd century military dress and belt-fittings.[109] The horse-archer graffiti from the town probably depict irregular Palmyrenes or Sassanids, not equites cohortales.[110]

Some other elements of equipment may have displayed oriental characteristics.[111] The putative conical helmet worn by the Housesteads archer corresponds roughly with two actual helmets in the Archaeological Museums of Zagreb and Sofia which also accord with some helmets on Trajan's Column (notably in Scene VIII).[112] It is possible, however, that the two surviving helmets, one very ornately decorated, were purely for parade use or belonged to officers. Normal auxiliary helmet types may have been everyday wear. Ethnic variation in military equipment is a vexed subject but units might be expected to gradually lose initial characteristics by moving away from their areas of recruitment, by a slow process of equipment replacement and by becoming inbedded in new supply systems. Moreover, surviving equipment from the Eastern Provinces does not differ markedly from pieces found in the West. The Zagreb and Sofia helmets are decorated with typically Roman mythological figures and the Housesteads archer's helmet may be crested rather than conical.

The bill-hook held by the Housesteads archer is unparalleled in military depictions and may be a specialist tool for the collection of arrow materials. Had it been an axe it would have been understandable as a side-arm because the man is otherwise only armed with a long knife. Axes are mentioned by Arrian in use by cavalry and in 6th century A.D. and later manuals they were associated with light troops.[113] Robertson suggested that for mural defence of their forts auxiliary sagittarii would have been equipped with shields and shafted weapons.[114]

There is no evidence for the use of shields by sagittarii in the first three centuries A.D. Vegetius specifically said that archers were unable to use them so should have had body armour.[115] The Mediterranean release would have precluded use whilst shooting. No weapons would have outranged foot-archers except artillery, against which shields would have been useless. In the event of a unit of archers being shot at it would reply in kind rather than use shields for protection. Some Palmyrene horsemen carried shields and bows but shafted weapons are also in evidence suggesting a melee role. Parthian horse-archers definitely did not use shields.[116]

Were later Roman sagittarii equipped with shields? In the Notitia Dignitatum 15 units of archers and 4 of ballistarii have shield-patterns. Grigg has demonstrated the arbitrary formulation of these emblems in the manuscripts and it is doubtful whether any weight may be placed on them.[117] It is true that the scola scutariorum sagittariorum has a shield and archer title but in the same chapter an equally unique scola scutariorum clibanariorum occurs.[118] Since all the ethnically designated clibanarii in the Notitia were 'Parthian' or 'Persian' they were very likely contus-armed, another weapon with which shields could not be used simultaneously. The small shield mentioned by Procopius, born by lance-armed and bow-armed cavalry, was to protect the face.[119] This has no bearing on the release question but is connected with the double-handed use of the contus. Even after the Avar introduction of stirrups Central Asian cavalry still used both hands rather than couching the lance.[120]

III. ARCHERS AND ARCHERY EQUIPMENT

1. BOW-ARMED UNITS

It would be an exaggeration to suggest that virtually all soldiers in the Roman Army could have practised archery at some point in their careers. However, archery equipment was much more widely employed than might be expected judging solely by regimental titles. Battlefield archery, siege-work and general military training presented a range of circumstances in which bows might have been used.

The archaeological material is the most informative source of information. The distribution of laths and arrow-heads in Britannia and along the Northern *limites* is important because in only a few cases may a direct connection be made between finds and known garrisons of *sagittarii*.

Laths from the Augustan legionary bases at Dangstetten (Catalogue No. 12) and Oberaden (Catalogue No. 13) may be explained by the recorded employment of *equites* and *pedites sagittarii* by Germanicus.[1] At Bar Hill (No. 1) the *cohors I Hamiorum sag.* was in residence at some point during the Antonine period and Ctraubing (No. 23) was garrisoned by the *cohors I Flavia Canathenorum mill. (eq.?)*. Depending upon their date the late laths from Intercisa (No. 25) might be attributed to the *equites sagittarii* of the *Notitia*. However, Bobertson has cautioned against a simple identification for the Bar Hill pieces and points to literary evidence (discussed below) suggesting archery practice outside units entitled *sagittariorum*. She suggested that the other unit in residence, *cohors I Baetasiorum c. R.*, could have been responsible for the laths and arrow-heads, thus the archery equipment is of no use in elucidating the garrison sequence.[2] Walke, on the other hand, takes the laths and heads from Straubing to indicate the presence of the *cohors sagittariorum*. [3] The difficulty in distinguishing 'Hunnic' laths from those found in indubitably Roman contexts makes any attributions for the Intercisa material quite insecure (No. 25).

All other laths in the Catalogue have no definite correlation with known *sagittarii*. At South Shields, Chesters, Chesterholm, Zugmantel, Heddernheim, Stockstadt and Buch none of the known garrisons were *sagittarii* (No. 2, 5, 15, 16, 17, 22). However, at Heddernheim the presence of elements of *cohors I Flavia Damascenorum mill. eq.* from Friedberg might offer an explanation.[4] Rißtissen (No. 21) had a mixed legionary-auxiliary garrison to judge from the *lorica segmentata* fittings and no known specialist *sagittarii*.[5] Garrisons of other forts in the Catalogue are unknown. Lath-finds in London (No.7) could be accounted for by this site being the provincial capital with military presence of singulares and administrators. At Silchester (No. 8) other small-finds suggest military presence of an unknown nature.[6]

Laths associated with the legionary fortresses of Mainz, Windisch and Carnuntum (Nos. 14, 20, 24) may be explained by analogy with the bow manufacture material at Caerleon (No. 9) and the presence of arcuarii as legionary immunes (Bow Construction above). Likewise the arrow-heads at Corbridge (No. 3) were produced in legionary Fabricae, presumably for campaign stocks. The laths may also have been of legionary manufacture.

In Britain the only known unit of sagittarii was the cohors I Hamiorum on the two walls.[7] Davies suggested that some Thracian units of the conquest period may have had bows but there is no proof of this.[8] Arrow-heads are very common on military sites with a wide distribution across the province. With a tolerably well-known auxiliary garrison and no Central Asian element in the archaeological record two logical explanations present themselves. Either the Hamii toured the province extensively discarding their bows and losing their arrows, or archery was widely practised by auxiliary units. The latter possibility seems preferable. The wide distribution of archery equipment on the German and Raetian limites led Erdmann to a similar conclusion for continental practice.[9]

Ox-skulls discovered at Chesterholm and Corbridge (Cat. No. 3, 5) were perforated with many punched holes made by an object which was square in section.[10] The smallness and multiplicity of these holes precludes association with animal slaughter. It seems most likely that the heads were mounted and used as targets for arrows or hand-hurled weapons. Square-sectioned ballista-bolts or pila would have either gone right through the skull or smashed it. Light javelins with appropriate head-sections would still have been too large. The neat, multiple holes are most likely to have been made by the 'bodkin' type of arrow-head (Arrows, types 3 - 4, above) found on Hadrian's Wall sites.[11] Neither Corbridge nor Chesterholm had known archer-garrisons.

The conclusion that archery may have played a part in general training for troops not in units entitled sagittariorum is quite reasonable on the archaeological evidence but has little to support it in the literary sources. Avidius Cassius is said to have exercised his troops in archery and the use of other weapons but this comes as no surprise in the Orient.[12] Arrian confines his detailed remarks on cavalry manoeuvres to javelin- and shield-armed cavalry.[13] Vegetius recommended mounted and foot practice for tirones shooting wooden bows at a palus target.[14] Elsewhere he says that other troops should also exercise using straw targets over a range of 600 feet.[15]

One reference suggests the use of archery by some Late Republican legionarii.[16] The construction of bows at Caerleon and by extension in legionary fortresses in general, might suggest the meeting of a need for practice weapons. Alternatively the bows were constructed for mural defence of the fortress, a supply of weapons being in store should the need arise. This is not as unlikely as it at first seems, considering the collection of caltrops found in the same building as the laths at Caerleon.[17] Another possibility is that such bows were for training or mural defence use by non-sagittarii auxiliaries, assuming that specialist units of archers manufactured their own weapons (above Bow Construction). Robertson suggested that a cohors sagittariorum would need shields and shafted weapons for the defence of its fort.[18] The reverse may have been true for non-sagittarii, greatly aided in defence by having a small complement of bows in store with a supply of arrows. By this token laths and arrow-heads cannot be used to identify the garrison of a given fort as specialist sagittarii with any degree of certainty.

One cannot imagine the Guard units in Rome being bow-armed for mural defence but surprisingly there is evidence confirming archery training for them. General doctores appear in the epigraphic record with campidoctores perhaps of superior rank and with an equestrian emphasis.[19] For the praetoriani there was a position of doctor sagittariorum[20] and in the equites singulares Augusti a campidoctor sagittariorum is attested.[21] Two funerary reliefs depict members of the latter body as horse-archers, suggesting that an element was permanently bow-armed (Fig.29).[22] The emperor Commodus received archery tuition from a Parthian who may have been a singularis.[23]

With regard to auxiliary regiments the title 'sagittariorum' is presumably a sure indication that a unit was bow-armed. Unfortunately the epithet is often omitted in inscriptions and on diplomata. Geographical titles without the epithet may be a guide and Davies suggested that all the cohortes Commagenorum, Petraeorum and Ituraeorum could have consisted of archers.[24] If most Eastern units were sagittarii then the Western units in the Orient assumed a greater tactical importance in providing spear- and shield-armed cavalry. Sometimes funerary reliefs provide the necessary information by depicting an archer, as with the ala Scubulorum or the ala Parthorum et Araborum discussed above.[25] Another problem is whether the cavalry of a cohors equitata sagittariorum were bow-armed. There are no definite depictions at Dura-Europos of equites cohortis XX Palmyrenorum[26] and no equites cohortales appear as tombstone figures. However, the cavalry element of cohors III Ulpia Petraeorum mill. sag. forming a screen on Arrian's march to meet the Alani consisted of horse-archers and those shooting over the heads of the _legionarii_ in the projected battle were presumably from the other _cohortes sagittariorum_ present.[27] Alternatively, comments addressed by Hadrian to the _equites of cohors VI Commagenorum_ would perhaps better suit men armed with javelins (though the unit is not definitely known to be _sagittariorum_).[28] The Straubing parade armour, presumably belonging to _equites_ of _cohors I Flavia Canathenorum mill._, suggests exercises no different from those involving practice javelins described by Arrian.[29]

It is probable, judging from the archaeological evidence, that some auxiliary regiments of non-_sagittarii_ were in part also bow-equipped for training and/or mural defence purposes. Indeed the finest recorded demonstration of archery skill was exhibited not by a member of an oriental _cohors sagittariorum_ but by Soranus, an _eques_ from a _cohors Batavorum mill. eq._[30] He swam the Danube on horseback under the eye of Hadrian, shot an arrow in the air and hit it with another before it came to earth. No units of Batavian archers are known. In small-scale engagements bow-armed elements of non-archer auxiliary regiments might give useful missile support. The distribution of arrow-heads might thus be compared with the incidence of _glandes_, there being no auxiliary units of _funditores_.[31]

A number of oriental 'national' _numeri_ were deployed, notably in Dacia, Africa and Mauritania. All may be assumed to have been composed of _sagittarii_ (most have the title) because this class of unit was formed to take advantage of particular skills of archery, light cavalry combat, or skills in forest and mountain warfare (see Deployment below). Quite why '_numeri_' of archers were employed rather than _alae_ or _cohortes_ from the same regions is unclear. The _numeri_ seem to have been internally organised along similar lines to the auxiliary units, however, with _centuriones_, _decuriones_, _optiones_, _librarii_, _signiferi_, _immunes_ etc. epigraphically attested.[32] Their overall size is unknown. It is probable that the _numerus Palmyrenorum Porolissensium c. R._ at least contained horsearchers because of the likely 3rd century formation of an _ala_ and a _cohors Palmyrenorum_ from it.[33] The _numerus Syrorum_ at Lalla Marnia in Mauretania Caesariensis may have had a mounted element because of its far-flung position and patrolling role.[34] Perhaps the _Palmyreni_ of 'Hyginus'' army were all infantry.[35]

Large numbers of irregular _symmachiarii_ were supplied by the Eastern client-states, the majority being light horse- and foot-archers. These were probably present in every Roman campaign from the Late Republic onwards and were transferred on occasion to the Rhine and Danube (Deployment below).

The best bow-armed troops in the Late Roman Army were provided by the field-army units. There was a predictable East-West imbalance in numbers and mounted and oriental units (assuming Persian-style clibanarii to be bow-armed): 13 vexillationes to 7 auxilia and legiones in the East; 10 vexillationes to 4 auxilia in the West.[36] Additional horse-archers were provided by 32 units of equites sagittarii indigenae in the East, 8 in the West, numbers again reflecting the cultural background of the Eastern provinces.[37] Moreover, some alae and cohortes had sagittariorum in their titles or appropriate geographical designations, but it is unclear whether they were still bow-armed.[38] Vegetius' description of an order of battle envisaged archers integral to the legiones and a late legio pseudocomitatensis has an archery title.[39] Presumably irregular contingents were used on campaign whenever available.

The impacts of the Huns and the Avars on Roman military practices made armoured, bow-armed cavalry the primary tactical arm, overlaying developments in the Roman-Sassanid frontier region. Infantry units continued to be bow-armed fulfilling the conventional skirmishing, screening and support roles. Hunnic horse-archer allies were also employed in significant numbers[40]

Archery had some permanent role in the Imperial Guard units. Most auxiliary regiments could have had archery training. Mural defence and limited field use could be other factors explaining the distribution of archaeological finds. The alae and cohortes sagittariorum probably supplied their own specialist equipment as would the 'national' numeri and more ad hoc bodies of symmachiarii. The legiones may have had some archery training but most likely manufactured bows for their own mural defence or for supply of weapons to non-archer auxiliary units. Field-use by legionarii would seem quite out of the question, at least before the Late Imperial period.

2. CALIBRE

There is no practical reason why the skills exercised by the best bowyers in Roman service should have been inferior to those outlined in the Islamic and Chinese sources. Details of construction, decoration. Stave proportions and mechanical properties obviously differed but a high degree of craftsmanship is likely in some contexts. The urban bowyers and fletchers in the Eastern Roman provinces and in the neighbouring Iranian empires might work for their cities' general supply of bows and arrows but their commercial business would have been dictated by their level of skill. Wealthy and noble customers would demand weapons commensurate with their social positions, for hunting purposes as much as for warfare. Bows and arrows made to an an individual's personal height and length of draw would be more expensive and given more skilful attention by the artisan than the bulk of the stock for general sale. The quality of the bowyer's work and his reputation might dictated price and demand.

Within the Roman forces such market factors might not have been applied except for the irregular symmachiarii and irregular units made up from noble exiles and their retainers already armed when they came into service.[41] The Levantine raising of auxiliary sagittarii and the continued recruitment of easterners to these units could, however, mean that the individuals would bring with them weapons that they already owned and which may have been manufactured to personal specifications. Moreover, the speculation that such regiments and the numeri sagittariorum would have specialised fabricae producing archery equipment (Bow Construction above) could also mean a high level of construction skill was exercised.

With regard to legionary bow construction the quality of the finished product would have depended upon the available personnel. If the arcuarii involved were easterners working with suitable materials then perfectly good composite bows could be produced. The laths from Mainz, Windisch and Carnuntum are not distinguished from other site-finds by inferior workmanship. The same cannot be said for the Caerleon laths which display ineptly cut nocks in sections of bone badly chosen with regard to curvature and coarseness of cellular structure.

Problems in constructing or obtaining good composite bows may be hinted at by Vegetius' mention of wooden practice bows. That composite bows were in wide use is demonstrated by the distribution of laths but wooden self-bows would have left no archaeological trace under normal western conditions. Arrow-types have no bearing on the types of bows which shot them. Self-bow archery was employed in Gaul and Free Germany,[42] but for Roman combat purposes the employment of orientals in the alae, cohortes and numeri indicates composite bow use. Physically the latter type is always superior to ancient or modern self-bows.

The practice of composite archery in the West was essentially a product of the political domination of the Mediterranean by one power. In Western climatic conditions the mechanical properties of the stave materials would have been more prone to the detrimental effects of cold and moisture than in the Levant. The time taken to construct bows and the stages of that process depended largely upon glue setting rates dictated by temperature. Moreover, the utmost care would be required in the storage of staves whilst out of use or in transit. Drying-frames and bow-cases would therefore have been extremely important, especially during winter months. In the hands of specialist troops archery equipment need not have been too adversely affected by these conditions. The framework of the Roman Army allowed for the presence of such troops in the West but after the eclipse of Western Roman power composite archery ceased to be widely employed despite periodic violent contacts with Asiatic peoples. The Roman Army was also the medium by which suitable horn for composite bellies was passed to the West. Not before the 10th century A.D. did composite construction appear again and then it was for shorter cross-bow staves, often substituting whale-bone for horn (Cross-Bows above).

The employment of Levantine sagittarii was a basic fact of Roman military policy from the Late Republican to the Late Imperial period. This and the continued recruitment of Easterners after Levantine units had been raised and moved away led Cheesman to the logical conclusion that "the reason for the adoption of these exceptional methods in the recruiting of the oriental auxilia was probably the purely military one that good archers were born in Syria, and could not be made elsewhere."[43] Good archers were in fact 'made' elsewhere, principally in Thrace, Cyrenaica, Numidia and on Crete, but it is clear that the Levantine cultural background produced the best archers at least before the impact of the Huns.[44] Commentators have emphasised the 'Eastern' elements within cohortes sagittariorum by studying personal names, origines and the tenacity of oriental religious cults.[45] The natural tendency has been to emphasise oriental elements especially in samples of personal names. Whilst no-one would deny the continuing oriental nature of important archer-units, the degree of local recruitment must not be played down. Even oriental numeri recruited locally and were capable of being culturally influenced by their location.[46] However, it is important to examine what exactly Cheesman's statement meant with regard to the calibre of archers in Roman service.

Ancient and modern observers of proficient archers often made a point of mentioning that skills were acquired at an early age.[47] This applies especially to horse-archery which virtually required the archer to be 'born in the saddle'. It was partly for this reason that the waves of steppe horse-archers made such impacts on settled peoples around the Asiatic fringes. Development of specific musculature early in life was probably very important. This is not to say that in a Roman context recruits from Pannonia, for example, entering a cohors sagittariorum at eighteen could not be trained and exercised to make them good archers. The basics of archery may be learnt fairly quickly and a moderate level of proficiency can be attained with constant practice. However, the higher levels of marksmanship, bow maintenance, good arrow-construction and horsemanship require years of experience and dedicated exercise.

Release of the string in shooting must be smoothly executed, the minutest variations in draw-length and finger-disengagement will affect the flight of the arrow. The mark of a skilled archer is constant accuracy under difficult circumstances so that he can function well in confused battlefield conditions. A regiment raised de novo from a population with no cultural background of archery, even if trained by oriental doctores, would be far inferior to a unit of Easterners already proficient with the bow on enlistment. Local, western recruits brought into this unit would benefit from the presence of oriental principales, doctores and arcuarii, not to mention living and working alongside their Eastern fellows. Therefore gradual induction of local recruits might not necessarily or immediately affect the efficiency of an oriental regiment's archery. It is unlikely that the strategically important auxiliary regiments would have been allowed to decline in skill. The unusual homeland recruitment which apparently continued after the unit had moved away was intended precisely to prevent degeneration.[48] Coupled with the circulation of Eastern doctores high standards would have been ensured.

The career of Barsemius Abbei may be of relevance in the last respect. He originated in Carrhae, served in a numerus Hosroruorum then in the ala firma Katafractaria as a decurio, before becoming a magister cohortis in the cohors I Hemensenorum mill. sag. eq. at Intercisa.[49] To judge from this series of units Barsemius must have been an equestrian trainer and/or an archery expert. If the former then it is likely that the numerus was mounted or part-mounted. If he was an archery instructor then it would be likely that the ala Katafractaria was bow-armed. For the cohors Hemensenorum, a unit of great numerical and strategic importance, it is reasonable to suppose that Barsemius was a skilled horse-archery campidoctor.[50]

A representation of a member of the ala I Augusta Parthorum at Cherchel suggests that Eastern regiments could degenerate if isolated from a continuous supply of oriental recruits.[51] By comparison with the Mainz ala Parthorum et Araborum tombstone horse-archers might reasonably be expected for the ala Parthorum but the African figure carries a round shield and a javelin. None of the known personnel of the ala Parthorum were of oriental origin and perhaps nearly two centuries of isolation in Mauretania resulted in a change of weaponry.[52]

A similar process of degeneration in skills may have affected the alae and cohortes in the Late Roman limitanei. None of those in the Notitia were given the suffix 'sagittariorum' although they included suitable geographical or ethnic titles and some pre-Severan units are known to have consisted of archers.[53] This may simply be because the full titles are not given in the

Notitia and, as all of these units occur in the East, they may have continued to be bow-armed. Alternatively reliance may have been put on the field-army units for the best archers (as in the West) and the limitanei have degenerated to spear- and shield-armed troops. The numerous units of equites sagittarii indigenae may thus represent a measure to compensate for this putative decline, or may have been a regularisation of the earlier symmachiarii. Although the emphasis, even with the limitanei, was on mounted troops,[54] on occasion it was possible for troops to be down-graded to infantry as in the case of the equites sagittarii recorded by Synesius.[55]

Some oriental regiments of sagittarii (oriental at least when raised) served in the Western field-armies but it seems unlikely that a supply of new, eastern recruits could have been kept up to these units. What level of archery skill they exercised may only be guessed at and some units of western origins with Gallic epithets also appear.[56] These troops, being 'mobile', may have relied entirely upon the Ticinum fabrica for their equipment. In the Eastern half of the empire supplies of materials and proficient personnel would not have been a problem. Contacts with Hunnic and Avar horse-archers and the increasing Roman bias towards mounted troops meant that the resulting 'composite cavalryman' did not meet his match in archery in the West, fighting Ostrogoths, Vandals, Alamanni and the Franks, and in the East, against the Sassanid Persians. The archery of these Late Roman cavalry with their Hunnic allies was of the highest calibre.

3. PERFORMANCE

The ranges of composite bows have been much discussed but with little consensus.[57] Vegetius quoted a surprisingly long practice range of 600 feet (274m).[58] This would certainly have to be worked up to be a novice starting at 200 to 300 feet (61-91m). Islamic treatises expected consistent accuracy at a range of 60 bows. With a side measurement of 114cm this gave a range of 75 yards (69m). An archer was capable, after a few years of practice of hitting a target of 3 feet (0.9m) diameter with every shot.[59] Late Roman and Byzantine tactical manuals used the bow-shot as a measure of intervals between troop-formations drawn up in the field but estimation of this 'unit' is not possible.[60] Bivar quoted a maximum of 250 yards (229m) and an effective range of 100 yards (91m). Collingwood and Richmond gave a modern range of 250 yards (229m) and an effective killing range of about 150 yards (137m).[61] Jobey used the last two figures when discussing the arrow-heads from Burnswark (Dumfries).[62] Along with artillery stones and glandes these arrows may have been loosed during manoeuvres. At the end of his very useful review of ancient literary sources McLeod concluded that the "Turkish target range of 160-190m, the Byzantine 'moderate bow-shot' and the Turkish extreme effective range of some 310m, and the sundry flight shots of 500m, and upwards all square with the ancient values postulated. They suggest that the ancient evidence, tenuous though it be, has not betrayed us".[63]

In fact many variables are at work within these figures. Range depended upon the type of bow involved. All figures pertaining to 'flight' bows must be discarded in a discussion of military practice because these bows were designed to shoot light arrows over a maximum distance without accuracy. As discussed above a bow designed for use on foot is likely to be more powerful than a horse-bow, thus infantry-archers would normally have outranged horse-archers. The introduction of new bow-types obviously meant variation (presumably improvement) in performance. McLeod put together figures from widely separated periods and contexts with the misguided belief that "the oriental bows used on the fringes of the Greco-Roman world underwent no startling improvement between 700 B.C. and A.D. 700".[64] If nothing else the present study demonstrates the erroneous nature of this statement.

The nature of the target, its size, vulnerability, rate of movement etc., also governs accuracy and effectiveness over various ranges. The Moroccan treatise gives instructions on how to shoot at still and moving targets, horsemen and lions.[65] More modern observers describe two target exercises carried out in common by various Asiatic nomads, the Ottoman Turks and the Egyptian Mamluks.[66] A tall mast was set up with a gourd (qabaq) or another suitable target on the top and horse-archers would shoot near-vertically at the target whilst galloping past. This exercise simulated the shooting of birds on the wing. A second target was a butt (qīghaj) at which the archer galloping by had to shoot downwards. Some of the most skilful shots could be practised from below the horse's neck and this was intended to simulate the archer coming across an unhorsed foe in battle and not having time to put his bow away and take up sword or mace. It developed accuracy at short range and at short notice. The Japanese still gallop past and shoot at targets in yabusame exercises. The exercises and targets of Roman auxiliaries, mentioned by Hadrian, Arrian and Vegetius, make it seem likely that targets similar to the qighaj at least were devised for archers in Roman service.

Quite different are the demands for accuracy made upon a body of archers shooting at a large troop-formation. Whilst it might be desirable to pick off chieftains/officers and standard-bearers, arrows dropping into the mass of troops need not be more than generally aimed and will be more likely to hit a target over the greater distance.[67]

The greater the range the less effective is the arrow's penetration on impact. The weight of the arrow and type of head are other contributory factors. The development of catafracti/clibanarii in the East was a direct response to archery. On the Danube heavily armoured Sarmatian cavalry would demand penetrative Roman archery. Light-armed, loose-formation targets such as skirmishing Moorish cavalry would require greater accuracy than penetration. However, in the West close-order, unarmoured Celtic or Germanic warbands would have formed slow-moving, soft targets.[68] British chariots would need only one horse to be killed for the whole vehicle to be neutralised.[69] Vegetius recommended that Roman foot-archers should be armoured to protect them from Gothic (foot-) archery.[70] The Armenian Tiridates with 1000 horse-archers offered to meet Corbulo if he was escorted by an unlimited number of unarmoured Roman troops. Corbulo refused because any number of unarmoured infantry would be useless against mounted bowmen.[71]

Taybughā's Ghunyah is most important as a treatise on horse-archery skills and he described a variety of shots from the saddle over a field of almost 180 degrees.[72] Some of these shots may have been difficult if not impossible, before the introduction of stirrups by the Avars. However, the performance of ancient horse-archers demonstrates that they were not markedly inferior to medieval horsemen. Stirrups are advantageous principally in providing firmer support for downward sword-slashes and for the couching of the lance, but some of Taybughā's shots below and across the front of the horse's neck demanded that all the rider's weight be put on one stirrup.

It must be said that there is nothing inherently 'Parthian' about the 'Parthian Shot'. The ability to shoot back over the horse's rump, and indeed all around, especially whilst retiring, was common to all good horse-archers from the Scythians to the Crimean Tatars.[73] Close-order, armoured cavalry might have shot behind but the shooting arc forward and to the sides would have been restricted by their formation.[74] The reason for the modern association of shooting backwards with the Parthians is the fact that the 'Parthian Shot' and Parthian archery generally became a Leitmotif in Roman literature.[75]

4. DEPLOYMENT

Vegetius recommended the use of archers and light infantry as a screen for an army on the march in potentially hostile country.[76] This is seen in practice on both Vespasian's invasion of Galilee and Arrian's projected advance to meet the Alani.[77] Arrian had horse-archers at the head of the column (from cohors III Petraeorum mill. sag. eq.) preceded by exploratores and backed by an ala.

In Roman battles foot-archers with javelinmen and slingers formed a screen to protect the close-order legionary infantry from enemy missiles. They served to provoke an enemy advance by skirmishing forward and retired through the main line as the enemy approached. They would then form up behind it or move to the flanks. Once the enemy was repulsed the light troops would move back through the line and pursue.[78] Archers might alternatively have been drawn up originally behind the legionary infantry, as Arrian, Vegetius and Julian stipulated giving missile support as the enemy charged in and during the ensuing melee.[79] The pedites sagittarii of Arrian's force were joined by horse-archers shooting over the heads of the legionarii.[80] Instead of, or in addition to this position, archers were often placed with the bulk of the cavalry on the flanks as in Caesar's North African battles or in Corbulo's order of battle in Armenia. Arrian placed his Armenian symmachiarii on his right flank. When Maximinus Thrax advanced to Emona his army was in a shallow square enclosing the baggage, with catafracti, Moors and bow-armed cavalry on the wings. Vegetius recommended light horse-archers be grouped on the wings. Julian placed catafracti and sagittarii on his flanks facing the Alamanni.[81] Arrian deepened his legionary formation to take the physical impact of the Alan cavalry charge but this impetus, it was hoped, would have been disrupted by an arrow storm projected from behind the line and from the flanks.[82]

In 6th century armies the armoured, bow-armed cavalry were the close-order battle-line. The tactics employed were intended to gain maximum effect of archery if the enemy was unable to reply in kind (in the West) or to fight an archery duel before charging in with the contus (in the East).[83] On one occasion Sarmatian cavalry with shorter-range bows attempted to lessen the effects of the enemy archery by discarding their bows, taking up contus and sword and charging in as soon as possible.[84] The greatest danger for an armoured, mounted army was to be drawn out and disordered by opposing light horse-archers. The Sassanid Peroz's army was fatally enticed by Ephthalites feigning flight and the main Byzantine line at Manzikert was stung into pursuing Selçuk horse-archers.[85] 'Feigned flight' was also employed by Roman horse-archers and on one occasion Ostrogothic cavalry were drawn forward by a horse-archer screen to make a fine, dense target.[86]

Unless armed with shafted-weapons and shields light horse-archers were at a great disadvantage if contacted by enemy troops with melee weapons.[87] However, their mobility enabled them to skirmish with both enemy cavalry and infantry formations. Facing infantry without missile-support they were extremely effective in harassment. If caught and routed by such troops the horsemen could easily outdistance pursuit and ensure that the defeat was indecisive. Precisely these tactics led to the Carrhae debâcle and to the savage mauling of Antonius' humiliatingly unsuccessful expedition.[88] Antonius realised the need for more cavalry and archers after the event. In the Imperial period this lack was rectified by the regular _auxilia_ and the wide use of client _symmachiarii_. Admittedly the invasions of Parthia by Trajan and Severus were aided by the weakness of the Arsacid monarchy[89] but Roman armies were well supplied with support troops and in no danger of suffering Crassus' fate on the battlefield. Even under Crassus the _legionarii_ could only be defeated in combat if detachments could be lured out and isolated. The Parthian _catafracti_ crushed the inferior Gallic cavalry but were too few to do more than hold the legionary front whilst horse-archers harassed the flanks.[90] Julian's advance down the Euphrates may have been ultimately disastrous but the Roman forces defeated the Sassanids in the major engagements, even when elephants were sent in to disorganise them. The Sassanids were superior in archers (horse, foot and elephant-mounted[91]) but the Roman troops were able to close with them quickly enough to lessen the effect of missiles.[92]

Infantry-archers were far more vulnerable than horsemen because they were more easily caught. At Idistaviso the Germans attempted to reach the _pedites sagittarii_ but these were protected by Gallic and Germanic _auxiliarii_ who repulsed the attack. Sabinus' archers in the Thracian war were effective at long range but were quickly routed by an unexpected barbarian attack.[93] Arrian protected his _Armeni_ on a hill by posting a _cohors Italica_ in front of and below them.[94] The auxiliary _sagittarii_, as in Vegetius' battle-order, were safe at the rear.

In less set-piece situations archers were used to support work-parties and river-crossings.[95] During sieges archery could be employed to drive the defenders from the parapets or archers with other missile troops could be put into siege-towers to support escalades.[96] The defensive value of archery is nicely demonstrated when an Ostrogothic, ox-drawn siege-tower being hauled up to the walls of Rome was neutralised by the simple expedient of shooting the oxen.[97] The mural provision of projecting stone towers, especially ones with cranellated tops, was primarily for archery, not the positioning of artillery-pieces, to give added range through height and command of the base of the curtain.[98]

Breeze has stated that Roman auxiliaries besieged in their forts were at a great disadvantage because they did not have suitable defensive weapons.[99] The wide distributions of laths and arrow-heads along the limites belies this view which is possibly a subconscious over-statement of the truism that the two British walls were not used as fighting platforms and that the army preferred to seek out the enemy in the field. Weapon-stores in forts must be envisaged as having had a few bows, bundles of arrows, bundles of light javelins, perhaps shaped stones for throwing and dropping, and a sack or two of caltrops to be scattered on causeways.

Caesar used ship-board archers, slingers and artillery to support troops making the contested landing of his first invasion of Britain. He had them in Africa also, and Antonius used archers at Actium. Vegetius recommended their employment and Belisarius hauled up archers to the tops of ships' masts.[100] A relief of two sagittarii from Koptos in Egypt demonstrates the use of Palmyrene archers on ships engaged in the Indian trade.[101] Fire-arrows would have been very useful in naval contexts, as in sieges.

Roman auxiliary archers were deployed all along the limites as an integral part of frontier armies. Units might be placed in positions of strategic importance, for example the cohors I Flavia Damascenorum mill. eq. at Friedberg in the neck of a frontier salient.[102] In more neutral positions regiments could support other troop formations with vexillationes in the field, as may have been the case with cohors I Hamiorum at corresponding medial positions on the British walls.[103] The cohors I Cilicum mill. certainly supplied a detachment to accompany Lower Moesian units in the Crimea, perhaps because of the Sarmatian enemy there.[104] However, concentrations of units of archers corresponded with areas where Roman armies faced peoples who were predominantly horsemen and/or archers.

In the Eastern theatre Armenian, Parthian, Jewish and Sassanid archery dictated that the Romans should raise units of oriental archers and employ local symmachiarii. Trajan's Parthian enterprises required the formation of specialist units of sagittarii and dromedarii, presumably aided by the annexation of Arabia.[105] The rump of the exercitus Cappadocicus on campaign with Arrian had a high proportion of archers amongst the auxilia, 5 out of 10 cohortes, and these were intended to play a very important part in the projected battle.[106] Knowledge of Cappadocia's entire auxiliary garrison is now fairly complete and out of 15 units 6 consisted of sagittarii with Armenian archers available in addition.[107] None of the alae were bow-armed and indeed Arrian's Techne Taktike only discusses Western cavalry in detail, mentioning horse-archers briefly.[108] With numerous infantry-archers to outrange enemy mounted bowmen perhaps greater need was felt for cavalry armed with melee weapons.

In Syria the native archers had a policing function in protecting the trading caravans between the 'caravan cities' from attack by desert nomads. The importance of this role is manifest in Palmyrene sculptural depictions of 'Caravan gods' armed and attired as light horse-archers and dromedary-bowmen (Fig.33).[109] Apart from Roman employment of these archers they were of wider importance in the 3rd century and were able to maul Sassanid troops retiring after plundering Syria.[110] The stationing of cohors I Flavia Chalcidenorum sag. eq. at Palmyra and of cohors XX Palmyrenorum mill. sag. eq. at Dura-Europos, probably respectively in association with the wars of Verus and Severus, suggest functions in the maintenance of desert communications.[111] Bodies of irregular Palmyrene archers were probably always present at Dura.[112]

The defence of the Danubian provinces involved conflict with three tactical classes of enemies. The Sarmatian peoples fielded armoured, bow-armed cavalry.[113] The Germanic tribes fought in close-order warbands.[114] The 'Free Dacian' tribes were probably skilled in mountain warfare and practised composite archery under Sarmatian influence.[115]

From the mid 1st century A.D. the North-South river line of Eastern Pannonia had a high proportion of cavalry facing the Iazyges as compared with the frontier of Northern Pannonia facing Germanic tribes.[116] After the Marcommanic Wars of Marcus in particular there was a marked concentration of auxiliary sagittarii.[117] At Szentendre (Ulcisia Castra) the cohors I mill. Aurelia Antonina Surorum (later on the cohors mill. Nova Surorum) was in garrison;[118] Nagyteteny (Campona) probably held the ala I Thracum veterana sag.;[119] Dunaujvaros (Intercisa) was garrisoned by the cohors I mill. Hemesenorum sag. eq.[120] In addition to Syrian archers units of North African troops versed in cavalry warfare were brought in.[121]

The Roman forces in Dacia flanked by Sarmatians and with Dacians to the north, contained a large number of cohortes sagittariorum centrally located or on the limes Alutanus facing the Roxolani.[122] These were supplemented by numeri sagittariorum, three of Palmyreni (Porolissensium, Optatianensium, Tibiscensium) and one of Suri (Malvensium).[123] These consisted of both cavalry and infantry and were presumably located to counter enemy archery in addition to fulfilling their frontier policing role.

Along the Danube, facing the Germanic tribes, units of sagittarii were markedly fewer except towards the 'knee' where

Iazyges were nearby and the ala III Augusta Thracum sag. was located at Almasfuzito (Azaum).[124] However, here and on the Rhine, greater use of archers, especially mounted ones, was favoured for armies in the field. An inscription from Carnuntum, apparently associated with Vespasianic forces in the Civil War, mentions a miles who was ex vexil(latione) sagit(tariorum) exer(citus) Syriaci.[125] A body of archers made up from various Eastern units might have seemed an effective expedient to supplement the Danubian defences whilst the legiones were active elsewhere. The army itemised by 'Hyginus' included 500 Palmyreni and similar, probably contemporary, troops are represented on the Marcus Column (Fig.24) for use primarily against the Germans.[126] Interestingly the only Western troops in the Notitia corresponding to the Eastern equites sagittarii indigenae all appear in the Pannonian commands.[127]

Further West Severus Alexander included Osrhoenians, Armenians and Parthians in the army assembled for his German campaign together with Moorish cavalry. "This force Alexander began to train to use against the Germans. An army of this kind is particularly harassing to them because the Moors with their javelin-throwing used their tactics of light-armed attack and withdrawal, and the archers found the Germans' bare heads and huge bodies an easy target for their arrows".[128] These were further augmented by Maximinus Thrax for his German campaign and later brought into Italy.[129] The sagittarii Osrhoeni were prominent enough to pose a danger to Maximinus through revolt.[130]

North-west Africa was another region where alae and cohortes sagittariorum were employed in some numbers.[131] The barbarian enemies consisted of skirmishing javelin-armed light cavalry (themselves employed widely by the Romans) and presumably javelin- or bow-armed light infantry.[132] The high proportion of part-mounted cohortes, especially in Numidia, reflected the tactical response to an elusive enemy in the field and the need to patrol vast areas. The composition of the Pannonian reinforcements sent in for the Mauretanian war of Antoninus Pius suggests a well-considered reaction to tactical problems posed by the enemy. The whole of ala I Ulpia contariorum mill. was joined by vexillationes of six other alae.[133] Of the latter three were from alae sagittariorum (III Augusta Thracum, I Thracum veterana and I Ituraeorum) and one may have been contus-armed (I Cannenefatium).[134] Evidently the vexillationes had to be mounted and were chosen with regard to their specialised weaponry. Perhaps experience in dealing with mounted adversaries was also a determining factor.[135]

Oriental numeri sagittariorum were also employed in Mauretania and Africa. In Numidia during the late 2nd to early 3rd century A.D. a numerus Palmyrenorum had its headquarters at

El-Kantara (Calceus Herculis), *vexillationes* of it appearing elsewhere, and a *numerus Hemesenorum* was also present.[136] In Mauretania Caesariensis a *numerus Osrhoenorum* was present at Sidi Ali ben Yub (Kaputtasaccora) and the *numerus Syrorum Malvensium* was in an extended position at Lalla Marnia (Numerus Syrorum) from the Severan period.[137] The latter was transferred from the *limes Alutanus*, may have been of milliarial strength and probably was part-mounted. The *numeri* perhaps primarily performed a policing function (as on other frontiers) but with the climate and terrain conditions being similar to those in Syria. This area and Dacia were the only occidental regions in which *numeri sagittariorum* were deployed with consistency.[138]

ACKNOWLEDGEMENTS

A great many people have advised the writer over several years on specific points. In addition to those acknowledged in the notes, a more general debt is owed to Prof. E. Birley, Mr G.C. Boon, Mr C.M. Daniels, Mr J.P. Gillam and Prof. J.J. Wilkes. The writer is particularly grateful to Dr B. Zouhdi, Dr K. As'sad and Dr A. Taha for granting access to the Damascus and Palmyra museum collections, and to Prof. A. la Regina and Arch. G. Martines for their tremendous help for studying the sculptural monuments of Rome. Much of this work would have been impossible without generous grants from the British School at Rome, the British School at Jerusalem, and the British Institute at Amman. Several fellow members of the Society of Archer-Antiquaries have been very helpful indeed, in particular, Lt-Cdr W.F. Paterson, Mr E. McEwen and Mr M. Anderssohn. Mr M.C. Bishop, Mr D.B. Boyson and Dr H. Dodge very kindly read the text and made many helpful suggestions and corrections. Lastly, Mr W. Hubbard and Mrs A. Gibson drew up the figures. The opinions expressed and any errors contained in this paper are purely the writer's and not necessarily held by the many people gratefully acknowledged here.

ADDENDUM

Two possible ear lath fragments have been found at St. Albans, Hertfordshire (Britannia) in Insula XIV. They are broken at both ends (3.6cm and 2.4cm in length) and appear to have back zone scoring on the convex faces. Their presence at a town site is perhaps similar to finds from Colchester and Silchester (Cat. Nos. 6 and 8) in their supposed civilian context - cf. S.S. Frere, *Verulamium Excavations*, Vol.III, (Oxford 1984), 75 No.293.

NOTES

INTRODUCTION

1. RICHMOND, 1935, 16.

2. Cf. especially DARKO, 1935; DE WEERD & LAMBRECHTS, 1938; CALLIES, 1964; SADDINGTON, 1970.

3. JONES, 1964, 655, 661-2.

4. LATHAM & PATERSON, 1970; FARIS & ELMER, 1945; KLOPSTEG, 1947.

5. CHIUNG, 1981.

I. BOWS

1. See in general ARMITAGE & CLUTTON-BROCK, 1976. The writer is very grateful to Dr S. Greep for discussing many aspects of bone, antler and horn materials.

2. MACDONALD & PARK, 1905-6, 524; NASH-WILLIAMS, 1932, 94, 96.

3. Ibid., 96.

4. VON GROLLER, 1901, 132; MACDONALD & PARK, 1905-6, 527.

5. WERNER, 1932, for the best summary.

6. RAUSING, 1967, 22-6; MAENCHEN-HELFEN, 1973, 223-5. Also EMENEAU, 1953, 79.

7. VON FLESCHENBERG, 1941.

8. For example, RAUSING, 1967, throughout.

9. WERNER, 1932, 48; POLASCHEK, 1932, 240, 256-7; ALFÖLDI, 1932, 18; WERNER, 1956, 48, Map 4.

10. MACDONALD & PARK, 1905-6, 523-8, Fig.44; MACDONALD, 1934, 283-4, Fig.39; ROBERTSON et al., 1975, 56, Fig.18.8. Hunterian Museum, Glasgow, Inv. F.1936.117(8-10). Dr L.J.F. Keppie very kindly discussed these pieces and allowed the writer to examine them.

11. ALLASON-JONES & MIKET, 1984, 2.16, 2.18; Museum of Antiquities, Newcastle upon Tyne, Inv. 1972.31; South Shields Museum, Inv. 1900.44. The writer is very grateful

to Miss L. Allason-Jones and Mr R. Miket for access to these pieces.

12. Corbridge Museum, Inv. 75.1219-23, 75.3630, 75.3634. Study of these pieces was kindly arranged by Mr J. Dore.

13. Inv. 75.3563.

14. BUDGE, 1903, No.339-40; Clayton Museum Inv. No.633-4.

15. BIRLEY, 1977, Pl.60. The writer is obliged to Mrs P. Birley for allowing examination of this lath.

16. CRUMMY, 1983, 138, Fig.160, Inv. 4245.

17. WEBSTER, 1958, No.156; GREEP, 1983, Fig.2.1-2. Museum of London, Inv. 13942, 18208, 20077. Mrs J. Hall kindly arranged for the study of these pieces.

18. WEBSTER, 1958, No.192; BOON, 1974, 68, Fig.8.7.

19. NASH-WILLIAMS, 1932, 94-6, Fig.42.

20. Ibid., Fig.42.1; BOON, 1972, Fig.30.12.

21. NASH-WILLIAMS, 1931, 131-3.

22. WEBSTER, 1965, 149, No.21, Fig.7.21.

23. GRÜNEWALD, 1981, 23.

24. STADE, 1933, 113-4, Fig.2; WERNER, 1956, 47.

25. BEHRENS, 1913-14, 72, No.23, Fig.6.20; KLUMBACH, 1968, 73,40, Pl.5.4; 1971, 231-2, Pl.96.9.

26. WERNER, 1932, 35, 57-8, Figs.2, 10; POLASCHECK, 1932, 256-7; RAUSING, 1967, 65.

27. O.R.L. Abt.B, No.8, 182, Pl.XX.80; ECKINGER, 1933, 289 n.[1].

28. FISCHER, 1973, Fig.48, No.4.

29. O.R.L. Abt.A, Str.6, 44, Fig.7.3-4; STADE, 1933, 111-2, n.[6], Fig.1.

30. ECKINGER, 1933, 289, n.[9].

31. STADE, 1933, 113.

32. ECKINGER, 1933, 289-90, Fig.1.

33. GRÜNEWALD, 1981, 23.

34. ULBERT, 1970, No.253-6, Pl.13.

35. PLANCK, 1983, 47-9, Fig.24.

36. WALKE, 1965, 55, 152, No.25-31, Pl.105.

37. VON GROLLER, 1901, 132, Pl.XXIV.22-5; WERNER, 1932, 33-5, 39, Fig.1; POLASCHECK, 1932, 241-2.

38. GRÜNEWALD, 1981, 23-4, Pl.16, 17. Mr M.C. Bishop has kindly drawn the writer's attention to the fact that the Waffenmagazin material is all commensurate with a 1st-2nd century A.D. date, despite disturbance. Perhaps all the laths are of this period therefore.

39. SALAMON, 1976, 48-50, Pl.24-6; LENGYEL & RADAN, 1980, 400, Fig.70.

40. Pers. comm. Mr S. James, contra ROSTOVTZEFF, 1935, 221-2.

41. BALFOUR, 1921, 306-7, Fig.14. The writer is very grateful to Miss L. Cheethan for allowing examination and photography of this piece.

42. Scene numbering follows CICHORIUS, 1896-1900.

43. For example on the 4th century B.C. Kul Oba electrum vase, MINNS, 1913, Fig.93-4; ROSTOVTZEFF, 1922, Pl.XXII; PHILLIPS, 1965, Fig.61; RAUSING, 1967, Fig.54.

44. WEBSTER, 1969, Pl.XXI. These lines are definitely on the stone itself and are not a product of the casting technique.

45. ROBINSON, 1975, 7, 83, 183.

46. KIESERITZKY & WATZINGER, 1909, No.501; MINNS, 1913, Fig.223; ROSTOVTZEFF, 1922, Pl.XXVII. For other bows with curved ears see Quivers and Bow-Cases, below n.58.

47. Scene numbering follows PETERSEN et al., 1896.

48. ORANGE & VAN GERKAN, 1939, 45-7, Pl.8-11.

49. FRESHFIELD, 1921-2, Pl.XVII, XX, XXIII.

50. G.R. III, Pl.IV,1; ÉSPÉRANDIEU, 1907-66, No.6136-7.

51. _Ibid._, No.6125.

52. _G.R._ III, Pl.IV,1; ÉSPÉRANDIEU, 1907-66, No.5861 (Monimus).

53. SMITH, 1968, Pl.XVI; WEBSTER, 1969, 151, Pl.XVI. The writer is very grateful to Dr D.J. Smith for his advice and patient discussion of this sculpture.

54. TOYNBEE, 1962, No.93, Pl.94.

55. PFEFFER, 1957, 121; INSTINSKY, 1958, Fig.1; HOLDER, 1980, Pl.3.B.

56. SPEIDEL, 1965, 86, n.540.

57. HOFMANN, 1905, Fig.23.

58. BARADEZ, 1954, Fig.10; BENSEDDIK, 1979, Fig.5.

59. CUMONT, 1942, Fig.25; _C.S.I.R._, _Ost._, I,5, No.9.

60. SEYRIG, 1937, Pl.IV; GHIRSHMAN, 1962, Fig.4,90; COLLEDGE, 1967, Pl.4.

61. CAMPBELL, 1968, 183-4, 191-6, 250, Pl.III-IV, XXII, XXVI, XXX, XXXVI-VIII, XL-XLI; ROSTOVTZEFF _et al._, 1939, Pl.XIV. For other deities see ÉSPÉRANDIEU, 1907-66, No.3540, 4916, 5233, 7804.

62. _A.R.B._, 55, No.9, Pl.XX.

63. For cross-sections see VON LUSCHAN, 1899, Fig.7; BALFOUR, 1890, Pl.VI; 1921, Fig.4-5; STONE, 1934, Fig.170; BROWN, 1937, Fig.2; RAUSING, 1967, Fig.3; LATHAM & PATERSON, 1970, Fig.10.

64. LEVI, 1947, Pl.XCa.

65. BALTY, 1969, Pl.VII.2, IX, XXXIV, XXXIX.

66. BRETT _et al._, 1947, 82, Pl.39.

67. GUSMAN, 1914, Pl.140.

68. BROWN, 1937, 5; DARKO, 1946-8, 87-9; RUBIN, 1955, 273-8; KAEGI, 1964, 98-9; BIVAR, 1972, 281-6; MAENCHEN-HELFEN, 1973, 228.

69. LATHAM, 1970; BIVAR, 1972, 290; LATHAM & PATERSON, 1970,

37-40; 1979, 79-80.

70. Dio, XL, 15 says exactly this. Cf. Julian, 63,C-D.

71. MEDINGER, 1933; DEBEVOISE, 1938, 86-7; ROSTOVTZEFF, 1943, 184-5; RUBIN, 1955, 273-5; BIVAR, 1972, 273-6.

72. BROWN, 1937, 1-3.

73. Ibid., 2.

74. RAUSING, 1967, 68.

75. Ibid., 100-1, 105, 150.

76. POPE, 1938, Pl.134a; GHIRSHMAN, 1962, Fig.340.

77. ROSTOVTZEFF, 1943, Pl.XVIII-XIX.

78. MAENCHEN-HELFEN, 1973, 224.

79. RAUSING, 1967, 105, n.17.

80. RAUSING, 1967, 105; PATERSON, 1969a, 29, Fig.1.3; MAENCHEN-HELFEN, 1973, 228-32. This is another inexact term because this bow-type occurs outside the Sassanid Empire and period. See EMENEAU, 1953, 86, Fig.10-14 for 4th-6th century Indian coins depicting it.

81. POPE, 1938, Pl.163a, 209-11, 213-4, 229, 239; GHIRSHMAN, 1962, Fig.236-8, 248, 252-4, 247, 314; BIVAR, 1972, Fig.23; MAENCHEN-HELFEN, 1973, Fig.7-8; HERRMANN, 1977, 15, 119.

82. BIVAR, 1972, 282, Fig.20-22; LATHAM & PATERSON, 1979, Fig.93.

83. GHIRSHMAN, 1962, Fig.343; BELENITZKY, 1968, Pl.34; PATERSON, 1969a, 31.

84. GHIRSHMAN, 1962, Fig.236-8; HERRMANN, 1977, 17, 132-3. Bows in the Tāq-i-Bustān game-park relief have angled ears.

85. No laths appear in MINNS, 1913; ROSTOVTZEFF, 1922; RUDENKO, 1970.

86. SULIMIRSKI, 1970, 32, 120. Suggested by the curled ears in Crimean art, Bow-Cases and Quivers, below, n.58.

87. Arrian, Ektaxis, 3,18.

88. DAVIES, 1977, 269-70.

89. SYME, 1929, 130-3; EADIE, 1967, 165-8 (for the evidence but his conclusions are worthless); SULIMIRSKI, 1970, 29; BIVAR, 1972, 274; MAENCHEN-HELFEN, 1973, 238-9; WILKES, 1983, 258.

90. Tacitus, Annals, VI,35.

91. RAUSING, 1967, 68-9, Fig.57.

92. BERGMAN, 1939, 121-4, Fig.30, Pl.18.10; HEDIN, 1940, 76-9.

93. BERGMAN, 1939, 122, Pl.XIII,a; HEDIN, 1940, 78. Composite bow in right background.

94. BERGMAN, 1939, 123.

95. Disribution map and discussion WERNER, 1956, 46-50, Map 4. See BROWN, 1937, 5; BERGMAN, 1939, 123-4; WERNER, 1939, 194, Fig.2a; FIELD & PROSTOV, 1940, 417; LÁSZLÒ, 1951, 99; RAUSING, 1967, 67, 123; MAENCHEN-HELFEN, 1973, 222.

96. SEBESTYEN, 1930; WERNER, 1932; ALFÖLDI, 1932.

97. WERNER, 1956, 48.

98. POLASCHEK, 1932, 239-40; ALFÖLDI, 1932, 18-22, Fig.1.

99. POLASCHEK, 1932, 241-2; WERNER, 1932, 33-5, Fig.1; ALFÖLDI, 1932, 23-4.

100. MAENCHEN-HELFEN, 1973, 255-6.

101. ALFÖLDI, 1932, Fig.2.

102. ALFÖLDI, 1932, 18-20, Fig.2; RAUSING, 1967, 68-9, 143; MAENCHEN-HELFEN, 1973, 222.

103. WERNER, 1932, Fig.5; ALFÖLDI, 1932, Fig.1; LÁSZLÒ, 1951, 99; WERNER, 1956, Pl.25.

104. ALFÖLDI, 1932, Pl.II; LÁSZLÒ, 1951, 93-8, Pl.XXI-II; HARMATTA, 1951, 143-4, 148; WERNER, 1956, 49-50, Map 4, Pl.16, 51-2, 61; PHILLIPS, 1965, 125-6, Fig.139; RAUSING, 1967, Fig.55; SULIMIRSKI, 1970, 192, Fig.72; ELMY, 1981.

105. DARKO, 1935, 465-9; 1946-8, 87-9; KAEGI, 1964, 98-9; BIVAR, 1972, 283-6; MAENCHEN-HELFEN, 1973, 227; HALDON, 1975, 11-13.

106. Sidonius Apollinaris, _Panegyric on Avitus_, 23; Gregory of Tours, _Historia Francorum_, II,8.

107. Zosimus, _Historia Nova_, 27, 34, 50; Procopius, _Wars_, V,xxii,5-6.

108. _Ibid._, I,i,8-15; III,viii,27-8; ix,11; V,xxvii,26-8; VIII,xxxii,7-10. See Agathius, _Historiae_, II,9.

109. WERNER, 1956, 48, 53; RAUSING, 1967, 67.

110. Procopius, _Wars_, I,xviii,31-4; BIVAR, 1972, 286.

111. Maurikios, _Strategikon_, i,2,2; DARKO, 1937, 119-22, 129, 134-7; 1946-48, 95; BIVAR, 1955, 62-5; 1972, 287; GROUSSET, 1970, 171-6; HALDON, 1975, 15, 21-2, 24. Other cavalry and archery equipment developed and spread by steppe peoples included _dracones_, _spangenhelme_, double-handed lances, types of quivers and bow-cases, the use of ear laths, sabres, sabretaches, pellises, thumb-rings, scabbard-slides and felt armour.

112. FETTICH, 1926, 46, 60. Fig.11,18; SEBESTYÉN, 1930, 206-19, Fig.1-6; HORVÁTH, 1935, Pl.I, III-V, IX, XVII-XVIII, XXXVI; MAROSI & FETTICH, 1936, 12, 14-16, 19, 46, 53, Pl.II-III, V-VI, VIII; VETTERS, 1948, 239, Pl.XX.F; WERNER, 1956, 35-6, 43; RAUSING, 1967, 69, Fig.30; GARAM _et al._, 1975, 63, 130, 215, 290-1, 308-9, 317-8; LENGYEL & RADAN, 1980, 410, Fig.75-6.

113. SEBESTYÉN, 1930, Fig.2.4; HORVÁTH, 1935, Pl.XVIII.7; MAROSI & FETTICH, 1936, Pl.III.7; RAUSING, 1967, Fig.30; GARAM _et al._, 1975, 215, 290-1, 317-8.

114. The next steppe bow-type, introduced by the Magyars, had the nock an even greater distance from the ear-tip and a reduced total of 6 laths. Dr G. Fabian has built working reconstructions with a moderately set-back handle and very long, stiff ears which angle sharply forward when the stave is reversed. Length 152.5cm. See SEBESTYÉN, 1930, Fig.9; RAUSING, 1967, 70, Fig.31; FABIAN, 1970, 12-16, Fig.1-3; DIENES, 1972, 38, Fig.10, Pl.18.

115. SEBESTYÉN, 1930, 211-12, Fig.8 for the clearest explanation.

116. _Ibid._, 214; HORVÁTH, 1935, 14, Pl.IV.11-12.

117. The writer is very grateful to Lt-Cdr W.F. Paterson for making available much information about recent Avar finds and Dr G. Fabian's reconstructions. See now FABIAN, 1984.

118. MACDONALD & PARK, 1905-6, 527.

119. CHIUNG, 1981, 172. Lt-Cdr W.F. Paterson, pers. comm.

120. RAUSING, 1967, 19. The term 'composite' was coined by General Pitt-Rivers (BALFOUR, 1890, 220).

121. BALFOUR, 1890, 227; 1921, 295-302; VON LUSCHAN, 1899, 228-33; MEDINGER, 1933, 228-30; BROWN, 1937, 4-5; KLOPSTEG, 1943, 185; EMENEAU, 1953, 78; FARIS & ELMER, 1945, 11-12; RAUSING, 1967, 147; MARSDEN, 1969, 9-10, Fig.2-3; LATHAM & PATERSON, 1970, xxv-viii; 1979, 79; MAENCHEN-HELFEN, 1973, 222.

122. KLOPSTEG, 1943, 178; PATERSON, 1966a, 80-1; 1969a, 29; RAUSING, 1967, 19-20, 144, 146; LATHAM & PATERSON, 1970, xxvii-viii.

123. PATERSON, 1966a, 85.

124. Ibid., 81, 83-4.

125. BROWN, 1937, 2, 5; PATERSON, 1966a, 72-3; LATHAM & PATERSON, 1970, 11; MCEWEN, 1979, 92.

126. KLOPSTEG, 1947, Fig.43; PATERSON, 1966a, Fig.3; RAUSING, 1967, Fig.62.

127. KLOPSTEG, 1943, 178; 1947, 145-7; PATERSON, 1966a, 78-9; 1966b, 19; RAUSING, 1967, 146-7; LATHAM & PATERSON, 1970, xxviii-ix; MCEWEN, 1979, 81.

128. PATERSON, 1966a, 79.

129. RAUSING, 1967, 17; MCEWEN, 1979, 81.

130. FABIAN, 1984.

131. Mr E. McEwen, pers. comm. See BROWN, 1937, 5.

132. LATHAM & PATERSON, 1970, 6, 16-17; FARIS & ELMER, 1945, 14.

133. STONE, 1934, 131, 682; MAENCHEN-HELFEN, 1973, 227.

134. ALFÖLDI, 1932, 18-20, Fig.2. Reproduced in MACDONALD, 1934, Fig.40; LENGYEL & RADAN, 1980, Fig.70.

135. BROWN, 1937, 2.

136. MAENCHEN-HELFEN, 1973, 226-7.

137. FARIS & ELMER, 1945, 160-1; LATHAM & PATERSON, 1970, 11.

138. FARIS & ELMER, 1945, 86.

139. LATHAM & PATERSON, 1970, 8.

140. MAENCHEN-HELFEN, 1973, 226.

141. VON LUSCHAN, 1899, 233; RAUSING, 1967, 157; LATHAM & PATERSON, 1970, 15-16; CHIUNG, 1981, 159-60.

142. FARIS & ELMER, 1945, 5; KLOPSTEG, 1947, 10; CHIUNG, 1981, 160.

143. LATHAM & PATERSON, 1970, 6-7.

144. FARIS & ELMER, 1945, 161; KLOPSTEG, 1947, 41-2, 121; PATERSON, 1966a, 70-1; RAUSING, 1967, 155; LATHAM & PATERSON, 1970, 11.

145. BALFOUR, 1890, Pl.VI; 1921, Fig.4-5; VON LUSCHAN, 1899, Fig.2,7-8; BROWN, 1937, Fig.2; PATERSON, 1966a, Fig.2; 1966b, 19; RAUSING, 1967, Fig.3; LATHAM & PATERSON, 1970, Fig.10; MCEWEN, 1979, 90.

146. MCEWEN, 1979, 91.

147. PATERSON, 1966a, 72, Fig.1; LATHAM & PATERSON, 1970, 11-12, Fig.12.

148. BROWN, 1937, 2.

149. MCEWEN, 1979, 92.

150. PATERSON, 1966a, 72.

151. Ibid., 70; MCEWEN, 1979, 91.

152. KLOPSTEG, 1947, 40-1; RAUSING, 1967, 155; CHIUNG, 1981, 165-6, 174.

153. VON GROLLER, 1901, 132; MACDONALD & PARK, 1905-6, 527.

154. MACGREGOR & CURREY, 1983, 75-6.

155. BALFOUR, 1879, Pl.IX; BROWN, 1937, 2; PATERSON, 1966a, Fig.1; LATHAM & PATERSON, 1970, Fig.9; CHIUNG, 1981, 177.

156. PATERSON, 1966a, 74.

157. RAUSING, 1967, 155.

158. BALFOUR, 1890, 229-30; 1897, 212; 1921, 291-301; ADLER, 1905, 12; FARIS & ELMER, 1945, 86, 161, 167-70; KLOPSTEG, 1947, 42, 121; EMENEAU, 1953, 79; PATERSON, 1966a, 81-2; RAUSING, 1967, 155; FABIAN, 1970, 12; LATHAM & PATERSON, 1970, 13; MCEWEN, 1979, 91.

159. BALFOUR, 1921, 292, Fig.1; BROWN, 1937, 1; MCLEOD, 1958, 401; 1962, 18.

160. *Pers. obs.*

161. PATERSON, 1969a, 30.

162. BALFOUR, 1897, 212; PATERSON, 1966a, Fig.2; RAUSING, 1967, Fig.3; CHIUNG, 1981, 183-4.

163. NASH-WILLIAMS, 1932, Fig.42.

164. PATERSON, 1966a, 74, Fig.1; LATHAM & PATERSON, 1970, Fig.7-8.

165. FARIS & ELMER, 1947, 162; KLOPSTEG, 1946, 46; PATERSON, 1966a, 75-6, Fig.1; LATHAM & PATERSON, 1970, 14-15, Fig.9.

166. CHIUNG, 1981, 185.

167. *Ibid.*, 159-60.

168. PATERSON, 1966a, 75; LATHAM & PATERSON, 1970, 14-15.

169. KLOPSTEG, 1947, 42; PATERSON, 1966a, 74-5; LATHAM & PATERSON, 1970, 14; FABIAN, 1970, 12; MCEWEN, 1979, 91-2. These are based on experimentation disproving the neck tendon identification of PAYNE-GALLWEY, 1907, 4; BROWN, 1937, 2; FARIS & ELMER, 1945, 186.

170. CHIUNG, 1981, 186.

171. KLOPSTEG, 1947, 47-9; PATERSON, 1966a, 75, Fig.1; RAUSING, 1967, 156; LATHAM & PATERSON, 1970.

172. *Ibid.*, Pl.2 and *pers. obs.*

173. PATERSON, 1966a, 76; RAUSING, 1967, 156; CHIUNG, 1981, 189.

174. KLOPSTEG, 1947, 50-1, Fig.21; PATERSON, 1966a, 77-8;

RAUSING, 1947, 156-7; LATHAM & PATERSON, 1970, 15, 177, Pl.18.3.

175. FABIAN, 1984, 30-1.

176. LATHAM & PATERSON, 1970, Fig.10.

177. FARIS & ELMER, 1945, 94-5; KLOPSTEG, 1947, 54-5; POPE, 1962, 35-6; LATHAM & PATERSON, 1970, xxix-xxxi; CHIUNG, 1981, 175, 192-3.

178. LATHAM & PATERSON, 1970, 14; MCEWEN, 1979, 92.

179. BALFOUR, 1890, 227-31; 1897, 212-3; 1921, 297; MCLEOD, 1958, 400; PATERSON, 1966b, 20; LATHAM & PATERSON, 1970, 15, Fig.2b-c; MCEWEN, 1979, 92. For leather and parchment see CHIUNG, 1981, 174, 191-2, Pl.IV,6-8. For other stave and grip materials see KLOPSTEG, 1947, 52; RAUSING, 1967, 157; FABIAN, 1970, 12; CHIUNG, 1981, 175.

180. EMENEAU, 1953, 79; PATERSON, 1966a, 77.

181. VERMASEREN, 1971, 7-8, Pl.III-V. For stave painting see BALFOUR, 1890, 232; ADLER, 1905, 15-16; KLOPSTEG, 1947, 52; EMENEAU, 1953, 79; NCLEOD, 1958, 400; PATERSON, 1966a, 77; MCEWEN, 1979, 92; CHIUNG, 1981, 175, 191.

182. GHIRSHMAN, 1962, Fig.62, 429; BELENITZKY, 1968, Pl.136-8; 1980, Pl.32.

183. KLOPSTEG, 1947, 47-51, Fig.13; CHIUNG, 1981, 162-73, Pl.I.

184. KLOPSTEG, 1947, 45, 51; RAUSING, 1967, 156-7; CHIUNG, 1981, 171, Pl.I.11.

185. KLOPSTEG, 1947, 45; CHIUNG, 1981, 162-3.

186. STONE, 1934, 678, Fig.868.

187. KLOPSTEG, 1947, 40-2, 53; CHIUNG, 1981, 162, 165-6, 172-3, 184-5, 188.

188. Ibid., 158-9, 161-2, 184-5.

189. Ibid., 173-5; KLOPSTEG, 1947, 40-2.

190. CHIUNG, 1981, 160.

191. KLOPSTEG, 1947, 52.

192. See MACMULLEN, 1960, 25-9; ROBINSON, 1975, 8; MANNING, 1976a, 152-3.

193. Digest, 50,6,7; WATSON, 1969, 181.

194. Vegetius, Epitoma rei militaris, II,11. A 2nd-3rd century A.D. papyrus includes arcus peracti amongst the products of a legionary fabrica (BRUCKNER & MARICHAL, 1979, No.409, Col.II,14). Mr M.C. Bishop kindly brought this reference to the writer's attention.

195. In addition to heads from pila, spears and javelins: RICHMOND & BIRLEY, 1940, 106, 112-3, Pl.XI. See CREDLAND, 1982, 16-18, for widely scattered suppliers of cross-bow quarrels for the Welsh wars of Henry III and Edward I.

196. SALAMON, 1976, 53-4.

197. MERRIFIELD, 1962, 38-9.

198. KLUMBACH, 1968, 36-7, Fig.1; 1971, 231-2.

199. Scriptores Historiae Augustae, Claudius, VII,5; Herodian, III,9,4.

200. MAENCHEN-HELFEN, 1973, 225.

201. BELENITZKY, 1968, 101.

202. CHIUNG, 1981, 156-7. In addition to blades Damascus was noted for its bows, FARIS & ELMER, 1945, 92.

203. PATERSON, 1966a, 81-2.

204. WERNER, 1932, 53; 1956, 48, 53; RAUSING, 1967, 67. Grave 86 in the Lombardic Nocera Umbra cemetery yielded ear and grip laths of Avar form (Museo dell'Alto Medio Evo, E.U.R., pers. obs.). An ivory plaque depicting Carolingian period cavalry suggests Avar or Magyar influence, see SCHINDLER, 1977, Fig.264.

205. STONE, 1934, 11; CREDLAND, 1981, 9; 1983, 12.

206. Notitia Dignitatum, Oc., IX,24, 28, 32.

207. HOFMANN, 1969, 63-5, 80, 112-13, 115.

208. PAYNE-GALLWEY, 1907, 60; STONE, 1934, 11-12; LATHAM & PATERSON, 1970, xxxi-ii, 18-19, Fig.3,a-b; CREDLAND, 1981, 9, 11; 1982, 17.

209. ÉSPÉRANDIEU, 1907-66, No.1679; MACGREGOR, 1975-6, 319.

210. ÉSPÉRANDIEU, 1907-66, No.1683; AYMARD, 1951, 334-5, Pl.XX.A.

211. STONE, 1934, 125.

212. MARSDEN, 1969, 7-8, 10-11, Fig.1,4.

213. BAATZ, 1978, 14, 16, contra MARSDEN, 1971, 209-10.

214. GUDEA & BAATZ, 1974, 66-7; BAATZ, 1978, 14-16.

215. BAATZ, 1983, Fig.123.

216. Vegetius, II,15; IV,21.

217. II,15; IV,21-2.

218. R.L.O., 10, 1909, 64, Fig.22; MACGREGOR, 1975-6, 319.

219. N.D., Or., VII,43, 57; VIII,46-7; IX,47; Oc., VII,97; XLI,23. MARSDEN, 1969, 196-7.

220. Ammianus Marcellinus, XVI,2, 5-6; WEBSTER, 1983, 118.

221. LATHAM & PATERSON, 1970, 9.

222. Vegetius, IV,9; BRENNEN, 1980.

223. Literary evidence collected by HALDON, 1970.

224. STONE, 1934, 13; CREDLAND, 1983, 15-16.

225. BERGMAN, 1939, 163-4, Pl.29.18.

226. GILBERT, 1975-6, MACGREGOR, 1975-6, 320.

227. STONE, 1934, 11; CREDLAND, 1983, 13-14; This is presumably what lies behind WEBSTER, 1969, 153.

228. LATHAM & PATERSON, 1970, 19.

II. OTHER EQUIPMENT

1. ERDMANN, 1976; 1982; DAVIES, 1977.

2. Ibid., 260.

3. For Asiatic heads see LE COQ, 1925, Fig.115; FETTICH, 1926,

Fig.12; WERNER, 1932, 42; POLASCHEK, 1932, 239; ALFÖLDI, 1932, 18, Pl.I,1-4, XXII.27; BERGMAN, 1939, 164, Pl.28, 42, 30.16; WERNER, 1956, 49, Pl.8, 11,46, 51, 53, 57; SULIMIRSKI, 1970, Fig.14, 16, 25, 42, 45, 49, 52, 60, 75; GARAM et al., 1975, 308-9, 317. GHIRSHMAN, 1946, Pl.XVI.17, XXI.13-19, XXXVI, XLVIII for Begram (Afghanistan) finds; OATES, 1959, 236, No.8, Pl.LIX, for examples from Ain Sinu and Hatra (Iraq). Mr J. Stewart and Mr A. Killick have kindly shown the writer heads from recent excavations at Jerash and Udruh (Jordan). See ROSTOVTZEFF et al., 1946, 57, 86, Pl.LI for Dura finds.

4. YADIN, 1963, 91, Fig.38-40; 1966, 57. Recent finds also from Gamala (Galilee).

5. DAVIES, 1977, 262-3.

6. WEBSTER, 1958, 86, No.94, 198; REED et al., 1964, 185; JOBEY, 1977-8, 89-90; POTTER, 1979, 223; pers. obs.

7. DAVIES, 1977, 264; ERDMANN, 1976, 78, except Krefeld-Gellep perhaps.

8. ROSTOVTZEFF, 1935, Fig.85; BAUR et al., 1932, Pl.XXXIV.4, XXXV.3-4; GHIRSHMAN, 1962, Fig.247, 252, 314.

9. RICHMOND & BIRLEY, 1940, 106, 112-13, Pl.XI.

10. NASH-WILLIAMS, 1932, 70, Fig.9; BOON, 1972, Fig.30.6.

11. ERDMANN, 1982, 9.

12. Ibid., 9-10; BUSHE-FOX, 1949, Pl.LIX.301.

13. POTTER, 1979, 222-3, Fig.88.105; ROSTOVTZEFF, 1936, 454. These do appear on the northern continental frontiers, e.g., WALKE, 1965, Pl.105.2-3.

14. BOSANQUET, 1904, 224-5, 290-1, Fig.16,27; MANNING, 1976b, 22-3, Fig.14.

15. BUSHE-FOX, 1949, Pl.LIX.294, 302.

16. Brit. X, 1979, 276, from the middle of the west ditch. The writer is very grateful to Dr D.J. Breeze for information about these heads.

17. Pers. obs., thanks to Mrs P. Birley.

18. JAMES, 1983, Fig.2-3.

19. MACDONALD & PARK, 1905-6, Fig.42; ROBERTSON et al., 1975, Fig.32.13; ATKINSON, 1942, 225, Pl.55. B.1.

20. Ammianus Marcellinus, XXIII,4,14-15; Vegetius, IV,18. The literary sources are well reviewed by BROK, 1978.

21. DAVIES, 1977, 262.

22. LATHAM & PATERSON, 1970, 144; COLLINS, 1975, 262.

23. JAMES, 1983, 142-3, Fig.2-4.

24. Pliny, Natural History, XVI,65,160.

25. ROSTOVTZEFF, 1936, 453, Pl.XXIV,1. See also BAUR & ROSTOVTZEFF, 1929, 18 and ROSTOVTZEFF, 1934, 166. Wooden ballista-bolts are repeatedly mistaken for arrows in the Dura reports starting with CUMONT, 1926.

26. ROSTOVTZEFF, 1936, 453-4, Pl.XXIV.1. For Egyptian dynastic reed stele with wooden fore-shafts see BALFOUR, 1897, 215-6, Pl.X.

27. YADIN, 1966, 96.

28. YADIN, 1963, 91, Fig.32, 40.

29. BERGMAN, 1939, 70, 79-82, 95, 102, 116, 164, Pl.7, 18, 28, 30. For painted Scythian stele with sinew whipping for the fletchings see RUDENKO, 1970, 143, 217-8.

30. BULANDA, 1913, Fig.41; SULIMIRSKI, 1970, Fig.42 from the Lower Volga region; LATHAM & PATERSON, 1970, xxix, Fig.55.

31. Pausanias, I,xxi,5-6; TODD, 1975, 175.

32. FABIAN, 1970, 16.

33. Ammianus Marcellinus, XXIII,4,14.

34. LATHAM & PATERSON, 1970, 26.

35. CHIUNG, 1981, 203-7.

36. PATERSON, 1966a, 80.

37. BALFOUR, 1897, 216; KLOPSTEG, 1943, 190. See POPE, 1938, Pl.229 for a crescent hunting arrow-head blade on a Sassanid dish.

38. MINNS, 1913, 68; RUDENKO, 1970, 143; TODD, 1975, 176; Pausanias, I,xxi,5-6; Ammianus Marcellinus, XXXI,2,9.

39. For experiments in penetration see POPE, 1962, 51, 54, 57-8, Pl.17; COLES, 1973, 123-8.

40. LATHAM & PATERSON, 1970, 25-6.

41. FARIS & ELMER, 1945, 107-9.

42. Plutarch, _Crassus_, 24-5.

43. Procopius, _Wars_, VI,ii,28. Cf. Celsus, _de Medecina_, VII,5,1-2 for the extraction of barbed missiles in this way.

44. I,i,12-15; xviii,32-5.

45. FARIS & ELMER, 1970, 109-10.

46. LATHAM & PATERSON, 1970, 25.

47. _N.D._, _Oc._, IX,24, 28, 32.

48. Such as _ballista_-bolts, ROSTOVTZEFF, 1936, Pl.XXIV.3.

49. KNOX _et al._, 1984, 99-100, Pl.11.A-B.

50. Plutarch, _Crassus_, 21, 25.

51. _Ibid._, 30; _Marcus Antonius_, 46, 49.

52. LÁSZLÒ, 1957, 158. See also Pliny _N.H._, XVI,65,159.

53. Frontinus, _Strategems_, IV,7,30; Dio, LVI,21,3; GROUSSET, 1970, 90; KAEGI, 1964, 100; LEYSER, 1965, 12. Dio comments that the Parthian lands were climatically good for bow-strings except in Winter (XL,15,4).

54. STONE, 1934, 619-21, Fig.793. For Chinese use on quivers 521, Fig.667. For Roman deities with caps see ÉSPÉRANDIEU, 1907-66, No.243, 3367, 3540, 4916, 5233, 5994, 6063, 6125, 6161, 6329, 7804, 8370, 9245.3.

55. See, for example, Dio, LVI,21,3 for soaked shields.

56. MINNS, 1913, 66-8, Fig.49, 53, 75, 93-4.

57. LATHAM & PATERSON, 1970, Fig.27, Pl.12; BIVAR, 1972, Fig.29.

58. KIESERITZKY & WATZINGER, 1909, No.559, 575, 587, 591, 593-4, 597, 599-600, 604, 606, 609, 618-9, 622, 624-5, 630, 633-5, 639-40, 642, 647, 650-52, 655, 657, 662, 664, 666, 669-71, 675, 680-1, 683, 686, 693, 703, 718; MINNS, 1913, Fig.213-6, 223; ROSTOVTZEFF, 1913, Pl.LI.6, LXXXIV.1-2, 4, XCIII.1-2; 1922, Pl.XXX.2, XXVIII.1.

59. ROSTOVTZEFF et al., 1935, Fig.40-1; ROSTOVTZEFF et al., 1936, Pl.XXX; GHIRSHMAN, 1962, Fig.13,86; COLLEDGE, 1976, Fig.26, Pl.33, 42-3.

60. SEYRIG, 1937, Pl.IV; GHIRSHMAN, 1962, Fig.4,90-1; COLLEDGE, 1967, Pl.4.

61. KIESERITZKY & WATZINGER, 1909, No.574, 626-7, 650; MAENCHEN-HELFEN, 1973, Fig.2; PHILLIPS, 1965, Fig.105 on a Sarmatian gold plaque.

62. GHIRSHMAN, 1962, Pl.69; COLLEDGE, 1977, Pl.20a.

63. SEYRIG, 1937, Pl.III.4; TAHA, 1982, Fig.V.

64. COLLEDGE, 1977, Pl.22; cf. Parthian terracotta on display in the British Museum, London, Inv.135684.

65. KRAELING, 1956, Pl.LXIV.

66. GHIRSHMAN, 1962, Fig.196-7, 220; BIVAR, 1972, Fig.10-11, 19, 21; HERRMANN, 1977, 88-9, 92-3, 100.

67. CUMONT, 1926, Pl.LIV; BAUR & ROSTOVTZEFF, 1931, Pl.XLIII.2; ROSTOVTZEFF, 1934, Pl.XXXV.4; ROSTOVTZEFF et al., 1939, Pl.LVI.2-3, XV; POPE, 1938, Pl.209-11, 213-4; GHIRSHMAN, 1962, Fig.223, 247, 253.

68. GHIRSHMAN, 1962, Fig.223-4; BIVAR, 1972, Fig.28; FUKAI & HORIUCHI, 1972, Pl.XLIX; HERRMANN, 1977, 135.

69. LE COQ, 1925, 21, Fig.33-24, 66, 94-100, 109; APPELGREN-KIVALO, 1931, Fig.82-3, 211, 307; GHIRSHMAN, 1962, Fig.363, 429; BELENITZKY, 1968, Pl.68, 74, 136-8; BIVAR, 1972, Fig.27; MAENCHEN-HELFEN, 1973, Fig.12A; BRINKER & GOEPPER, 1980, 258-62, Fig.157, 172-3; BELENITZKY, 1980, Pl.25, 30-1, 34.

70. FETTICH, 1926, Fig.18; MAROSI & FETTICH, 1936, 49, Pl.III.14; LENGYEL & RADAN, 1980, 410, Fig.75-6.

71. LÁSZLÓ, 1957, Fig.6; DIENES, 1972, Fig.5, Pl.20.

72. BRETT *et al.*, 1947, Pl.39.

73. SADLER & BAGATTI, 1949, 71, Pl.25.1.

74. GHIRSHMAN, 1962, Fig.343; BELENITZKY, 1968, Pl.34.

75. FUKAI & HORIUCHI, 1972, Pl.XL.

76. LE COQ, 1925, 21, Fig.32-3, 65, 93, 100-2; LÁSZLÒ, 1957, Fig.8; BELENITZKY, 1968, Pl.74, 110, 136-8; 1980, 11, 83, Pl.26-8, 30-1, 41, 44; BRINKER & GOEPPER, 1980, Fig.157, 172-3.

77. BULANDA, 1913, Fig.40; MAENCHEN-HELFEN, 1973, Fig.9.

78. LÁSZLÒ, 1957, 172-82, Fig.3-7.

79. HALDON, 1975, 21.

80. MORSE, 1886, 12-16, Fig.8-9; FARIS & ELMER, 1945, 43-4: "This is indeed a corrupt draw used by the ignorant" according to the Moroccan treatise!

81. Freeze-frame photography revealed the arrow's path, KLOPSTEG, 1943, 185-6, Fig.12-13; 1947, 161-2.

82. MORSE, 1886, 17, Fig.11-12; BULANDA, 1913, 40, Fig.25; FARIS & ELMER, 1945, 45-6; LATHAM & PATERSON, 1970, xxv, 34-6, 51-3, Fig.1, 17, 22.a-e; BIVAR, 1972, 284-5.

83. MORSE, 1886, 17-20, Fig.3,18; STONE, 1934, 14-17, Fig.22-3; PATERSON, 1963; LATHAM & PATERSON, 1970, Pl.8; WIGGINS, 1973.

84. PATERSON, 1969b, 28.

85. MORSE, 1886, 4. Japanese archers, especially in the Heian and Kamakura periods, often had a tight left sleeve and a baggy right one. For a bound Mongolian sleeve see WEBB, 1982, Fig.

86. ROBINSON, 1976, 29, 39; BUDGE, 1903, No.146; Clayton Museum Inv. No.474.

87. BAUR & ROSTOVTZEFF, 1931, 73-4.

88. British Museum, London, Inv.135684.

89. GHIRSHMAN, 1962, Fig.125; HERRMANN, 1977, Frontispiece.

90. PUGACHENKOVA, 1971, 70-1.

91. FARIS & ELMER, 1945, 43-5. Cf. PATERSON, 1966, 30, Fig.2; LATHAM & PATERSON, 1970, 53, Fig.23; BIVAR, 1972, 285-6.

92. Ammianus Marcellinus, XXV,1,13. For points of draw see FARIS & ELMER, 1945, 49-52. For the angle of the string at the hand see RAUSING, 1967, 143, 146-7; PATERSON, 1966a, 78; 1966b, 19; 1969, 30.

93. No thumb-rings occur in funerary contexts or artistic representations of Mongolian releases.

94. LE COQ, 1925, Fig.107; BELENITZKY, 1980, 80, Pl.34 (?). Cf. LATHAM & PATERSON, 1970, Pl.8, 12-13.

95. ERDÉLYI, 1966, 44.

96. BIVAR, 1972, 286, Fig.29.

97. Ibid., 284-5.

98. Ibid., 282, Fig.20-2; LATHAM & PATERSON, 1979, Fig.93.

99. POPE, 1938, Pl.163A; GHIRSHMAN, 1962, Fig.236-8; HERRMANN, 1977, 17, 132-3.

100. FUKAI & HORIUCHI, 1972, Pl.XL; HERRMANN, 1977, 135.

101. ÉSPÉRANDIEU, 1907-66, No.6125.

102. No.5801.

103. SMITH, 1968, 286.

104. ROBINSON, 1975, 29; TOYNBEE, 1964, 189; SMITH, 1968, 286.

105. ÉSPÉRANDIEU, 1907-66, No.3001.

106. SMITH, 1968, 286.

107. GHIRSHMAN, 1962, Fig.59.

108. CICHORIUS, 1896-1900, II, 310, 328-9; III, 189, 238.

109. CUMONT, 1926, Pl.XLIX-L.

110. ROSTOVTZEFF, 1935, 264.

111. WEBSTER, 1969, 153-4 evidently thought so.

112. ROBINSON, 1975, 85, Pl.237; 1976, 29.

113. Arrian, Ektaxis, 31; HALDON, 1975, 33.

114. ROBERTSON et al., 1975, 25.

115. Vegetius, I,20; II,15. See Julian, 57, C-D, partly contradictory, but in accord over lack of shields.

116. Dio, XL,15,2.

117. GRIGG, 1979, 111-12; 1983, 140-1.

118. N.D., Or., XI,7-8.

119. Procopius, Wars, I,i,13; HALDON, 1975, 18, n.33.

120. BELENITZKY, 1968, Pl.132, 136-8; 1980, 54, 80, 83, 116, Pl.23-4.

III. ARCHERS AND ARCHERY EQUIPMENT

1. Tacitus, Annals, I,56; II,16.

2. ROBERTSON et al., 1975, 26. The writer is very grateful to Dr L.J.F. Keppie for discussing this question.

3. WALKE, 1965, 85.

4. SCHÖNBERGER, 1973.

5. ULBERT, 1970, 12-16. It is difficult to imagine auxiliary troops wearing this type of armour.

6. BOON, 1974, 66-70, Fig.8, mixed garrison presence.

7. ROBERTSON et al., 1975, 24-6.

8. DAVIES, 1977, 265. See Dio, LXII,12,3-4 and JARRETT, 1969, 224.

9. ERDMANN, 1976, 10.

10. FORSTER & KNOWLES, 1911, 51, Fig.40; BIRLEY, 1977, Pl.58.

11. ERDMANN, 1982, 9.

12. S.H.A., Avidius Cassius, VI,3.

13. Arrian, Techne Taktike, 43,1.

14. Vegetius, I,15.

15. Ibid., II,23; III,4.

16. Suetonius, Divus Iulius, 68, surely unreliable.

17. NASH-WILLIAMS, 1932, 71, Fig.22. Legionary weapons such as pila and ballista-bolts suggest that the deposit was intended for legionary use, not to be sent out to auxiliary units.

18. ROBERTSON et al., 1975, 25.

19. DOMASZEWSKI, 1908, 26, No.39-40.

20. Ibid., No.41.

21. SPEIDEL, 1965, 45.

22. Ibid., 86, n.540. The second is perhaps identical with a fragmentary tombstone relief seen by the writer in the catacombs of SS Marcellino e Pietro.

23. Herodian, I,15,2. See SPEIDEL, 1975, 212, n.39.

24. DAVIES, 1977, 270.

25. WAGNER, 1938, 64-5; DAVIES, 1977, 269.

26. ROSTOVTZEFF, 1938, 96.

27. Arrian, Ektaxis, 1,21; DAVIES, 1977, 754.

28. C.I.L., VIII, 18042 = I.L.S., 2487.

29. GARBSCH, 1978, 33-4; cf. Arrian, T.T., 34,1-8.

30. C.I.L., III, 3672 = I.L.S., 2558. Cf. Dio, LXIX,9,6.

31. JOBEY, 1977-78, 87-9; WATSON, 1969, 60-1; KORFMANN, 1972, 6. See Arrian, T.T., 43,1; C.I.L., VIII, 18042 = I.L.S., 2487; Vegetius, II,23. Scenes, LXVI, LXX, LXXII, CVIII and CXIII on Trajan's Column depict irregular slingers. See N.D., Or., VII, 52 for late funditores.

32. DOMASZEWSKI, 1908, 59-61; R.E.[1], XVII, s.v. 'numerus', cols 1327-41; MANN, 1954; CALLIES, 1964, 182-209.

33. WAGNER, 1938, 56, 174-5, 212.

34. SPEIDEL, 1973, 170-1; BENSEDDIK, 1979, 75-8.

35. Liber de munitionibus castrorum, 29-30.

36. N.D., Or., V, 29-30, 40, 54-6; VI, 31-2, 40, 54-6; VII, 31-4, 56; VIII, 30-1; IX, 19; Oc., V, 170, 174, 193, 211; VI, 67-73, 77, 83-4.

37. N.D., Or., XXXI, 25-9; XXXII, 22-6, 29; XXXIII, 18-22; XXXIV, 25-9; XXXV, 20-3; XXXVI, 25-8; XXXVII, 20; XXXVIII, 11-12; HOFMANN, 1969, 240-1, 255, 264-5.

38. N.D., Or., XXVIII, 40, 44, 46; XXXI, 49, 60; XXXII, 33, 38, 44; XXXIV, 32, 46-7; XXXV, 30, 33; XXXVI, 34-5; XXXVII, 34; XXXVIII, 27-8.

39. Vegetius, III,14; N.D., Or., VII, 56.

40. DARKO, 1935, 461-9; 1937, 119-25; KAEGI, 1964, 98-9; BIVAR, 1972, 281-6; HALDON, 1975, 11-25.

41. KENNEDY, 1977, 527-30.

42. Caesar, de Bello Gallico, VII, 31, 36, 80; TODD, 1975, 175.

43. CHEESMAN, 1914, 84.

44. For lists of oriental units see CHEESMAN, 1914, 178, 180-3; DE WEERD & LAMBRECHTS, 1938, 230-4; DAVIES, 1977, 269-70; HOLDER, 1980, 227-8, 230-2; R.E.[1], I, s.v. 'ala', cols 1223-70; IV, 'cohors', cols 231-56; R.E.[2], I.A, s.v. sagittarius, cols 1743-57 (esp. 1744-6).

45. CHEESMAN, 1914, 82-4; CATACUZÈNE, 1927, 159-71; DE WEERD & LAMBRECHTS, 1938, 234-9; FITZ, 1972, 149-58.

46. MANN, 1974, 260; SPEIDEL, 1973, 171, 176-7; BENSEDDIK, 1979, 73, 75-8.

47. For example, Herodotus, I,136 of the Achaemenid Persians; Dio, XL,15,3 of the Parthians; Ammianus Marcellinus, XXV,1,13 of the Sassanids; Strabo, VII,4,6 of Sarmatian horsemanship. Cf. LATHAM & PATERSON, 1970, xxv, 78.

48. CHEESMAN, 1914, 82-4; CATACUZÈNE, 1927, 171; DE WEERD & LAMBRECHTS, 1938, 237; CALLIES, 1964, 206; MÓCSY, 1974, 154, 195; ROXAN, 1976, 59, n.1; HOLDER, 1980, 115.

49. C.I.L., III, 10307 = I.L.S., 2540; DOMASZEWSKI, 1908, 59; CATACUZÈNE, 1927, 165, 169; WAGNER, 1938, 145-6, 211; CALLIES, 1964, 193; FITZ, 1972, 127; SPEIDEL, 1975, 228-9.

50. Cf. Vegetius, I,15; II,15.

51. BENSEDDIK, 1979, 38-40, Fig.11.

52. KENNEDY, 1977, 526-7; BENSEDDIK, 1979, 40.

53. Especially N.D., Or., XXXVIII. XXVIII, 40 is a partial exception.

54. ROXAN, 1976, 61.

55. Synesius, Epistulae, 131, quoted by JONES, 1964, 646.

56. N.D., Oc., V, 170, 174, 211.

57. For good summaries see MCLEOD, 1965; RAUSING, 1967, 31; KORFMANN, 1972, 17-19.

58. Vegetius, II,23.

59. PATERSON, 1966a, 83-4, 86; FARIS & ELMER, 1945, 77, 167.

60. MCLEOD, 1965, 9-12.

61. BIVAR, 1972, 283; COLLINGWOOD & RICHMOND, 1969, 306. See also MARSDEN, 1969, 12, c.150-200 yards (137-83m).

62. JOBEY, 1977-78, 89-91.

63. MCLEOD, 1965, 13.

64. Ibid., 3.

65. FARIS & ELMER, 1945, 145-9.

66. PATERSON, 1966a, 84; LATHAM & PATERSON, 1970, 73-9, Pl.13-15.

67. PATERSON, 1966a, 83. See Plutarch, Crassus, 24.

68. Herodian, VI,7,8; Procopius, Wars, VIII,xxix,17; Agathius, II,9.

69. Dio, LXII,12,3.

70. Vegetius, I,20.

71. Tacitus, Annals, XIII,38.

72. LATHAM & PATERSON, 1970, 71-5, 80-2, Fig.28.

73. Plutarch, Crassus, 24; Trajan's Column, Scene XXXVII; ÉSPÉRANDIEU, 1907-66, No.5270; Procopius, Wars, I,i,14; Anna Comnena, Alexiad, XV,3; MINNS, 1913, Fig.9; DARKO, 1935, 450; ROSTOVTZEFF, 1943, 180-7; BRETT et al., 1947, Pl.39; SULIMIRSKI, 1952, 455-9, Pl.I; GHIRSHMAN, 1962, Fig.248, 285, 289, 363, 422, 437, 444-5, 449; COLLINS, 1965, 264; BALTY, 1969, Pl.IX, XII, XXXIV; LATHAM & PATERSON, 1970, 74, 82, Fig.27, Pl.15; BRINKER & GOEPPER, 1980, Fig.127; BELENITZKY, 1980, Pl.71.

74. Cf. Procopius, Wars, V,xxix,47.

75. DEBEVOISE, 1938, 209.

76. Vegetius, III,6.

77. Josephus, Jewish War, III,116; Arrian, Ektaxis, 1. Also Josephus, V,47 in Titus' advance to Jerusalem.

78. Onasander, Strategos, XVII; Vegetius, III,14; Arrian, Ektaxis, 29. See Plutarch, Marcus Antonius, 41-2.

79. Arrian, Ektaxis, 18; Vegetius, III,14; Julian, 57D. See Tacitus, Annals, II,16 and Josephus, Jewish War, V,131.

80. Arrian, Ektaxis, 21.

81. Caesar, de bello Africano, 60, 81; Tacitus, Annals, XIII,38, 40; Arrian, Ektaxis, 12, 14; Herodian, VIII,1,3; Vegetius, III,16; Ammianus Marcellinus, XVI,12,7. Also see Julian, 57B-C, for Constantius II's army similarly arrayed.

82. Arrian, Ektaxis, 26. See Dio LXXV,7,2 for Niger's deployment at Issos.

83. Procopius, Wars, V,xxvii,27-9; I,xviii,32-5.

84. Tacitus, Annals, VI,35.

85. Procopius, Wars, I,iv,8-14; KAEGI, 1964, 102, 104-7.

86. Procopius, Wars, V,xxvii,5.

87. Plutarch, Marcus Antonius, 44; Dio, XLIX, 29; Herodian,

IV,15,2-3.

88. Crassus had some archers but they were lost with Publius (Plutarch, _Crassus_, 25). Antonius had light infantry who did much to save his army from Crassus' fate (_Marcus Antonius_, 41-2).

89. DEBEVOISE, 1938, 218, 258-9.

90. Plutarch, _Crassus_, 24-5, 27-8. See Tacitus, _Annals_, XIII,40 for the refusal of Corbulo's troops to be drawn out. See COLLINS, 1975, 271 for Tatar ineffectiveness against formed infantry.

91. Ammianus Marcellinus, XXIV,2,5; XXV,1,13; Julian, 63B, 65B; Vegetius, III,24.

92. Ammianus Marcellinus, XXIV,6,9; XXV,1,16.

93. Tacitus, _Annals_, II,17; IV,47.

94. Arrian, _Ektaxis_, 13.

95. Caesar, _de bello Gallico_, II,19; Tacitus, _Annals_, I,56; Dio, LXVIII,21,2; LXXI,3.

96. Caesar, _de bello Gallico_, VIII,40; Josephus, _Jewish War_, II,535-6; III,168, 258, 285, 486; V,263, 281, 296; Vegetius, IV,6, 17, 21.

97. Procopius, _Wars_, V,xxii,7-9. See Vegetius, IV,9.

98. Vegetius, IV,29; BAATZ, 1983, 137-8.

99. BREEZE, 1982, 84-5.

100. Caesar, _de bello Gallico_, IV,25; _de bello Africano_, 20; Plutarch, _Marcus Antonius_, 64; Vegetius, IV,44; Procopius, _Wars_, V,v,13.

101. INGOLT, 1970-71, 198-9, Pl.VI.1.

102. BAATZ, 1975, 135-6; SCHÖNBERGER, 1973; WELLES _et al._, 1959, 25.

103. ROBERTSON _et al._, 1975, 24-6.

104. WAGNER, 1938, 119-20.

105. CALLIES, 1964, 164-5; BIRLEY, 1966, 56. For the title _Ulpia_

in Syrian diplomata see HOLDER, 1980, 206.

106. Arrian, Ektaxis, 26.

107. RITTERLING, 1902, 361-72; PELHAM, 1911, 227-30; CHEESMAN, 1914, 159-60; HOLDER, 1980, 201; SPEIDEL, 1983, 16-17.

108. Arrian, T.T., 44,1.

109. ROSTOVTZEFF, 1932a, 805-8; 1932b, 112; SEYRIG, 1941. See Bow-Cases and Quivers, above, n.59.

110. Zosimus, Historia Nova, I,39.

111. SEYRIG, 1933, 167; WELLES et al., 1959, 25.

112. Ibid., 24.

113. Trajan's Column scenes XXXI and XXXVII; KIESERITZKY & WATZINGER, 1909, No.650; ROSTOVTZEFF, 1913, Pl.LXIV.1, LXXIX, LXXVIII.1, LXXXVIII.2, CXXXIV.3; 1922, 169; SULIMIRSKI, 1970, 28-9; WILKES, 1983, 258.

114. Tacitus, Germania, 6; TODD, 1975, 174, 178.

115. Trajan's Column, scenes XXIV, XXXII, CXIII, CXXXIV. RICHMOND, 1935, 38 was mistaken in seeing an arrow held by Trajan in XXV. It is the lower half of a spear, depicted in both stone and bronze.

116. SZILAGYI, 1952, 191.

117. FITZ, 1972, 59; MÓCSY, 1974, 194.

118. SZILÁGYI, 1952, 196; FITZ, 1976, 77.

119. WAGNER, 1938, 69-71; FITZ, 1976, 93.

120. WAGNER, 1938, 142-6; FITZ, 1971, 46-52; 1976, 101; MÓCSY, 1974, 195.

121. SAXER, 1967, No.74; MÓCSY, 1974, 194.

122. WAGNER, 1938, 123-8, 157-9, 173-4, 182-3; GUDEA, 1979, 84-5.

123. MANN, 1954, 502, 504, 506-7; CALLIES, 1964, 182-3, 185, 205, 207; MACREA, 1964, 155; SPEIDEL, 1973, 1-2; 1975, 207.

124. SZILÁGYI, 1952, 200; FITZ, 1972, 39.

125. C.I.L., III, 13483a = I.L.S., 9168; SAXER, 1967, 119, No.33; MANN, 1954, 502. See also SAXER, 1967, No.43, possibly also sagittarii (?).

126. 'Hyginus', Liber de munitionibus castrorum, 29-30.

127. N.D., Oc., XXXII, 32, 35; XXXIII, 38, 44; XXXIV, 17, 21, 32-3.

128. Herodian, VI,7,8. Also S.H.A., Alexander Severus, LXI,8.

129. S.H.A., Maximini Duo, XI; Herodian, VII,2,1-5; VIII,1,3.

130. S.H.A., Maximini Duo, XI; Herodian, VII,1,9-11. See also S.H.A., Tyranni Triginti, 32 and C.I.L., XIII, 6677a (for a possible damnatio memoriae following revolt). Osrhoeni were also employed by Caracalla, Dio LXXVIII,14,1.

131. BENSEDDIK, 1979, 51-3, 65-6; HOLDER, 1980, 210-14.

132. Trajan's Column scene LXIV; SPEIDEL, 1975, Pl.2; HORN & DE RUGER, 1979, Pl.107.

133. BARADEZ, 1956; SPEIDEL, 1977; BENSEDDIK, 1979, 27-30, 35-6, 43, 47.

134. From a funerary relief, BARADEZ, 1954, Fig.12; 1956, Pl.II.5; BENSEDDIK, 1979, Fig.2. However, this is so similar to the ala contariorum tombstone figure that they are most likely the products of the same sculptor, thus an unreliable guide to the armament of the Canenafates.

135. BARADEZ, 1956, 8, 11.

136. CAGNAT, 1913, 205-6; CHEESMAN, 1914, 88, 165; CARCOPINO, 1925, 43-4, 119-34; SAXER, 1967, No.328-31; FENTRESS, 1979, 87-8, 91, 116-7.

137. CAGNAT, 1913, 251; WAGNER, 1938, 214-5; SPEIDEL, 1973, 170-1; BENSEDDIK, 1979, 73, 75-8.

138. See WAGNER, 1938, 216, for another numerus Syrorum, in Moesia Inferior. The writer is very grateful to Mr A. Rushworth for discussing the North African dispositions.

FIGURE CAPTIONS

Unless otherwise acknowledged the photographs are from the writer's collection.

Fig.1: Bow terminology. The key is the same for all the drawings.

Fig.2: Yrzi bow (after BROWN, 1937).

Fig.3: Belmesa bow ear.

Fig.4: Qum Darya bow fragments. Left Piece A, right Piece B (after BERGMAN, 1937)

Fig.5: Stages of stave construction.

Fig.6: Ch'engtu workshop plan. A = drying cabinet with brazier; B = work table and two benches; E = 'big board bench'; F = areas for the storage of materials which are also hung from the rafters above (after CHIUNG, 1981).

Fig.7: Stave positions - I Yrzi bow; II 'Sassanid' bow: a = full draw, b = strung or braced, at rest, c = unstrung and reversed (after BROWN, 1937 and PATERSON, 1969a).

Fig.8: Arrow terminology. A = general arrow, B = Dura type with wooden pile, C = Bar Kochva Cave of Letters with two part, wood and reed stele.

Fig.9: Bar Hill ear laths. Convex faces (by permission of the Hunterian Museum, Glasgow).

Fig.10: Bar Hill ear laths. Flat faces (by permission of the Hunterian Museum, Glasgow).

Fig.11: Caerleon ear laths. Concave faces of the more complete examples (by permission of the National Museum of Wales).

Fig.12: Caerleon ear laths. Flat faces, note cellular structure (by permission of the National Museum of Wales).

Fig.13: Caerleon grip laths. Convex faces (by permission of the National Museum of Wales).

Fig.14: Caerleon grip laths. Flat faces (by permission of the National Museum of Wales).

Fig.15: Belmesa bow ear. Sinister face, note score lines (by permission of the Pitt Rivers Museum, Oxford).

south-east of the 'Funerary Temple', Palmyra.

Fig.35: Detail of statue of Atys. London, British Museum.

Fig.36: 'Worcester Hunt' mosaic from Daphne, Syria (after LEVI, 1947).

Fig.37: Detail of 'Triclinos', hunt mosaic. Apamea Syriae, Musées Royaux, Brussels.

Fig.38: Hunt mosaic detail, Great Palace, Istanbul (after BRETT et al., 1947).

Fig.39: Parthian terracotta horse-archer. Berlin Museum (after GHIRSHMAN, 1962).

Fig.40: Aquatic hunt scene, Tāq-i-Bustān, Iran (after GHIRSHMAN, 1962).

Fig.41: Mounted Sassanid king (Chosroes II?), Tāq-i-Bustān, Iran (after GHIRSHMAN, 1962).

Fig.42: Sven Hedin with finds from Mass Grave 1, beside Lop Nor. Complete Qum Darya bow in bottom right (after BERGMAN, 1939).

Fig.43: Detail of horse-archer fresco from Khasr el-Hayr el-Gharbi, National Museum in Damascus.

Fig.44: Detail of Sassanid Persian silver dish (after GHIRSHMAN, 1962).

Fig.45: Stele with fletchings and wooden piles from Dura-Europos, Syria (after ROSTOVTZEFF et al., 1936).

Fig.46: Tanged, trilobate arrow-heads and stele from Masada, Palestine (after YADIN, 1966).

Fig.47: Barbed, flat, tanged arrow-heads from Bearsden, Dumb. (by permission of the Scottish Development Department, Ancient Monuments).

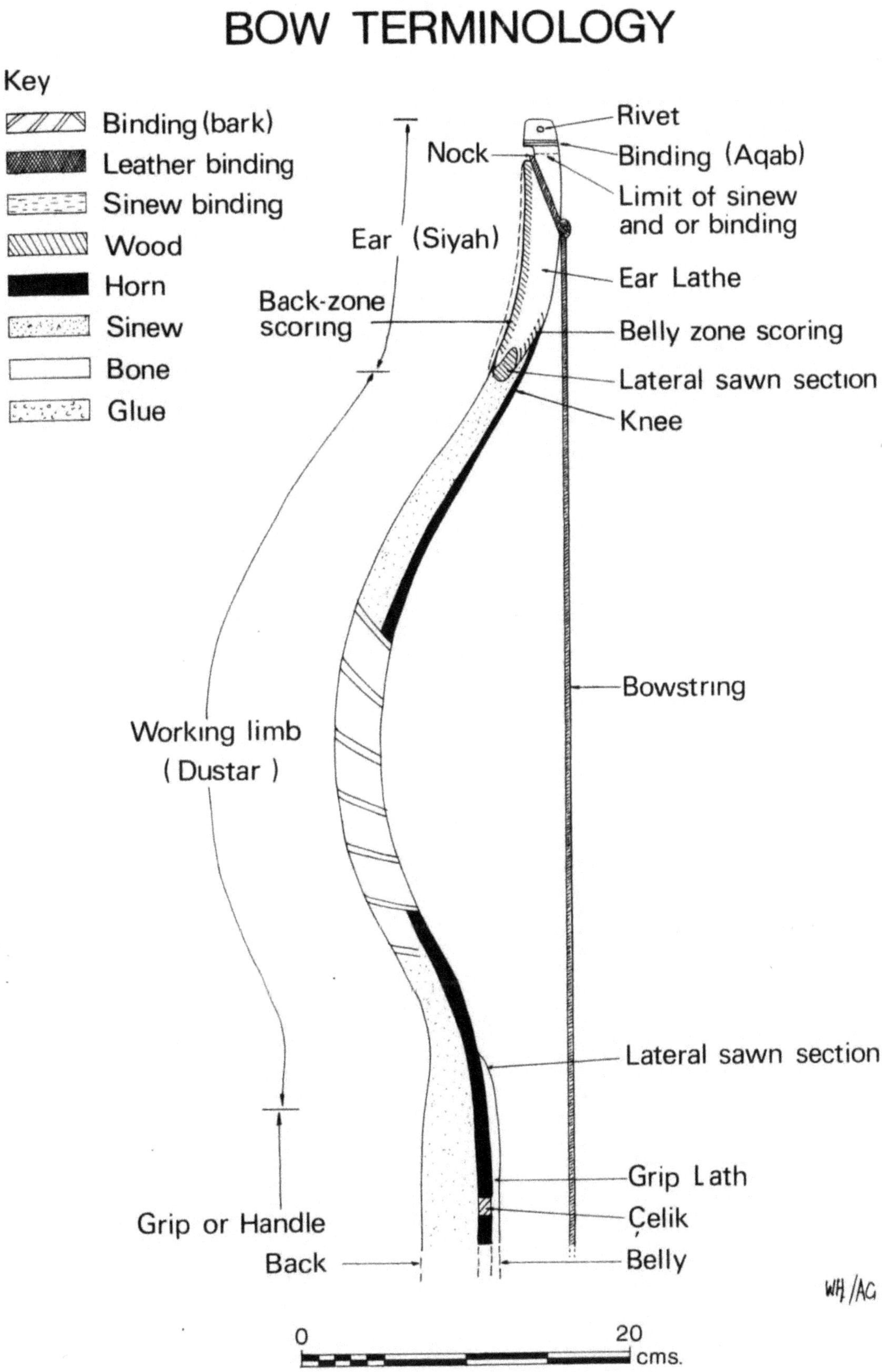
BOW TERMINOLOGY
Key
Binding (bark)
Leather binding
Sinew binding
Wood
Horn
Sinew
Bone
Glue
Nock
Ear (Siyah)
Back-zone scoring
Working limb (Dustar)
Grip or Handle
Back
Rivet
Binding (Aqab)
Limit of sinew and or binding
Ear Lathe
Belly zone scoring
Lateral sawn section
Knee
Bowstring
Lateral sawn section
Grip Lath
Çelik
Belly
WH/AG
0
20
cms.

Fig.1

YRZI BOW

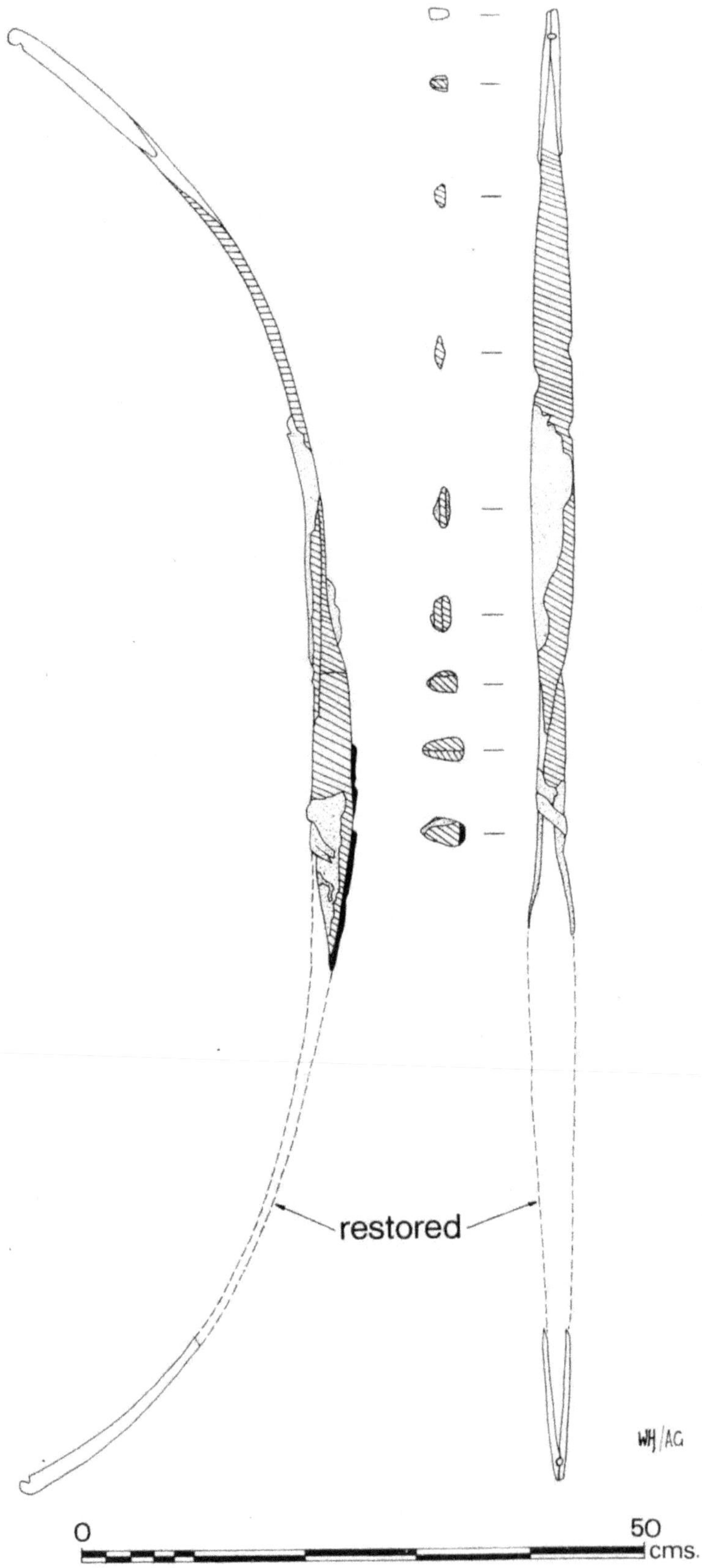

Fig.2

BELMESA EAR

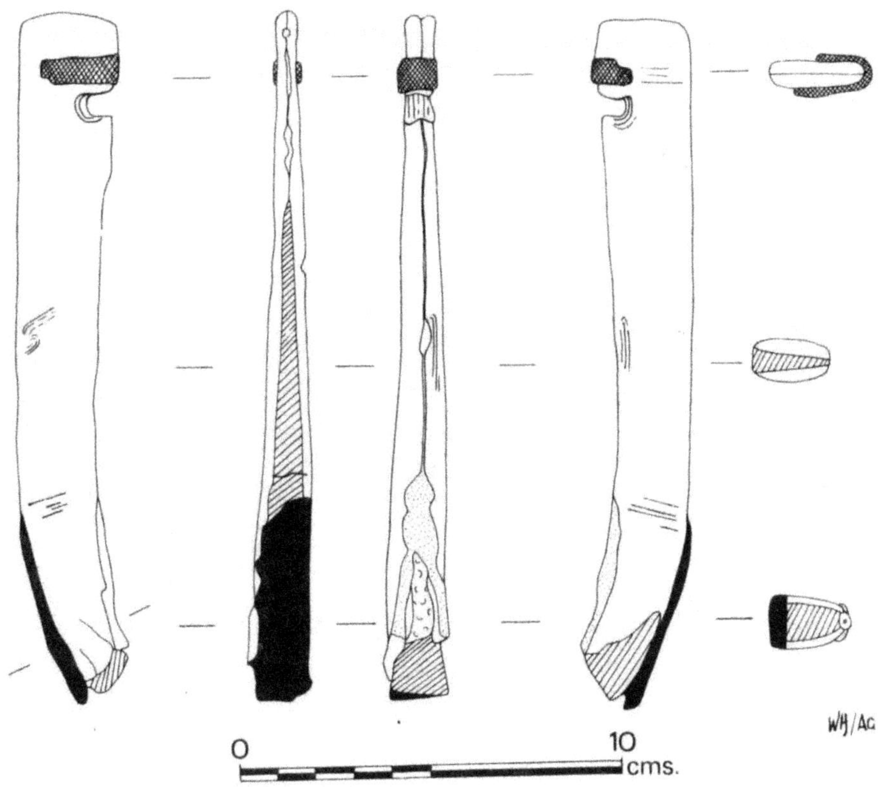

Fig.3

QUM DARYA BOW

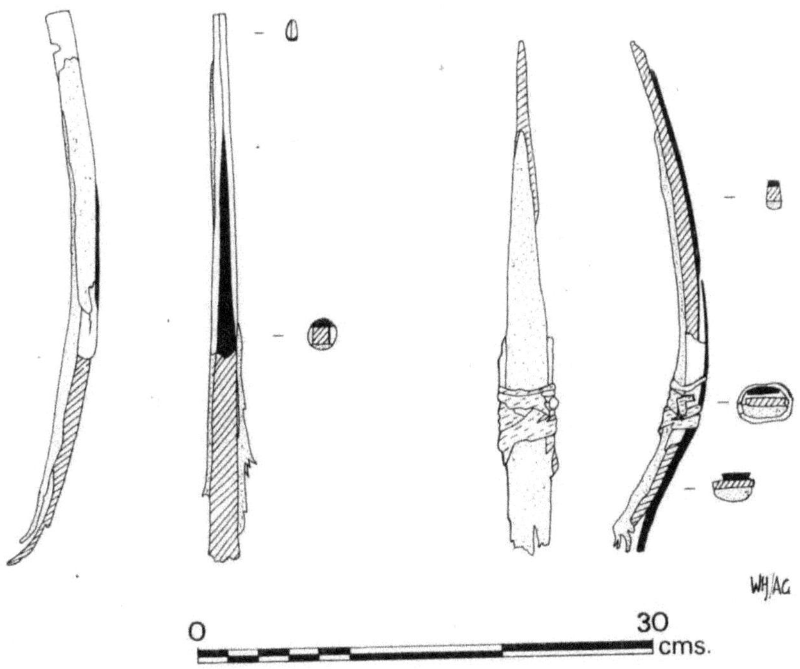

Fig.4

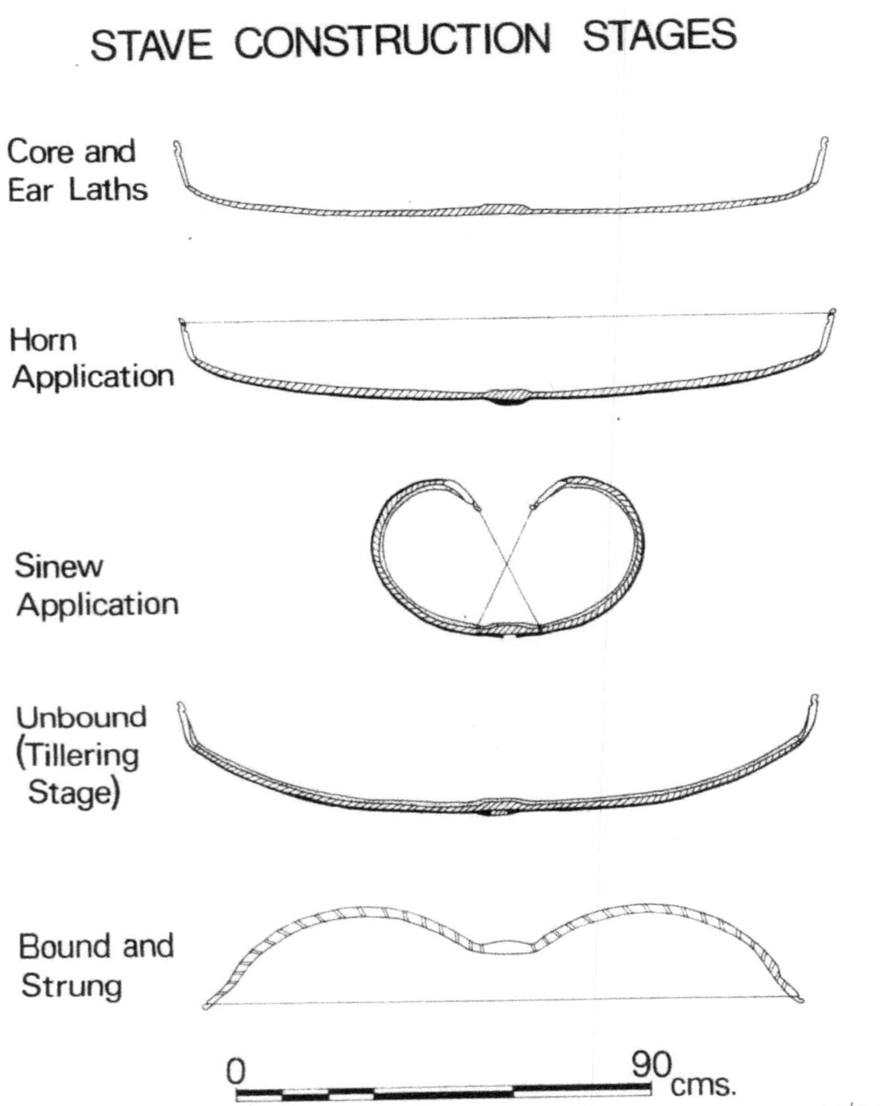
STAVE CONSTRUCTION STAGES
Core and Ear Laths
Horn Application
Sinew Application
Unbound (Tillering Stage)
Bound and Strung
0
90 cms.
WH/AG

Fig.5

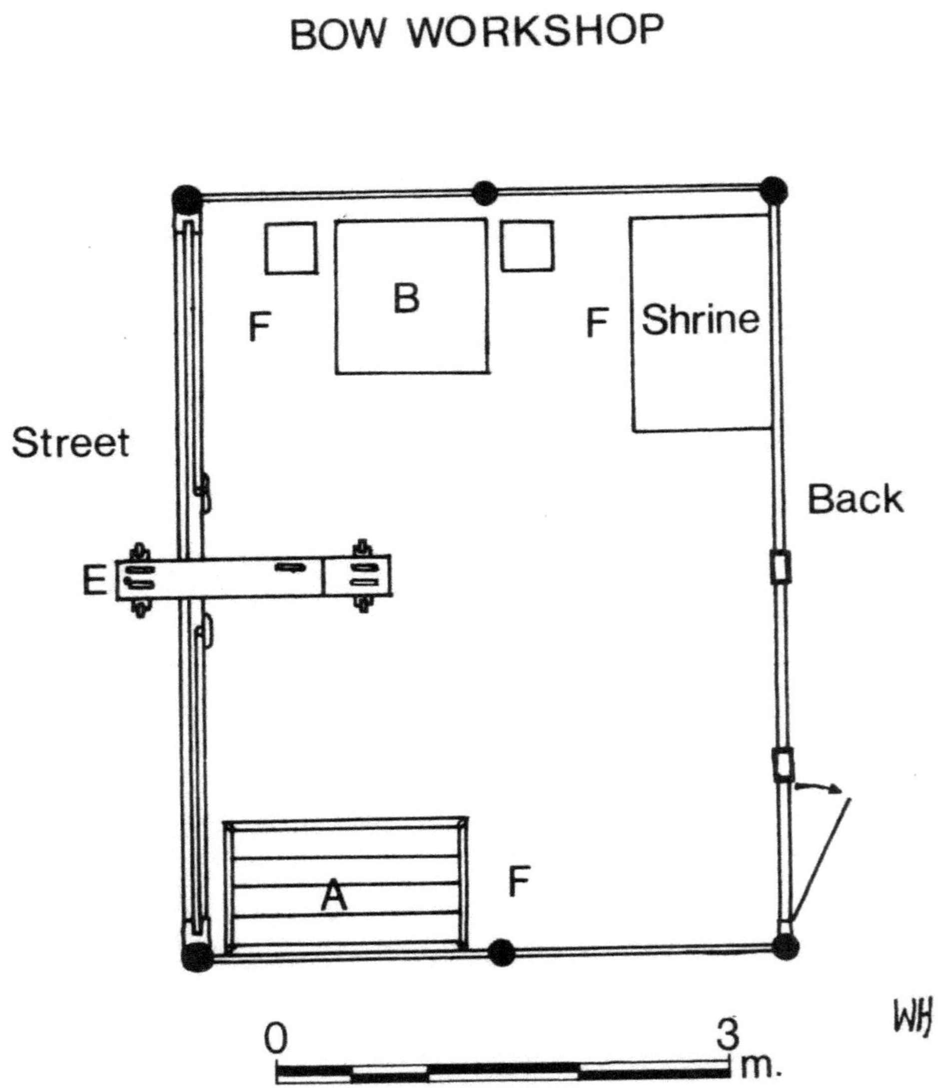
BOW WORKSHOP
Street
Back
B
F
F
Shrine
E
A
F
0
3 m.
WH

Fig.6

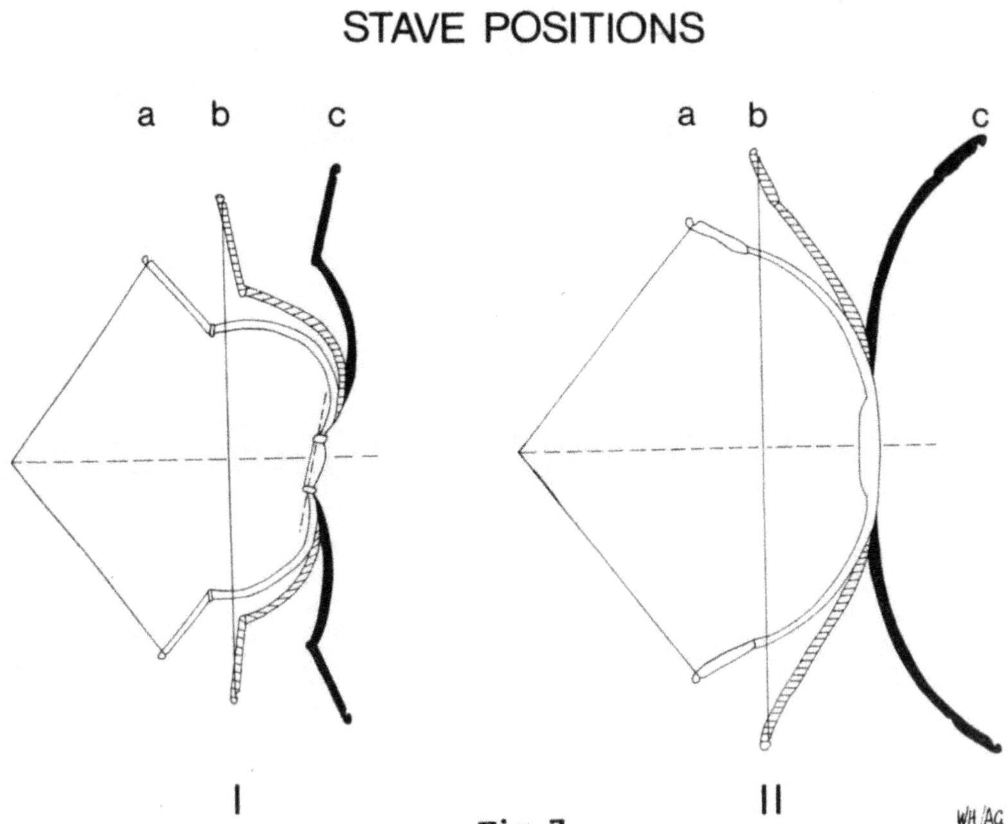
STAVE POSITIONS
a
b
c
a
b
c
I
II
WH/AG

Fig.7

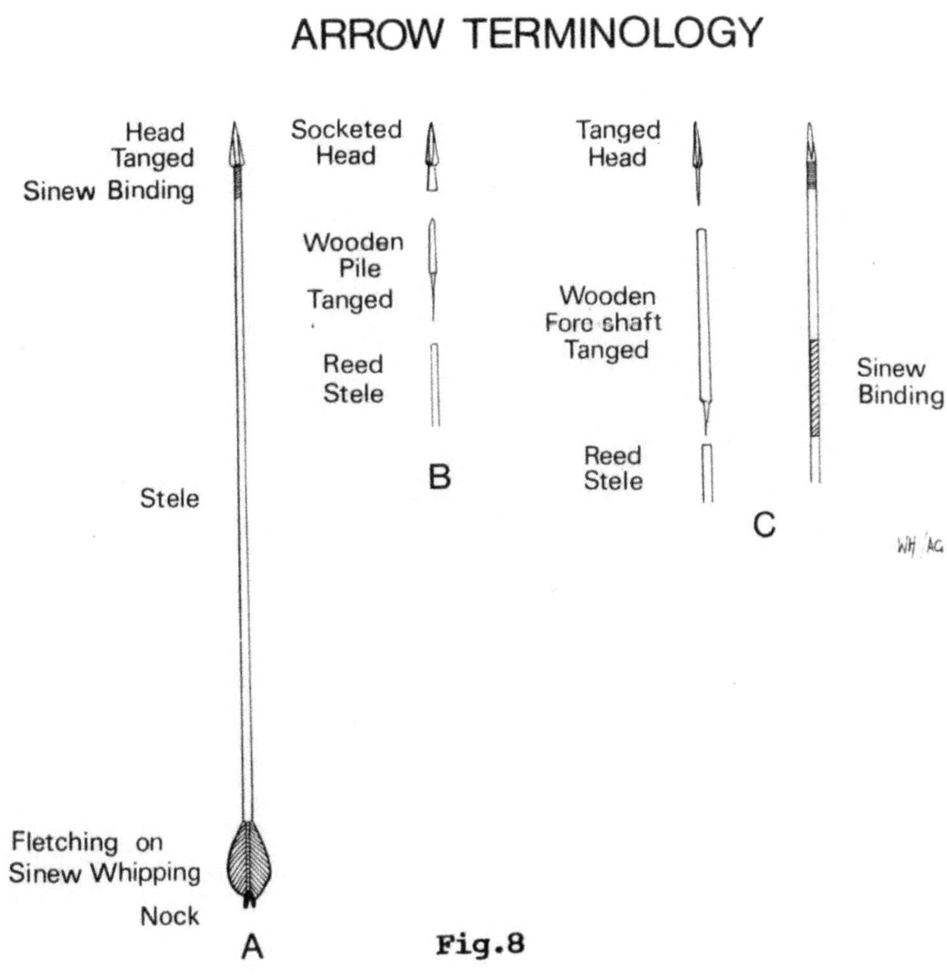
ARROW TERMINOLOGY
Head
Tanged
Sinew Binding
Stele
Fletching on
Sinew Whipping
Nock
A
Socketed
Head
Wooden
Pile
Tanged
Reed
Stele
B
Tanged
Head
Wooden
Fore shaft
Tanged
Reed
Stele
C
Sinew
Binding
WH/AG

Fig.8

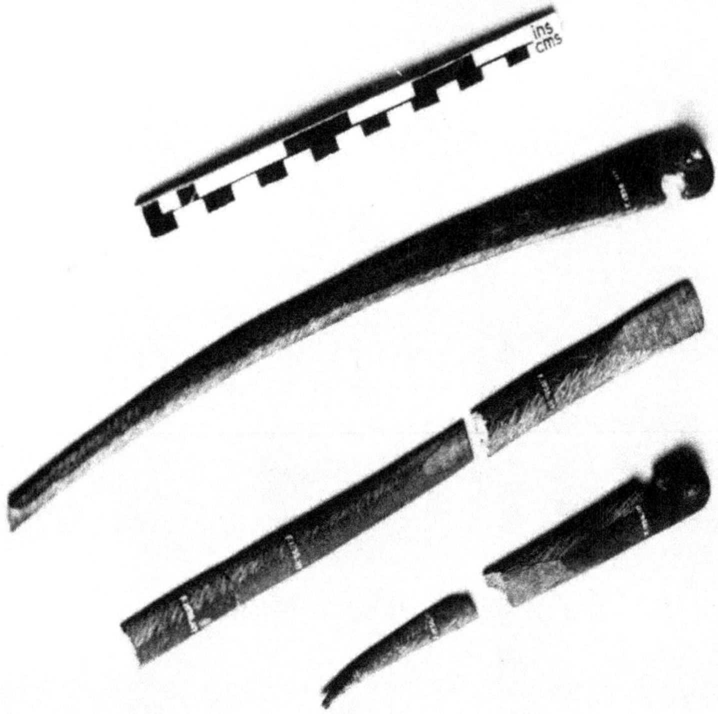

Fig.9

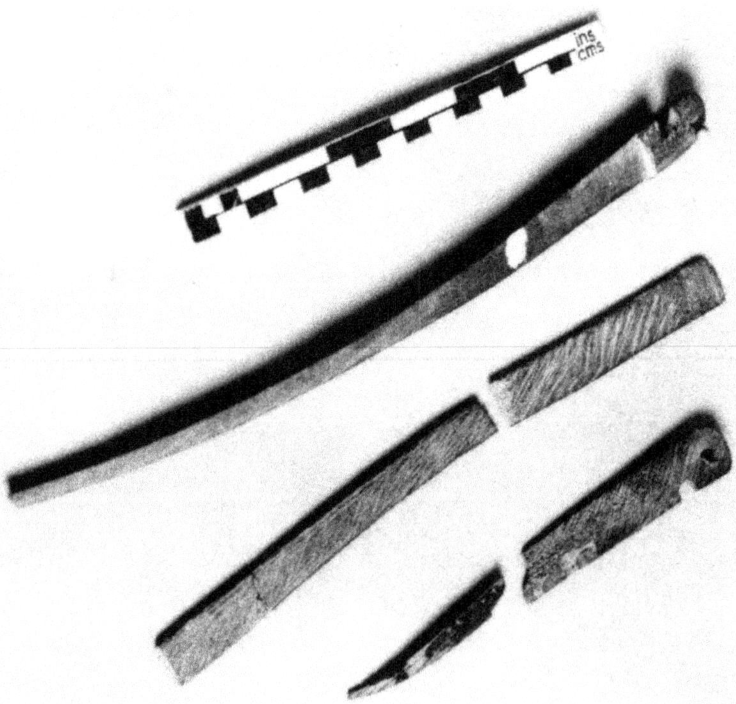

Fig.10

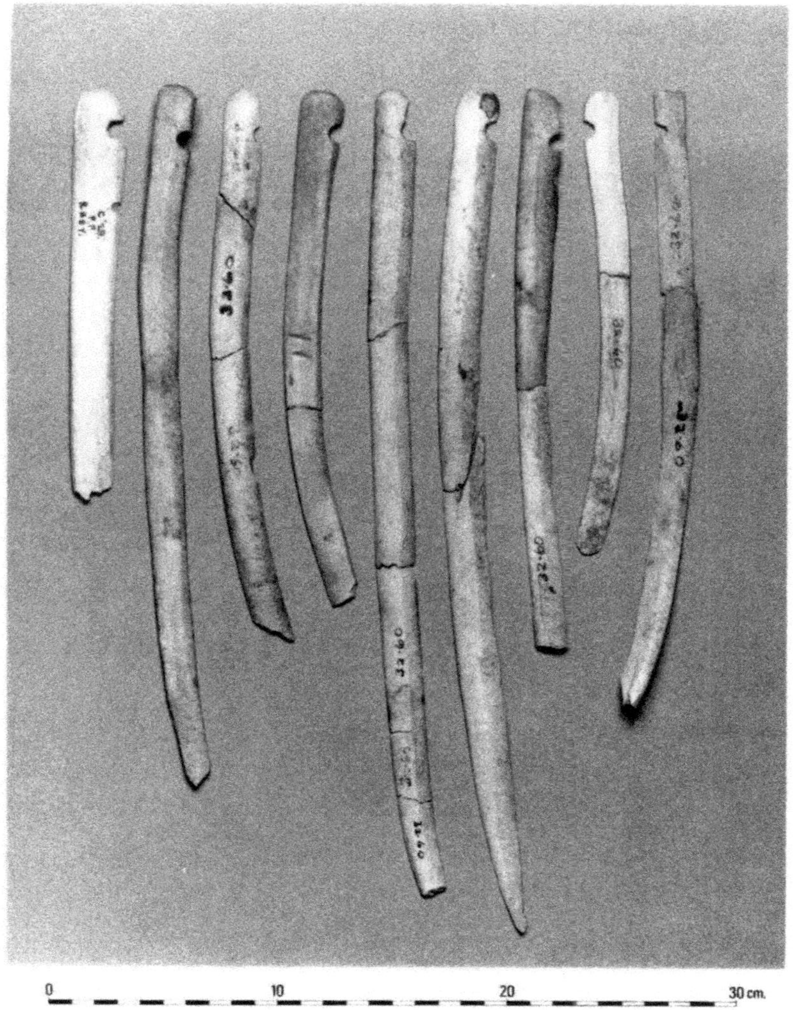

Fig.11

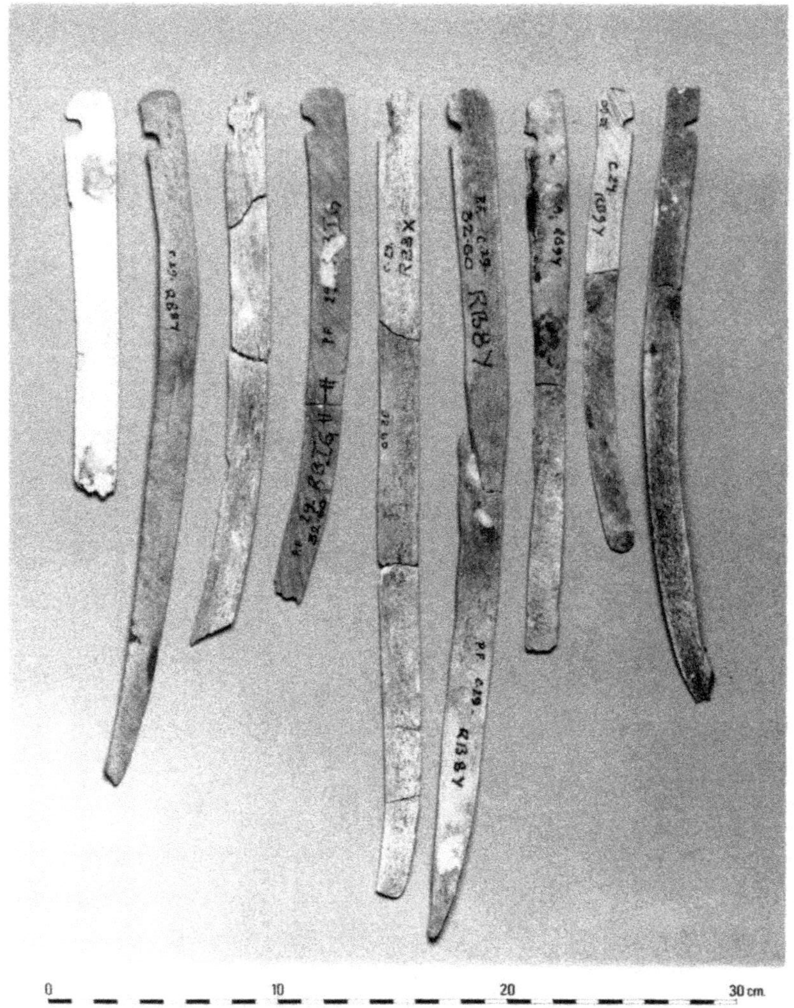

Fig.12

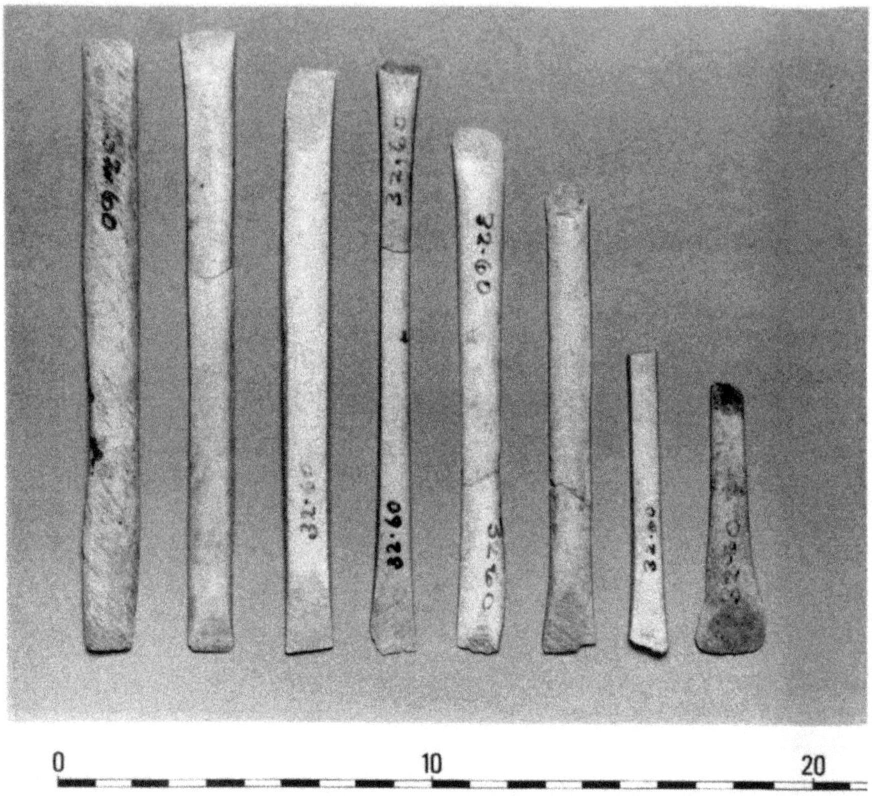

Fig.13

Fig.14

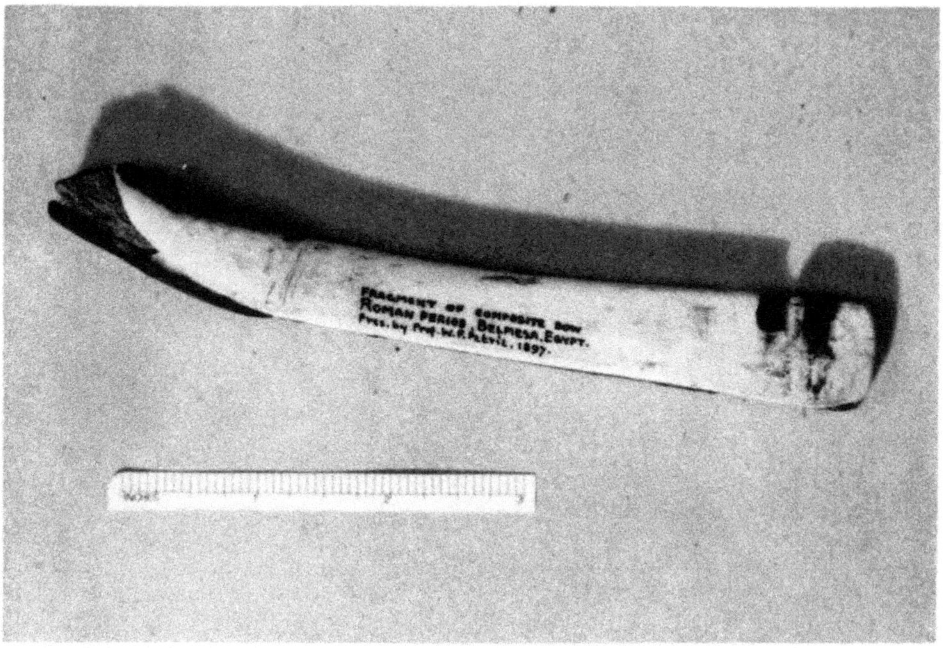
FRAGMENT OF COMPOSITE BOW
ROMAN PERIOD, BELMESA, EGYPT.

Fig.15

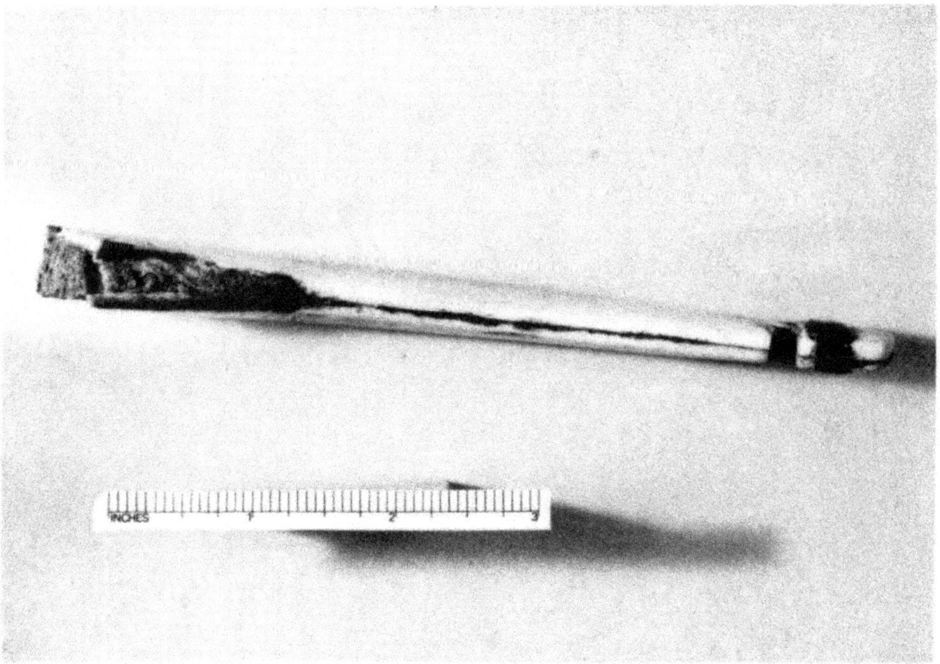
INCHES 1 2 3

Fig.16

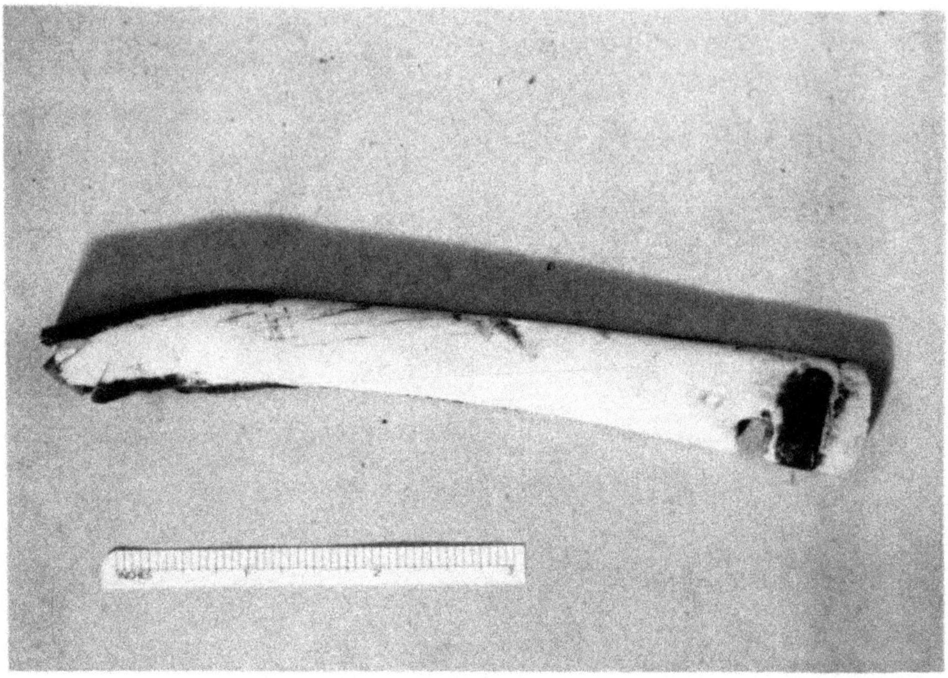

Fig.17

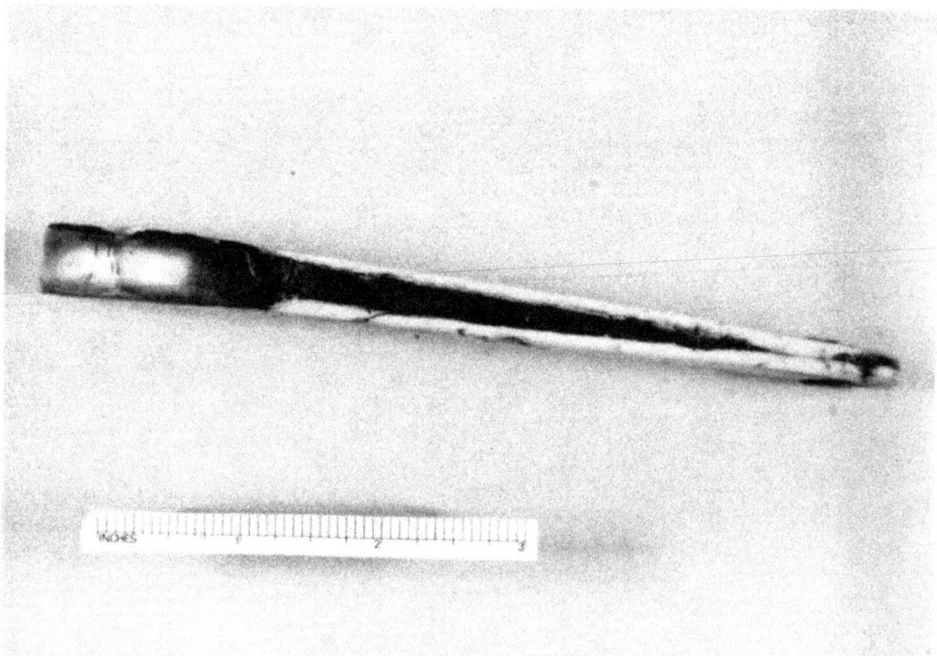

Fig.18

Fig.19

Fig.20

Fig.21

Fig.22

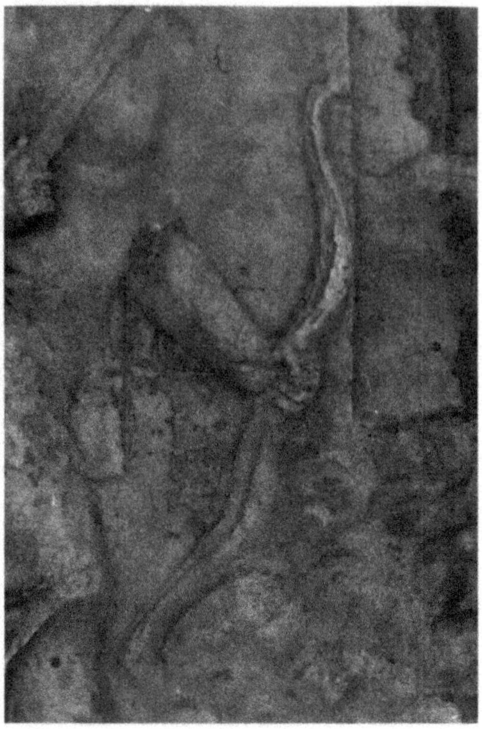

Fig.23

Fig.24

Fig.25

Fig.26

Fig.27

Fig.28

Fig.29

Fig.30

Fig.31

Fig.32

Fig.33

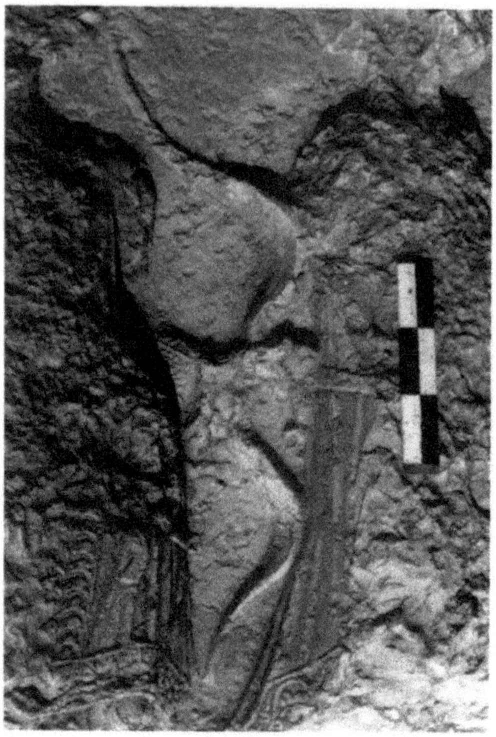

Fig.34

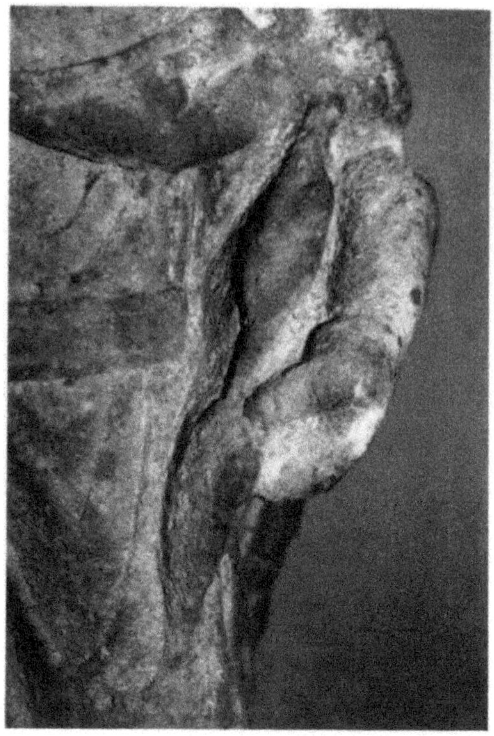

Fig.35

Fig.36

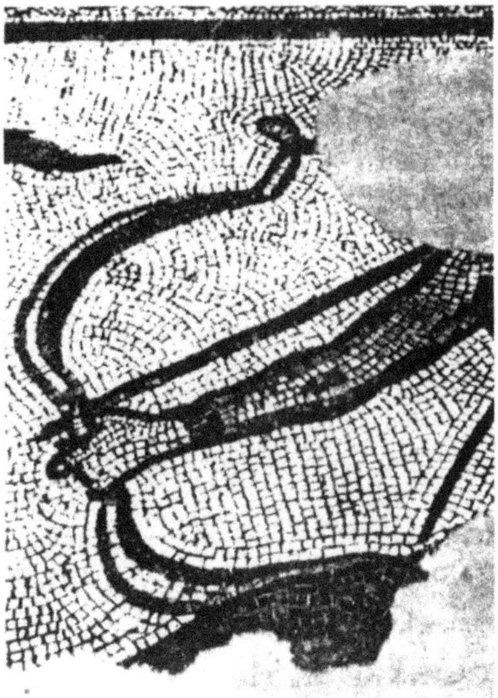

Fig.37

Fig.38

Fig.39

Fig.40

Fig.41

Fig.42

Fig.43

Fig.44

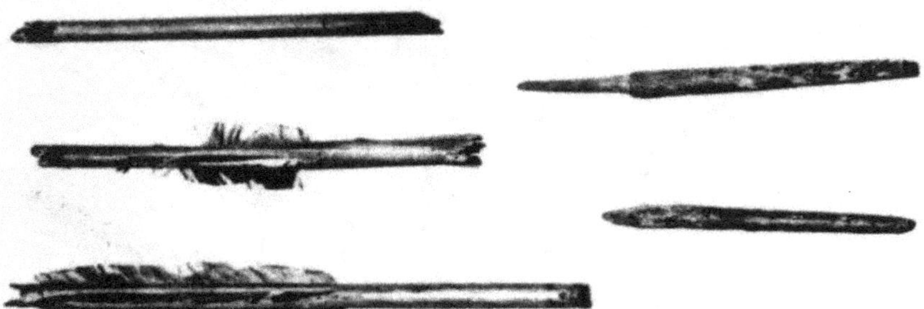

Fig.45

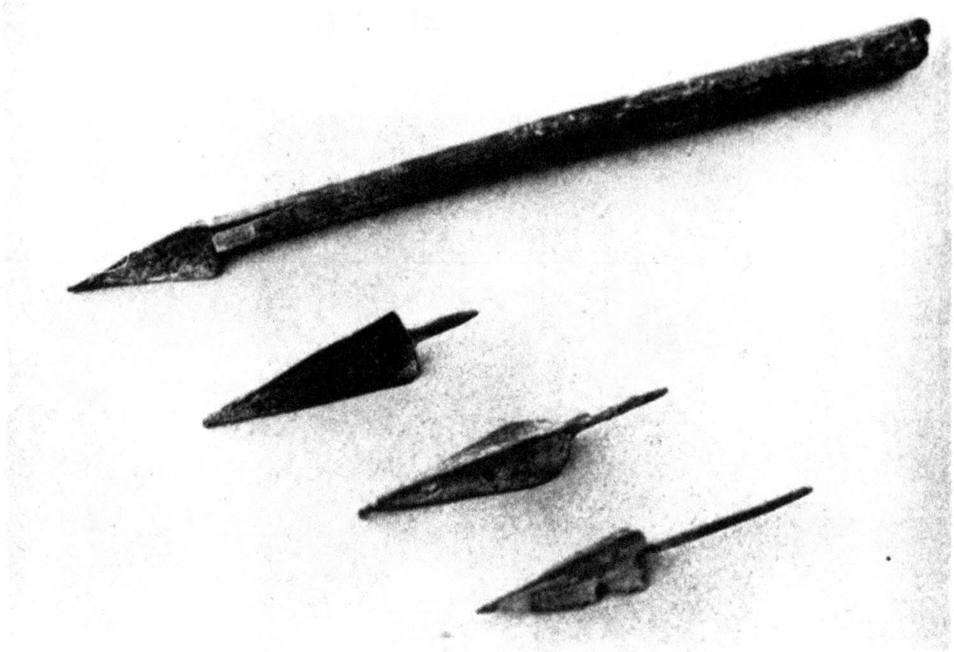

Fig.46

Fig.47

ABBREVIATIONS

A.A.	Archaeologia Aeliana
A.H.	Archaeologia Hungarica
A.J.A.	American Journal of Archaeology
A.R.B.	Guide to the Antiquities of Roman Britain, 3rd ed., British Museum, (London 1964)
Arch. Camb.	Archaeologia Cambrensis
A.S.H.	Acta Archaeologia Academiae Scientiarum Hungaricae
Brit.	Britannia
B.R.G.K.	Bericht der Römisch-Germanischen Kommission
C.I.L.	Corpus Inscriptionum Latinarum
C.S.I.R.	Corpus Signorum Imperii Romani
Ger.	Germania
G.R., III	Germania Romana: ein Bilder-Atlas, III, (Bamberg 1926)
I.E.J.	Israel Exploration Journal
I.L.S.	Inscriptiones Latinae Selectae
J.A.S.	Journal of Archaeological Science
J.R.A.I.	Journal of the Royal Anthropoligical Institute
J.R.S.	Journal of Roman Studies
J.S.A.A.	Journal of the Society of Archer-Antiquaries
M.Z.	Mainzer Zeitschrift
O.R.L.	Obergermanisch-raetische Limes
P.S.A.S.	Proceedings of the Society of Antiquaries of Scotland
R.E.	Paulys Real-Encyclopädie der Classischen Altertums
R.L.O.	Der römische Limes in Österreich

S.J. Saalburg Jahrbuch

BIBLIOGRAPHY

ADLER 1902: B. Adler, 'Die Bogen Nordasiens', International Archiv für Ethnographie, 15, 1902, 1-27

ALFOLDI 1932: A.Alfoldi, 'Funde aus der Hunnerzeit und ihre ethnische Sonderung', A.H., 9, 1932

ALLASON-JONES & MIKET 1984: L. Allason-Jones & R. Miket, The Catalogue of Small Finds from South Shields Roman Fort, (Newcastle 1984)

APPELGREN-KIVALO 1931: H. Appelgren-Kivalo, Alt-Altaische Kunstdenkmäler, (Helsinki 1931)

ARMITAGE & CLUTTON-BROCK 1976: P.L. Armitage & J. Clutton-Brock, 'A system for classification and description of the horn cores of cattle from archaeological sites', J.A.S., 3, 1976, 329-48

ATKINSON 1942: D. Atkinson, Report on the Excavations at Wroxeter, 1923-27, (Oxford 1942)

AYMARD 1951: J. Aymard, Essai sur les chasses romaines des origines a la fin di siècle des Antonins (Cynogetia), (Paris 1951)

BAATZ 1975: D. Baatz, Der römische Limes, (Berlin 1975)

BAATZ 1978: D. Baatz, 'Recent finds of ancient artillery', Brit., IX, 1978, 1-17

BAATZ 1983: D. Baatz, 'Town walls and defensive weapons' in MALONEY & HOBLEY, 1983, 136-40

BALFOUR 1890: H. Balfour, 'The structure and affinities of the composite bow', J.R.A.I., XIX, 1890, 220-50

BALFOUR 1897: H. Balfour, 'On a remarkable ancient bow and arrows', J.R.A.I., XXVI, 1897, 210-20

BALFOUR 1921: H. Balfour, 'The archer's bow in the Homeric poems', J.R.A.I., LI, 1921, 289-309

BALTY 1969: J. Balty, La grande mosaïque de chasse du Triclinos: fouilles de Apamee de Syrie, miscellanea, (Brussels 1969)

BARADEZ 1954: J. Baradez, 'Fouilles de Tipasa', *Libyca*, II, 1954, 89-148

BARADEZ 1956: J. Baradez, 'L'Algeria romaine et la Pannonie' in SWOBODA, 1956, 4-11

BAUR & ROSTOVTZEFF 1929: P.V.C. Baur & M.I. Rostovtzeff, *The Excavations at Dura-Europos, Preliminary Report of the First Season of Work, Spring, 1928*, (New Haven 1929)

BAUR & ROSTOVTZEFF 1931: P.V.C. Baur & M.I. Rostovtzeff (eds.), *The Excavations at Dura-Europos, Preliminary Report of the Second Season of Work, Oct. 1928 - Apr. 1929*, (New Haven 1931)

BAUR *et al.* 1932: P.V.C. Baur, M.I. Rostovtzeff, A.R. Bellinger, *The Excavations at Dura-Europos, Preliminary Report of the Third Season of Work, Nov. 1929 - Mar. 1930*, (New Haven 1932)

BEHRENS 1912-14: G. Behrens, 'Dritter Bericht über Funde aus dem Kastell Mainz', *M.Z.*, VIII-IX, 1913-14, 65-94

BELENITZKY 1968: A.M. Belenitzky, *Central Asia*, (Geneva 1968)

BELENITZKY 1980: A.M. Belenitzky, *Mittelasien Kunst der Sogden*, (Leipzig 1980)

BENSEDDIK 1979: N. Benseddik, *Les troupes auxiliaires de l'armée romaine en Maurétanie Césarienne sous le Haut-Empire*, (Algiers 1979)

BERGMAN 1939: F. Bergman, *Archaeological Researches in Sinkiang*, (Stockholm 1939)

BIRLEY 1966: E. Birley, '*Alae* and *cohortes milliariae*', *Corolla Memoriae Erich Swoboda Dedicata*, (Graz 1966), 54-67

BIRLEY 1977: R. Birley, *Vindolanda: A Roman Frontier Post on Hadrian's Wall*, (London 1977)

BISHOP 1983: M.C. Bishop (ed.), *Roman Military Equipment*, (Sheffield 1983)

BIVAR 1955: A.D.H. Bivar, 'The stirrup and its origins', *Oriental Art*, 1.2, 1955, 61-5

BIVAR 1972: A.D.H. Bivar, 'Cavalry tactics and equipment on the Euphrates', *Dumbarton Oaks Papers*, 26, 1972, 273-91

BOON 1972: G.C. Boon, *The Roman Legionary Fortress at Caerleon,*

Mon., (Cardiff 1972)

BOON 1974: G.C. Boon, Silchester: The Roman Town of Calleva, (London 1974)

BOSANQUET 1904: R.C. Bosanquet, 'Excavations on the line of the Roman Wall in Northumberland: the Roman camp at Housesteads', A.A., ser.2, XXV, 1904, 193-300

BREEZE 1982: D.J. Breeze, The Northern Frontiers of Roman Britain, (London 1982)

BRENNEN 1980: P. Brennen, 'Combined legionary detachments as artillery units in Late Roman Danubian bridgehead dispositions', Chiron, 10, 1980, 553-67

BRETT et al. 1947: G. Brett, W.J. Macauley, R.B.K. Stevenson, The Great Palace of the Byzantine Emperors, (Oxford 1947)

BRINKER & GOEPPER 1980: H. Brinker & R. Goepper, Kunstschätze aus China, (Zurich 1980)

BROK 1978: F.A. Brok, 'Ein spätrömischer Brandpfeil nach Ammianus', S.J., XXXV, 1978, 57-60

BROWN 1937: F.E. Brown, 'A recently discovered composite bow', Seminarius Kondakovianum, IX, 1937, 1-10

BRUCKNER & MARICHAL 1979: A. Bruckner & R. Marichal (eds.), Chartae Latinae Antiquiores, X, Germany, 1, (Zurich 1979)

BUDGE 1903: E.A.W. Budge, An Account of the Roman Antiquities Preserved in the Museum at Chesters, Northumberland, (London 1903)

BULANDA 1913: A. Bulanda, Bogen und Pfeil bei den Völkern der Altertums, (Vienna-Leipzig 1913)

BUSHE-FOX 1949: J.P. Bushe-Fox, Fourth Report on the Excavations of the Roman Fort at Richborough, Kent, (Oxford 1949)

CAGNAT 1913: R. Cagnat, L'armée romaine a'Afrique et l'occupation militaire de l'Afrique sous les empereurs, (Paris 1913)

CALLIES 1964: H. Callies, 'Die fremden Truppen im römischen Heer des Principats und die sogennanten nationalen Numeri. Beiträge zur Geschichte des römischen Heeres', B.R.G.K., 45, 1964, 130-227

CAMPBELL 1968: L.E. Campbell, Mithraic Iconography and Ideology,

(Leiden 1968)

CANTACUZÈNE 1927: G. Cantacuzène, 'Le recruitment de quelques cohortes syriennes', Musée Belges, 31, 1927, 5-18

CARCOPINO 1925: J. Carcopino, 'Le Limes de Numidie et sa garde syrienne', Syria, VI, 1925, 30-57, 118-49

CHEESMAN 1914: G.L. Cheesman, The Auxilia of the Roman Army, (Oxford 1914)

CHIUNG 1981: T'an Tan-Chiung, 'Investigative report on bow and arrow manufacture in Ch'engtu', Soochow University Journal of Chinese Art History, XI, 1981, 143-216

CICHORIUS 1896-1900: C. Cichorius, Die Reliefs der Traianssäule, (Berlin 1896-1900)

COLES 1973: J.M. Coles, Archaeology by Experiment, (London 1973)

COLLEDGE 1967: M.A.R. Colledge, The Parthians, (London 1967)

COLLEDGE 1976: M.A.R. Colledge, The Art of Palmyra, (London 1976)

COLLEDGE 1977: M.A.R. Colledge, Parthian Art, (London 1977)

COLLINGWOOD & RICHMOND 1969: R.G. Collingwood & I.A. Richmond, The Archaeology of Roman Britain, (London 1969)

COLLINS 1975: L.J.D. Collins, 'The military organisation and tactics of the Crimean Tatars, 16th-17th centuries' in PARRY & YAPP, 1975, 257-76

LE COQ 1925: A. von Le Coq, Bilderatlas zur Kunst- und Kulturgeschichte Mittel Asiens, (Berlin 1925)

CREDLAND 1981: A.G. Credland, 'Crossbow remains, 2', J.S.A.A., 24, 1981, 9-16

CREDLAND 1982: A.G. Credland, 'Crossbow remains, 3', J.S.A.A., 25, 1982, 16-21

CREDLAND 1983: A.G. Credland, 'The crossbow in the far north', J.S.A.A., 26, 1983, 12-23

CRUMMY 1983: N. Crummy, Colchester Archaeological Report 2: The Roman Smallfinds from Excavations in Colchester, 1971-79, (Colchester 1983)

CUMONT 1926: F. Cumont, Fouilles de Doura Europos, (1922-23),

(Paris 1926)

CUMONT 1942: F. Cumont, Recherches sur le symbolisme funéraire des romains, (Paris 1942)

DARKO 1935: E. Darko, 'Influences touraniennes sur l'evolution de l'art militaire des Grecs, des Romains et des Byzantines', Byzantion, X, 1935, 443-69

DARKO 1937: E. Darko, 'Influences touraniennes sur l'evolution de l'art militaire des Grecs, des Romains et des Byzantines', Byzantion, XII, 1937, 119-47

DARKO 1946-48: E. Darko, 'Le rôle des peuples nomades cavaliers dans la transformation de l'empire romain aux premiers siècles du moyen âge', Byzantion, XVIII, 1946-48, 85-97

DAVIES 1971: R.W. Davies, 'Cohortes equitatae', Historia, 20, 1971, 751-63

DAVIES 1977: J.L. Davies, 'Roman arrowheads from Dinorben and the sagittarii of the Roman army', Brit., VIII, 1977, 257-70

DEBEVOISE 1938: N.C. Debevoise, A Political History of the Parthians, (Chicago 1938)

DIENES 1972: I. Dienes, The Hungarian Cross the Carpathians, (Budapest 1972)

DOMASZEWSKI 1908: A. Domaszewski, Die Rangordnung der römischen Heeres, (Bonn 1908)

EADIE 1967: J.W. Eadie, 'The development of Roman mailed cavalry', J.R.S., 57, 1967, 161-73

ECKINGER 1933: K. Eckinger, 'Bogenversteifungen aus römischen Grabern', Germ., 17, 1933, 289-90

ELGOOD 1979: R. Elgood (ed.), Islamic Arms and Armour, (London 1979)

ELMY 1981: D. Elmy, 'Golden bows', J.S.A.A., 24, 1981, 3

EMENEAU 1953: M.B. Emeneau, 'The composite bow in India', Proceedings of the American Philosophical Society, 97, 1953, 77-87

ERDÉLY 1966: I. Erdély, The Art of the Avars, (Budapest 1966)

ERDMANN 1976: E. Erdmann, 'Dreiflügelige Pfeilspitzen aus Eisen von der Saalburg', S.J., XXXVIII, 1982, 5-11

ÉSPÉRANDIEU 1907-66: E. Éspérandieu, Recueil des bas-reliefs, statues et bustes de la Gaule romaine, (Paris 1907-66)

FABIAN 1970: G. Fabian, 'The Hungarian composite', J.S.A.A., 13, 1970, 12-16

FABIAN 1984: G. Fabian, 'An Avar bow', J.S.A.A., 27, 1984, 30-1

FARIS & ELMER 1945: N.A. Faris & R.P. Elmer, Arab Archery, an Arabic Manuscript of about A.D.1500, (Princeton 1945)

FENTRESS 1979: E.W.B. Fentress, Numidia and the Roman Army, (Oxford 1979)

FETTICH 1926: N. Fettich, 'Das Kunstgewerbe der Avarenzeit in Ungarn', A.H., I, 1926

FETTICH 1953: N. Fettich, 'L'arc hunnique en or', A.H., 32, 1953, 171-7

FIELD & PROSTOV 1940: H. Field & E. Prostov, 'Archaeology in the Soviet Union', Antiquity, XIV, 1940, 404-26

FITZ 1972: J. Fitz, Les Syriens à Intercisa, (Brussels 1972)

FITZ 1976: J. Fitz (ed.), Der römische Limes in Ungarn, (Székesfehérvar 1976)

FITZ 1977: J. Fitz (ed.), Akten des XI Internationalen Limeskongress, 1976, (Budapest 1977)

FISCHER 1973: U. Fischer, Grabungen im römischen Steinkastell von Heddernheim, 1957-59, (Frankfurt 1973)

VON FLESCHENBERG 1964: O.S. von Fleschenberg, 'Spätantike Anleitung zum Bogenschiessen', Wiener Studien, 59, 1964, 110-24

FORSTER & KNOWLES 1911: R.H. Forster & W.H. Knowles, 'Corstopitum: report on the excavations in 1910', A.A., ser.3, VII, 1911, 143-267

FRESHFIELD 1922: E.H. Freshfield, 'Notes on a Vellum Album containing some original sketches of public buildings and monuments, drawn by a German artist who visited Constantinople in 1574', Antiquity, LXXII, 1922, 87-104

FUKAI & HORIUCHI 1972: S. Fukai & K. Horiuchi, Tāq-i-Bustān, II,

(Tokyo 1972)

GARAM et al. 1975: E. Garam, I. Kovrig, J.G. Szabó, G. Török, Avar Finds in the Hungarian National Museum, (Budapest 1975)

GARBSCH 1978: J. Garbsch, Römische Paraderüstungen, (Munich 1978)

GHIRSHMAN 1946: R. Ghirshman, Begram, récherches archéologiques et historiques sur les Kouchans, (Cairo 1946)

GHIRSHMAN 1962: R. Ghirshman, Iran: Parthians and Sassanids, (London 1962)

GILBERT 1975-6: J.M. Gilbert, 'Crossbows on Pictish stones', P.S.A.S., 1975-6, 316-17

GOODBURN & BARTHOLOMEW 1976: R. Goodburn & P. Bartholomew (eds.), Aspects of the Notitia Dignitatum, (Oxford 1976)

GREEP 1983: S. Greep, 'Approaches to the study of bone, antler and ivory military equipment' in BISHOP, 1983, 16-21

GRIGG 1979: R. Grigg, 'Portrait-bearing codicils in the illuminations of the Notitia Dignitatum', J.R.S., 73, 1983, 132-42

GRIGG 1983: R. Grigg, 'Inconsistency and lassitude: the shield emblems of the Notitia Dignitatum', J.R.S., 73, 1983, 132-42

VON GROLLER 1901: M. von Groller, R.L.O., 2, (Vienna 1901)

GROUSSET 1970: R. Grousset, The Empire of the Steppes: a History of Central Asia, (New Brunswick, New Jersey 1970)

GRÜNEWALD 1981: M. Grünewald, Die Kleinfunde des Legionslagers von Carnuntum mit Ausnahme und Gefässkeramik (Grabungen 1968-74), R.L.O., XXXI, (Vienna 1981)

GUDEA 1974: N. Gudea & D. Baatz, 'Teile spätrömischer Ballisten aus Gornea und Orsova (Rumänien)', S.J., XXXI, 1974, 50-72

GUDEA 1979: N. Gudea, 'The defensive system of Roman Dacia', Brit., X, 1979, 63-87

GUSMAN 1914: P. Gusman, L'art décoratif de Rome de la fin de la republique au IVe siècle, (Paris 1914)

HALDON 1970: J.F. Haldon, 'Solenarion - the Byzantine crossbow?', University of Birmingham Historical Journal, XII.2, 1970, 155-7

HALDON 1975: J.F. Haldon, 'Some aspects of Byzantine military technology from the 6th to 10th centuries', Byzantine and Modern Greek Studies, I, 1975, 11-47

HARMATTA 1951: J. Harmatta, 'The golden bow of the Huns', A.S.H., I, 1951, 107-49

HAVERFIELD 1911: F. Haverfield (ed.), Essays in Roman History, (Oxford 1911)

HEDIN 1937: S. Hedin, The Wandering Lake, (London 1940)

HERRMANN 1977: G. Herrmann, The Iranian Revival, (Oxford 1977)

HOFMANN 1905: A. Hofmann, Römische Militärgrabsteine der Donauländer, (Vienna 1905)

HOFMANN 1969: D. Hofmann, Das spätrömische Bewegungsheer, (Köln 1969)

HOLDER 1980: P.A. Holder, The Auxilia from Augustus to Trajan, (Oxford 1980)

HORN & RÜGER 1979: H.G. Horn & C.B. Rüger (eds.), Die Numider, (Bonn 1979)

HORVÁTH 1935: T. Horváth, 'Die avarischen Gräberfeld von Üllo und Kiskörös', A.H., 19, 1935

INGOLT 1970-71: H. Ingolt, 'The sarcophagus of Be'elai and other sculptures from the tomb of Malkû, Palmyra', Mélanges de l'Université Saint-Joseph, XLVI-VII, 1970-71, 173-200

INSTINSKY 1958: H.U. Instinsky, 'Grabstein eines berittenen Bogenschützen der ala Parthorum et Araborum', Ger., 36, 1958, 72-7

JAMES 1983: S. James, 'Archaeological evidence for Roman incendiary projectiles', S.J., 39, 1983, 142-3

JARRETT 1969: M.G. Jarrett, 'Thracian units in the Roman army', I.E.J., XIX, 1969, 215-24

JOBEY 1977-78: G. Jobey, 'Burnswark Hill', Transactions of the Dumfriesshire and Galloway Natural History and Antiquarian Society, LIII, 1977-78, 57-108

JONES 1964: A.H.M. Jones, The Later Roman Empire, 284-602, (Oxford 1964)

KAEGI 1964: W.E. Kaegi, 'The contribution of archery to the conquest of Anatolia', Speculum, XXXIX, 1964, 96-108

KENNEDY 1977: D.L. Kennedy, 'Parthian regiments in the Roman army', in FITZ, 1977, 521-31

KIESERITZKY 1909: G.K. Kieseritzky & C. Watzinger, Griechische Grabreliefs aus Südrussland, (Berlin 1909)

KLOPSTEG 1943: P.E. Klopsteg, 'The physics of bows and arrows', American Journal of Physics, II, no.4, 1943, 175-92

KLOPSTEG 1947: P.E. Klopsteg, Turkish Archery and the Composite Bow, (Evanston, Illinois 1947)

KLUMBACH 1968: H. Klumbach, 'Orientalisches Rollsiegel vom Mainzer Legionslager', Ger., 46, 1968, 36-40

KLUMBACH 1971: H. Klumbach, 'Gerätegriff aus Hirschgeweih vom Mainzer Legionslager', Jahrbuch des Römisch-Germanischen Zentralmuseum Mainz, 18, 1971, 226-32

KNOX et al. 1984: R. Knox, R. Maddin, J.D. Muhly, T. Stech, 'Iron objects from Masada: metallurgical studies', I.E.J., 33, 1984, 97-107

KORFMANN 1972: M. Korfmann, Schleuder und Bogen in Sudwestasien von den frühesten Belegen bis zum Begin der historischen Stadtstaaten, (Bonn 1972)

KRAELING 1956: C.H. Kraeling, The Synagogue, the Excavations at Dura-Europos conducted by Yale University and the French Academy of Inscriptions and Letters. Final Report VIII, Pt.1, (New Haven 1956)

LÁSZLÒ 1951: G. Lászlò, 'The significance of the Hun golden bow', A.S.H., I, 1951, 91-104

LÁSZLÒ 1957: G. Lászlò, 'Contribution à l'archèologie de l'epoque des migrations, ii: le carquois d'arc des Hongrois conquérants', A.S.H., VII, 1957, 165-98

LATHAM 1970: J.D. Latham, 'The archers of the Middle East: the Turco-Iranian background', Iran, VIII, 1970, 97-103

LATHAM & PATERSON 1970: J.D. Latham & W.F. Paterson, Saracen Archery, (London 1970)

LATHAM & PATERSON 1979: 'Archery in the lands of Eastern Islam' in ELGOOD, 1979, 78-88

LENGYEL & RADAN 1980: A. Lengyel & G.T.B. Radan, The Archaeology of Roman Pannonia, (Budapest 1980)

LEVI 1947: D. Levi, Antioch Mosaic Pavements, II, (Oxford 1947)

LEYSER 1965: K. Leyser, 'The battle of the Lech, 955, a study in tenth century warfare', History, L, 1965, 1-25

LUSCHAN 1899: H. von Luschan, 'Zusammengesetzte und verstärkte Bogen', Zeitschrifte für Ethnologie, 31, 1899, 221-39

MACDONALD & PARK 1905-6: G. MacDonald & A. Park, 'The Roman forts on the Bar Hill', P.S.A.S.[4], IV, 1905-6, 403-546

MACDONALD 1934: G. MacDonald, The Roman Wall in Scotland, (Oxford 1934)

MACGREGOR 1975-6: A. MacGregor, 'Two antler crossbow nuts and some notes on the early development of the crossbow', P.S.A.S., 1975-6, 317-21

MACMULLEN 1960: R. MacMullen, 'Inscriptions on armour and the supply of arms in the Roman Empire', A.J.A., 64, 1960, 23-40

MACREA 1964: M. Macrea, 'Exercitus Daciae Porolissensis et quelques considérations sur l'organisation de la Dacie romaine', Dacia, VIII, 1964, 1-16

MAENCHEN-HELFEN 1973: O. Maenchen-Helfen, The World of the Huns, (Berkeley 1973)

MANN 1954: J. Mann, 'A note on the numeri', Hermes, 82, 1954, 501-6

MANN 1974: J. Mann, Review of FITZ 1972, J.R.S., 64, 1974, 259-60

MANNING 1976a: W.H. Manning, 'Blacksmithing' in STRONG & BROWN, 1976, 143-53

MANNING 1976b: W.H. Manning, Catalogue of Romano-British Ironwork in the Museum of Antiquities, Newcastle upon Tyne, (Newcastle 1976)

MALONEY & HOBLEY 1983: J. Maloney & B. Hobley (eds.), Roman Urban Defences in the West, (London 1983)

MAROSI & FETTICH 1936: A. Marosi & N. Fettich, 'Trouvailles Avares de Dunapentele', A.H., 18, 1936, 9-105

MARSDEN 1969: E.W. Marsden, Greek and Roman Artillery: Historical

Development, (Oxford 1969)

MARSDEN 1971: E.W. Marsden, Greek and Roman Artillery: Technical Treatise, (Oxford 1971)

MCEWEN 1979: E. McEwen, 'The Chahār Kham or "four curved" bow in India' in ELGOOD, 1979, 89-96

MCLEOD 1958: W.E. McLeod, 'An unpublished Egyptian composite bow in the Brooklyn Museum', A.J.A., 62, 1958, 397-401

MCLEOD 1962: W.E. McLeod, 'Egyptian composite bows in New York', A.J.A., 66, 1962, 13-19

MCLEOD 1965: W.E. McLeod, 'The range of the ancient bow', Phoenix, 19, 1965, 1-14

MEDINGER 1933: P. Medinger, 'L'arc turquois et les archers parthes à la bataille de Carrhes', Revue Archéologique, 2, 1933, 227-34

MERRIFIELD 1962: R. Merrifield, 'Coins from the bed of the Walbrook and their significance', Antiquaries Journal, XLII, 1962, 38-52

MINNS 1913: E.H. Minns, Scythians and Greeks, (Cambridge 1913)

MITCHELL 1983: S. Mitchell (ed.), Armies and Frontiers in Roman and Byzantine Anatolia, (Oxford 1983)

MÓCSY 1974: A. Mócsy, Pannonia and Upper Moesia, (London 1974)

MORSE 1885: E.S. Morse, 'Ancient and modern methods of arrow-release', Bulletin of the Essex Institute, XVII, 1885, 3-56

NASH-WILLIAMS 1931: V.E. Nash-Williams, 'The Roman legionary fortress at Caerleon in Monmouthshire: report on the excavations carried out in the Prysg Field, 1927-29, Pt. I', Arch. Camb., 86, 1931, 99-157

NASH-WILLIAMS 1932: V.E. Nash-Williams, 'The Roman legionary fortress at Caerleon, II', Arch. Camb., 87, 1932, 48-108

OATES 1959: D. & J. Oates, 'Ain Sinu: a Roman frontier post in northern Iraq', Iraq, XXI, 1959, 207-42

L'ORANGE & VON GERKAN 1939: H.P. L'Orange & A. von Gerkan, Der spätantike Bildschmuck des Konstantinsbogens, (Berlin 1939)

PARRY & YAPP 1975: V.J. Parry & M.E. Yapp (eds.), War, Technology and Social Change in the Middle East, (London 1975)

PATERSON 1963: W.F. Paterson, 'Thumb-rings', J.S.A.A., 6, 1963, 14-15

PATERSON 1966a: W.F. Paterson, 'The archers of Islam', Journal of the Economic and Social History of the Orient, 9, 1966, 69-87

PATERSON 1966b: W.F. Paterson, 'A bow from the Far East', J.S.A.A., 9, 1966, 19-22

PATERSON 1969a: W.F. Paterson, 'The Sassanids', J.S.A.A., 12, 1969, 29-32

PATERSON 1969b: W.F. Paterson, 'Shooting under a shield', J.S.A.A., 12, 1969, 27-8

PAYNE-GALLWEY 1907: R. Payne-Gallwey, Projectile-Throwing Weapons of the Ancients with a Treatise on Turkish and Other Oriental Bows, (London 1907)

PELHAM 1911: H.F. Pelham, 'Arrian as legate of Cappadocia' in HAVERFIELD, 1911, 212-33

PETERSEN et al. 1896: E. Petersen, A. von Domaszewski, G. Calderini, Die Marcus-Säule auf der Piazza Colonna in Rom, (Monaco 1896)

PFEFFER 1957: W.V. Pfeffer, 'Neugefundener Grabstein eines arabisch-parthischen Reiters', M.Z., 52, 1957, 120-2

PHILLIPS 1965: E.D. Phillips, The Royal Hordes: Nomad Peoples of the Steppes, (London 1965)

PLANCK 1983: D. Planck, Das Freilichtmuseum am rätischen Limes im Ostalbkreis, (Stuttgart 1983)

POLASCHECK 1932: E. Polascheck, 'Wiener Grabfunde aus der Zeit des untergehenden römischen Limes', Wiener Prähistorische Zeitschrift, 19, 1932, 239-66

POPE 1938: A.U. Pope, A Survey of Persian Art, I, (Oxford 1938)

POPE 1962: S.T. Pope, Bows and Arrows, (Berkeley 1962)

POTTER 1979: T.W. Potter, Romans in North-West England, (Kendal 1979)

PUGACHENKOVA 1971: G.A. Pugachenkova, Sculptura Khaltchayana,

(Moscow 1971)

RAUSING 1967: G. Rausing, The Bow: Some Notes on its Origins and Development, (Lund 1967)

REED et al. 1964: A. Reed, R. Harper, W. Dodds, 'Excavations at Ebchester in 1962-3', A.A., ser.4, XLII, 1964, 173-85

RICHMOND 1935: I.A. Richmond, 'Trajan's Army on Trajan's Column', Papers of the British School at Rome, XIII, 1935, 1-40

RICHMOND & BIRLEY 1940: I.A. Richmond & E. Birley, 'Excavations at Corbridge, 1938-40', A.A., ser.4, XVII, 1940, 85-115

RITTERLING 1902: E. Ritterling, 'Zur Erklärung von Arrians ektaxis kat Alanon', Wiener Studien, XXIV, 1902, 359-72

ROBERTSON et al. 1975: A. Robertson, M. Scott, L. Keppie, Bar Hill: a Roman Fort and its Finds, (Oxford 1975)

ROBINSON 1975: H.R. Robinson, The Armour of Imperial Rome, (London 1975)

ROBINSON 1976: H.R. Robinson, What the Soldiers Wore on Hadrian's Wall, (Newcastle 1976)

ROSTOVTZEFF 1913: M.I. Rostovtzeff, Antichnaya Dekorativnaya Zheevopeec na Yuge Rossiye, (Petrograd 1913)

ROSTOVTZEFF 1922: M.I. Rostovtzeff, Iranians and Greeks in South Russia, (Oxford 1922)

ROSTOVTZEFF 1932a: M.I. Rostovtzeff, 'Les inscriptions caravanières de Palmyre', Mélanges G. Glotz, II, 1932, 793-811

ROSTOVTZEFF 1932b: M.I. Rostovtzeff, 'The caravan-gods of Palmyra', J.R.S., XXII, 1932, 109-16

ROSTOVTZEFF 1934: M.I. Rostovtzeff (ed.), The Excavations at Dura-Europos: Preliminary Report of the 5th Season of Work, Oct. 1932 - Mar. 1933, (New Haven 1934)

ROSTOVTZEFF 1935: M.I. Rostovtzeff, 'Dura and the problem of Parthian art', Yale Classical Studies, V, 1935, 157-305

ROSTOVTZEFF et al. 1936: M.I. Rostovtzeff, A.R. Bellinger, C. Hopkins, C.B. Welles, The Excavations at Dura-Europos: Preliminary Report of the 6th Season of Work, Oct. 1932 - Mar. 1933, (New Haven 1936)

ROSTOVTZEFF 1938: M.I. Rostovtzeff, Dura Europos and its Art, (Oxford 1938)

ROSTOVTZEFF et al. 1939: M.I. Rostovtzeff, F.E. Brown, C.B. Welles (eds.), The Excavations at Dura-Europos: Preliminary Report of the 7th and 8th Seasons of Work: 1933-34 and 1934-35, (New Haven 1939)

ROSTOVTZEFF 1943: M.I. Rostovtzeff, 'The Parthian shot', A.J.A., XLVII, 1943, 174-87

ROSTOVTZEFF et al. 1946: M.I. Rostovtzeff, A.R. Bellinger, F.E. Brown, C.B. Welles, The Excavations at Dura-Europos: Preliminary Report of the 9th Season of Work, 1935-36, II, (New Haven 1946)

ROXAN 1976: M.M. Roxan, 'Pre-Severan auxilia named in the Notitia Dignitatum' in GOODBURN & BARTHLOMEW, 1976, 59-80

RUBIN 1955: B. Rubin, 'Die Entstehung der Kataphraktenreiterie im Lichte der chorezmischen Ausgrabungen', Historia, IV, 1955, 264-83

RUDENKO 1970: S.I. Rudenko, Frozen Tombs of Siberia: the Pazyryk Burials of Iron Age Horsemen, (London 1970)

SADDINGTON 1970: D.B. Saddington, 'The Roman auxilia in Tacitus, Josephus and other early imperial writers', Acta Classica, 13, 1970, 89-124

SALAMON 1967: A. Salamon, 'Archäologische Angaben zur spätrömischen Geschichte des pannonischen Limes - Geweihmanufaktur in Intercisa', Mittheilungen des Archäologischen Instituts der Ungarischen Akademie der Wissenschaften, 6, 1967, 47-54

SALLER & BAGATTI 1949: S.J. Saller & B. Bagatti, The Town of Nebo (Khirbet el-Mekhayyat) with a Brief Survey of Other Ancient Christian Monuments in TransJordan, (Jerusalem 1949)

SAXER 1967: R. Saxer, Untersuchungen zu den Vexillationen des römischen Kaiserheeres von Augustus bis Diokletian, (Köln 1967)

SCHINDLER 1977: B. Schindler, Führer durch das Landesmuseum Trier, (Trier 1977)

SCHÖNBERGER 1973: H. Schönberger, 'Die 1. Damascenerkohorte aus Friedberg in zwei Heddernheimer Inschriften', Ger., 51, 1973, 146-51

SEBESTEYÉN 1930: K.S. Sebesteyén, 'Ratselhafte Beinplatten in den Gräbern der Völkerwanderungenzeit', Dolgazatok, 6, 1930, 178-220

SEYRIG 1933: H. Seyrig, 'Textes relatifs à la garrison romaine à Palmyre', Syria, XIV, 1933, 152-68

SEYRIG 1937: H. Seyrig, 'Armes et costumes iraniens de Palmyre', Syria, XVIII, 1937, 4-31

SEYRIG 1941: H. Seyrig, 'Archers et méharistes palmyréniens', Syria, XXII, 1941, 170-80

SMITH 1968: D.J. Smith, 'The archer's tombstone from Housesteads', A.A., ser.4, XLVI, 1968, 184-91

SPEIDEL 1965: M. Speidel, Die Equites Singulares Augusti, (Bonn 1965)

SPEIDEL 1973: M.P. Speidel, 'Numerus Syrorum Malvensium. The transfer of a Dacian army unit to Mauretania and its implications', Dacia, XVII, 1973, 169-77

SPEIDEL 1975: M.P. Speidel, 'The rise of ethnic units in the Roman imperial army', Aufstieg und Niedergang der Römischen Welt II,3, 1975, 202-31

SPEIDEL 1977: M.P. Speidel, 'Pannonian troops in the Moorish war of Antoninus Pius' in FITZ, 1977, 129-35

SPEIDEL 1983: M.P. Speidel, 'The Roman army in Asia Minor, recent epigraphical discoveries and research' in MITCHELL, 1983, 7-34

STADE 1933: K. Stade, 'Beinplatten zur Bogenversteifung aus römischen Waffenplätzen', Ger., 17, 1933, 110-14

STONE 1934: G.C. Stone, A Glossary of the Construction, Decoration and Use of Arms and Armour, (New York 1934)

STRONG & BROWN 1976: D. Strong & D. Brown (eds.), Roman Crafts, (London 1976)

SULIMIRSKI 1952: T. Sulimirski, 'Les archers à chéval, cavallerie légère des anciens', Revue Internationale d'Histoire Militaire, III, 1952, 447-61

SULIMIRSKI 1970: T. Sulimirski, The Sarmatians, (London 1970)

SWOBODA 1956: E. Swoboda (ed.), Carnuntina, (Graz-Köln 1956)

SYME 1929: R. Syme, 'The Argonautica of Valerius Flaccus', Classical Quarterly, 23, 1929, 129-37

SZILÁGYI 1952: J. Szilágyi, 'Roman garrisons stationed at the northern Pannonian-Quadian frontier-sectors of the Empire', A.S.H., 2, 1952, 189-200

TAHA 1982: A. Taha, 'Men's costume in Palmyra', Les Annales Archéologiques Arabes Syriennes, XXXII, 1982, 117-32

TODD 1975: M. Todd, The Northern Barbarians, 100 B.C.-A.D.300, (London 1975)

TOYNBEE 1962: J.M.C. Toynbee, Art in Roman Britain, (Oxford 1962)

TOYNBEE 1964: J.M.C. Toynbee, Art in Britain under the Romans, (Oxford 1964)

ULBERT 1970: G. Ulbert, Das römische Donau-Kastell Rißtissen, I, (Stuttgart 1970)

VERMASEREN 1971: M.J. Vermaseren, Mithraica I, the Mithraeum at S.M. Capua Veteres, (Leiden 1971)

VETTERS 1948: H. Vetters, 'Austria' (Archaeological News), A.J.A., 52, 1948, 230-41

WAGNER 1938: W. Wagner, Die Dislokation der römischen Auxiliarformationen in den Provinzen Noricum, Pannonien, Mosien und Dakien von Augustus bis Gallienus, (Berlin 1938)

WALKE 1965: N. Walke, Das römische Donaukastell Straubing-Sorviodunum, (Berlin 1965)

WATSON 1969: G.R. Watson, The Roman Soldier, (Bristol 1969)

WEBB 1982: A. Webb, 'Nadon', J.S.A.A., 25, 1982, 12

WEBSTER 1958: G. Webster, 'The Roman military advance under Ostorius Scapula', The Archaeological Journal, 115, 1958, 49-98

WEBSTER 1965: G. Webster, 'Further investigations on the site of the Roman fort at Waddon Hill, Stoke Abbott, 1960-62', Proceedings of the Dorset Natural History and Archaeological Society, 86, 1965, 135-49

WEBSTER 1969: G. Webster, The Roman Imperial Army, (London 1969)

WEBSTER 1983: G. Webster, 'The functions and organisation of Late

Roman civil defences in Britain' in MALONEY & HOBLEY, 1983, 118-20

DE WEERD & LAMBRECHTS 1938: H. van de Weerd & P. Lambrechts, 'Note sur le corps d'archers au haut empire', Dissertations Pannonicae, 10, 1938, 229-42

WELLES et al. 1959: C.B. Welles, R.O. Fink, J.F. Gilliam, The Excavations at Dura-Europos: Final Report V,1. The Parchments and Papyri, (New Haven 1959)

WERNER 1932: J. Werner, 'Bogenfragmente aus Carnuntum und von der unteren Wolga', Eurasia Septentrionalis Antiqua, 7, 1932, 33-58

WERNER 1939: J. Werner, 'Ein hunnisches Lager der Hanzeit in Transbaikalien', Sinica, XIV, 1939, 193-6

WERNER 1956: J. Werner, Beiträge zur Archäologie des Attila-Reiches, (Munich 1956)

WIGGINS 1973: J. Wiggins, 'The oriental thumb-ring', J.S.A.A., 24, 1973, 215-21

WILKES 1969: J.J. Wilkes, Dalmatia, (London 1969)

WILKES 1983: J.J. Wilkes, 'Romans, Dacians and Sarmatians in the first and early second centuries' in B. Hartley & J.S. Wacher (eds.), Rome and her Northern Provinces, (Gloucester 1983), 255-89

YADIN 1963: Y. Yadin, Judaean Desert Studies, the Finds from the Bar Kokhba Period in the Cave of Letters, (Jerusalem 1963)

YADIN 1966: Y. Yadin, Masada: Herod's Fortress and the Zealots' Last Stand, (London 1966)

INDEX

www.ingramcontent.com/pod-product-compliance
Ingram Content Group UK Ltd.
Pitfield, Milton Keynes, MK11 3LW, UK
UKHW061844041026
15591UKWH00015B/900

9 780860 54347